Researchers in both the brain and cognitive sciences are attempting to understand the mind. Neuroscientists and cognitive psychologists should be natural allies, but tend to work in isolation of one another. *Brain and Mind* represents a pioneering attempt to bring these two fields closer together. The editors' objective was to force scientists who are working on the same problem but from different perspectives to address each other. Through a series of written dialogues on four topics – attention, perception, memory, and emotion – leading researchers sought to discover similarities and differences in their varied approaches. The dialogues demonstrate compellingly that neural and cognitive scientists have much to gain by increasing their interactions.

The eight main chapters, which cover cognitive and neuroscientific approaches to attention, perception, memory, and emotion, are detailed, up-to-date, scholarly reviews of the major methodological, empirical, and theoretical issues in each field. The accompanying dialogues assess these issues across disciplinary boundaries.

This book is bound to stimulate fruitful debate among all who study mind and brain. The editors and authors have striven to make the presentation accessible to advanced students in a wide range of fields: cognitive, perceptual, and physiological psychology; psychobiology; neurobiology; physiology; and medicine. Together, they have given the brain and cognitive sciences an invaluable resource.

Joseph E. LeDoux is Associate Professor of Psychology and Neurobiology, Cornell University Medical College. William Hirst is Assistant Professor of Psychology, Princeton University.

Mind and brain

Mind and brain

Dialogues in cognitive neuroscience

Edited by

JOSEPH E. LeDOUX
Cornell University Medical College

and

WILLIAM HIRST
Princeton University

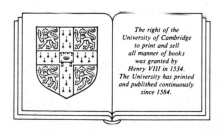

*The right of the
University of Cambridge
to print and sell
all manner of books
was granted by
Henry VIII in 1534.
The University has printed
and published continuously
since 1584.*

CAMBRIDGE UNIVERSITY PRESS

Cambridge
London   New York   New Rochelle
Melbourne   Sydney

Published by the Press Syndicate of the University of Cambridge
The Pitt Building, Trumpington Street, Cambridge CB2 1RP
32 East 57th Street, New York, NY 10022, USA
10 Stamford Road, Oakleigh, Melbourne 3166, Australia

First published 1986

Printed in the United States of America

Library of Congress Cataloging-in-Publication Data
Main entry under title:
Mind and brain.
Bibliography: p.
Includes index.
1. Neuropsychology.   2. Cognition.   3. Mind and body.
I. LeDoux, Joseph E.   II. Hirst, William.
QP360.M523   1986   152   85–22350

British Library Cataloguing in Publication Data
Mind and Brain : dialogues in cognitive neuroscience.
1. Neuropsychology
I. LeDoux, Joseph E.   II. Hirst, William
152   QP360

ISBN 0 521 26756 0 hard covers
ISBN 0 521 31853 X paperback

To Michael S. Gazzaniga

Contents

Part III   Memory

Part IV   Emotion

Editors' preface

Several years ago over lunch, we had an extended discussion of the relation between cognitive psychology and neuroscience. One of us defended neurobiological studies of mental processes as fundamental to psychology, while the other argued that such an approach was tolerable but unnecessary for the enterprise of psychology to advance. We each, in other words, stood faithfully by our party lines. The impasse led us to embark on a grueling journey through the philosophy of mind literature. After several months of reading we were convinced that the "in principle" solutions to the mind–body problem offered by philosophers from Descartes onward had few if any implications for practicing experimentalists, regardless of their academic background. Our experiments were not going to change even if philosophers came to a consensus about the mind–body problem.

Our reading and discussions were useful to us, however. We found that when we got down to specific issues, we agreed that neuroscientists and psychologists had something to offer each other. When, in other words, we strayed away from the question of how mind relates to brain and instead considered how specific mental processes are studied by neuroscientists and psychologists, we found much room for fruitful interchange. We decided to put our personal experience to the test and see whether other cognitive psychologists and neuroscientists could find anything to discuss. This book was the result.

We are grateful to Susan Milmoe and her associates at Cambridge University Press for their support throughout this project. We also appreciate the thoughtful and punctual contributions of the authors.

Joseph LeDoux
William Hirst

New York, N.Y.
November 1985

ix

Contributors

Ross Buck
Department of Communication
  Sciences
University of Connecticut

Michael Gabriel
Department of Psychology
University of Illinois

William Hirst
Department of Psychology
Princeton University

James E. Hoffman
Department of Psychology
University of Delaware

Joseph E. LeDoux
Laboratory of Neurobiology
Department of Neurology
Cornell University Medical College

Richard T. Marrocco
Department of Psychology
University of Oregon

Steven E. Petersen
National Eye Institute
National Institutes of Health

David Lee Robinson
National Eye Institute
National Institutes of Health

Daniel L. Schacter
Department of Psychology
University of Toronto

Steven P. Sparenborg
Department of Psychology
University of Illinois

Neal Stolar
Department of Psychology
University of Illinois

1 Cognitive neuroscience: an overview

Joseph E. LeDoux and William Hirst

Recent years have been exciting for researchers in the brain and cognitive sciences. Both fields have flourished, each spurred on by methodological and conceptual developments, and although understanding the mechanisms of mind is an objective shared by many workers in these areas, their theories and approaches to the problem are vastly different. In spite of common goals, two fields that are natural allies are proceeding independently.

The purpose of this book is to encourage neuroscientists and cognitive scientists to take stock, while their disciplines are still young, and consider whether mutual independence is the most prudent course. It may take considerable soul searching to find the substantive links, but they are there and, in our opinion, well worth pursuing.

The separation of cognitive and brain sciences today has as much to do with circumstance as substance. Early experimental psychologists, such as Wundt and James, were as interested in and knowledgeable about the anatomy and physiology of the nervous system as about the young science of the mind. However, the experimental study of mental processes was short-lived, being eclipsed by the rise of behaviorism early in this century. It was not until the late 1950s that the signs of a new mentalism first appeared in scattered writings of linguists, philosophers, computer enthusiasts, and psychologists.

In its new incarnation, the science of mind had a specific mission: to challenge and replace behaviorism. In the meantime, brain science had in many ways become allied with a behaviorist approach, if not with behaviorism itself. Behavioral psychology provided many practical techniques that researchers interested in brain function could readily apply to animal studies. Instrumental and classical conditioning techniques are staples today in neuroscience. Moreover, it is relatively straightforward for a biologist to think about how physiological response systems underlying the expression of behavior are represented in the brain. Thoughts, memories, and feelings, however, are a different matter. Studies of the neural substrates of psychological processes therefore focused on understanding behavioral expression, devoid of any mental content. Behaviorism made life easy for the brain scientist interested in function.

Equally important to the affinity of brain science and behaviorism is a philosophical link. Both behaviorism and neuroscience encompass reductionist approaches to psychology. While behaviorism sought to reduce the mind to statements about bodily action, brain science seeks to explain the mind in terms of physicochemical events occurring in the nervous system. These approaches contrast with contemporary cognitive science, which tries to understand the mind as it is, without any reduction, a view sometimes described as functionalism (Fodor, 1975).

The cognitive revolution is now in place. Cognition is *the* subject of contemporary psychology. This was achieved with little or no talk of neurons, action potentials, and neurotransmitters. Similarly, neuroscience has risen to an esteemed position among the biological sciences without much talk of cognitive processes. Do the fields need each other? Is there something to be gained by cross-disciplinary interchanges? Are interactions even possible? The answers are all yes because the problem of understanding the mind, unlike the would-be problem solvers, respects no disciplinary boundaries. It remains as resistant to neurological as to cognitive analyses.

In the last few years, there has been talk of a "cognitive neuroscience," an interdisciplinary blend of researchers concerned with brain and cognition. Edited volumes have been compiled on the subject (Gazzaniga, 1984), meetings have been held (see LeDoux, Barclay, and Premack, 1980), journals have appeared (*Brain and Cognition*), and the Society for Neuroscience now often includes a cognitive neuroscience section in the program of its annual meeting. These are important steps aimed at bringing the brain and cognitive sciences together. We hope that this volume further contributes to this effort.

Organization of the book

Our objective in organizing this book was to provide a direct form of interchange, a written dialogue, between psychologists and neuroscientists. From experience, we knew that unstructured interactions between cognitive and brain scientists tend to fail miserably on the scientific level. The language of discourse differs. Discussions often drift into hopeless generalities. To overcome these problems, as much as they could be overcome, we selected topics that are currently being worked on by both cognitive scientists and neuroscientists and organized the book so that interactions take place only between persons working on the same topic. In this way, we maximize the overlap of interest and experience of the interacting scientists.

We selected scientists from four areas of investigation: attention, perception, memory, and emotion. The first three of these need no justification. Attention, perception, and memory are central to any cognitive theory, and significant progress has been made by both neuroscientists and cognitive psychologists in understanding these processes.

Emotion, however, is not traditionally in the cognitive camp. Like motivation and intention, feeling and affect are usually viewed as noncognitive processes, things to be controlled in a good experiment (Miller and Gazzaniga, 1984). Nevertheless, the interplay between cognitive and emotional processes has become an exciting area of research. Moreover, recent approaches viewing emotion from an information-processing point of view have opened new possibilities for interactions between psychologists and brain scientists. Interdisciplinary discussion of emotion is timely.

For each of the four topic areas, a psychologist and a neuroscientist were each asked to write an up-to-date overview of the field, as seen through his eyes, focusing on the important issues being studied, the approaches taken to answer the questions, and the possible future direction of the field. The two contributors were also asked to discuss each other's chapters, emphasizing existing and possible points of overlap, specifying areas of divergence, considering whether they constitute real or semantic differences, and in general commenting on the possibility of a true interdisciplinary approach to their field. Finally, each was given the opportunity to respond to the other's discussion.

As is apparent from the format selected, our strategy was to provide intensive, detailed discussions of the real issues in a small number of areas rather than general overviews of many aspects of cognitive psychology and neuroscience. In so doing, some basic topics have been left out. For example, the reader will no doubt quickly notice the absence of work on linguistics. Although language has been a major area of interest to students of cognition since the earliest years of experimental psychology, work on the brain substrates of language has lagged significantly behind the progress made in psycholinguistics. The limited range of experimental techniques available to the neuroscientist for studying humans has restricted work on the biology of language to analyses of language disorders in brain-damaged patients. Aphasiology has seen major advances in the last few years (Zurif and Blumstein, 1978), but mainly in characterizing functional deficits. An understanding of the underlying neural mechanisms of language, in terms of circuits and transmitters, is still only a hope.

This book is an experiment, and in a sense we stacked the deck in our favor by choosing topics conducive to interdisciplinary discussion and by choosing contributors who were willing to interact. However, our reasoning was that if hand-picked cognitive and brain scientists cannot work out differences, then we are not ready for a cognitive neuroscience. Happily, we feel that the chapters to follow demonstrate that we are ready. At the same time, they make it painfully obvious how far we have to go. While a unified science of mind is not likely to be easily or quickly achieved, we hope that the present work at least takes us one step closer.

Part I

Perception

James E. Hoffman

It is easy to take perception for granted. A glance around our visual world seems to provide us with an immediate "understanding" of objects: their shapes, sizes, colors, and relations to neighboring objects. This understanding also seems to be purchased rather cheaply, the effort is certainly less than solving most other cognitive problems, such as puzzles or mathematical proofs. Yet, perception, as well as language and thinking, remain the exclusive province of the biological machine and beyond the reach of current computers.

Perception invites a multidisciplinary approach. The electrophysiologist seeks to open the black box and observe the "hardware" of the perceptual system. The computer scientist provides a test of the *sufficiency* of various theories of perception by finding out whether a particular set of algorithms allows a camera and computer system to "see" (Marr, 1982). The experimental psychologist starts with the behavior of a perceiving organism and hopes that careful manipulation of the stimulus together with changes in the task to be performed by the observer will allow insights into the underlying representations and processes of perception.

We are currently a long way from the ultimate goal of specifying the information-processing architecture in sufficient detail to build a seeing machine. At present, we can only sketch pieces of the puzzle and identify the basic problems that need solution. To see why perception is difficult, consider the problems involved in understanding a photograph of a common scene such as a city street. The human perceiver quickly understands such scenes and can identify both the objects that are present and their interrelationships (Biederman, 1981). The representation of the scene could conceivably benefit from two different kinds of processes: a "bottom–up" analysis, which takes data present in the input image and constructs a representation of the component objects, and "top–down" knowledge about which objects and interrelationships usually occur together. As we will see, the perceptual system appears to use both of these processes in a seamless interaction to arrive at an interpretation.

Feature analysis

The initial stages in image analysis appear to have the status of what Fodor (1983) calls a "module." Modules are hard-wired, innate, specialized systems that do not share general "system resources," such as memory and attention, with other cognitive systems. As such, they are also "cognitively impenetrable" (Pylyshyn, 1981) and cannot be influenced by higher-level processes, such as expectations. Their workings are not available to conscious awareness, and, therefore, relatively indirect methods must be used to study them.

Almost all current theories make the following strong assumptions about the initial input module in vision. Complex images are initially analyzed into components by a set of independent, feature-analyzing systems. Perceptual experience occurs late in the processing sequence after features have been reassembled. This counterintuitive theory has by now accumulated a remarkably large and varied set of supporting data. The main disagreements concern the nature of the features extracted at the initial stages and how the later assembly into objects is achieved.

The enthusiasm for a feature-based approach stems from several sources. First, the ability to recognize objects must depend on matching a description of the image with a representation in memory. What form should this memory representation take? A *template* approach is to store the collection of two-dimensional views of known objects and attempt to find the best match with the current input image. This approach is sure to fail because objects generally do not ever present the same exact view twice. Changes in viewer perspective, distance, and so forth produce changes in an object's shape that doom the template approach to failure.

Templates are too literal to serve as the basis for recognizing objects. What is needed is a somewhat more abstract, symbolic representation. Features are one possibility. For example, the letter "A" can be described as two diagonal lines meeting at the top with a horizontal line connecting the diagonals. Notice that features such as a diagonal line are abstract: They make no mention of brightness, length, or retinal position. Thus, once the features are abstracted, matching to memory can proceed independently of incidental aspects of the shape such as its size.

There have been four major approaches to characterizing the feature-analysis stage: single-unit recording, spatial frequency analysis, "texton theory" (Julesz and Bergen, 1983), and feature-integration theory (Treisman, 1979; Treisman and Gelade, 1980). The first three will only be briefly described. Treisman's theory will be described in some detail both because of its theoretical importance and because it is a good example of the use of converging operations (Garner, Hake, and Eriksen, 1956) as a method for studying cognitive processes.

Single-unit recording

Recording neural activity in single cells at various points in the visual system has suggested the presence of a complex feature-analyzing system. These cells generally have receptive fields that are sensitive to lines of specific size, orientation, and direction of movement. In addition, higher-level cells have been discovered that respond to quite specific properties of the input, such as the angle of intersection of two lines and even the shape of a "monkey's hand" (Gross, Rocha-Miranda, and Bender, 1972).

Some neurophysiologists have been quite explicit about the role of these detectors in the recognition of objects. Barlow (1972), for example, suggests that one can create neurons capable of detecting objects by suitably combining the outputs of earlier detectors. The "monkey hand" detector could be taken as an instantiation of Barlow's "neuron doctrine for perceptual psychology."

Spatial frequency analysis

This approach suggests that visual images are analyzed by a set of more or less independent "channels" that are sensitive to orientation and spatial frequency. The nature of these channels has been determined primarily from psychophysical experiments investigating the sensitivity of human observers to threshold-level presentations of sine-wave gratings. There appears to be approximately four channels, each maximally tuned to a different range of spatial frequencies (Wilson and Bergen, 1979). Thus, these four channels simultaneously analyze an image into four representations with different size scales. Low-frequency channels can be used to identify "large" details in the image, whereas high-frequency channels represent small details and sharp edges.

This model of early visual analysis is attractive because it provides an explicit, quantitative description of visual patterns. In addition, the idea that several representations of the image exist at different grain sizes is valuable heuristically for modeling the hierarchical analysis of images, as we will see later.

Texton theory

Consider the two texture pairs shown in Figure 2.1. Each pattern consists of a center and surround that contain different shapes. In the left half, the disparate quadrant is immediately apparent, whereas in the right half, it can be discovered only through a slow process of "scrutiny" in which visual attention is directed to each individual element in the texture. These two cases are thought to represent the operation of two different levels in the visual

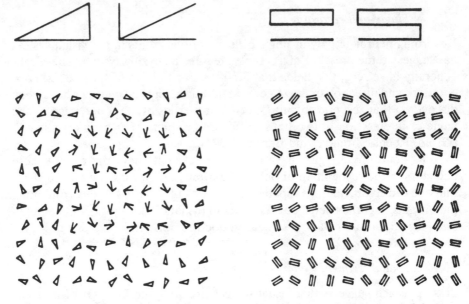

Figure 2.1. The center and surrounding areas of the left texture pattern are composed of elements belonging to two different classes of textons that differ in the number of line terminators. These different regions are immediately discriminable by preattentive vision. In the right panel, the two regions can only be discriminated by a slow process of focal attention. The two regions here are composed of members of the same texton class even though these elements look quite different in focal attention. (From Julesz and Bergen, 1983.)

system. Immediate texture segregation is provided by the preattentive system. Patterns that cannot be picked up by this parallel system can be detected by a serial process of focal attention.

What kinds of differences between two regions can be picked up by the preattentive system? If a relatively small number of such differences can be catalogued, then texture segregation experiments may provide an insight into the "perceptual alphabet" that forms the building blocks of pattern perception. In fact, Julesz and Bergen (1983) have identified three basic elements they call *textons*. Textons are "1. Elongated blobs – e.g., rectangles, ellipses, line segments with specific colors, angular orientations, widths, and lengths. 2. Terminators – ends-of-line segments. 3. Crossings of line segments" (Julesz and Bergen, 1983, p. 1622).

The left texture pattern in Figure 2.1 can be rapidly discriminated because it has two regions composed of different textons. The arrow shape has three terminators, while the triangle has none. The right-hand pattern consists of regions differing in two elements that look quite different in focal attention, as the reader can verify, but, in fact, are drawn from the same texton class. Each element has two line endings that are not connected to any other lines, and, therefore, they do not differ in number of terminators. Consequently, these patterns cannot be discriminated by preattentive vision.

Feature-integration theory

Treisman (1979), Treisman and Gelade (1980), and Treisman and Schmidt (1982) have recently proposed a feature-integration theory of attention that accounts for many of the findings in the areas of attention and visual search and, in addition, makes several surprising predictions that have been experimentally confirmed. The basic claim is that early visual processes analyze the image into separate "feature maps." Thus the letter "T" printed in red ink would be initially represented in three separate maps: a hue map for the color and two separate shape maps for the horizontal and vertical lines. The perceptual awareness of a single object possessing these three attributes only occurs when focal attention is directed to the letter's location, which then allows information in the three maps to be integrated into a composite representation.

Treisman's characterization of preattentive vision is clearly quite similar to that of Julesz and Bergen (1983). One difference between them is the "discovery criteria" used for deciding the nature of the separate features. Julesz and Bergen have used a combination of statistical criteria and Monte Carlo generation of texture patterns. Treisman starts with an important distinction made by Garner (1974) between separable and integral features. According to Garner, separable and integral features can be distinguished by a battery of diagnostic tests. Two dimensions are integral if one dimension cannot exist independently of the other. An example would be the hue and saturation of a colored patch. Any hue must necessarily be associated with some value of saturation. In contrast, the angle of a line would be a separable dimension from the size of the circle that surrounds it. Each dimension can exist independently of the other.

Integral dimensions act as if they come packaged together by the visual system and can only be "pulled apart" with considerable time and effort. Support for this claim is provided by selective attention experiments that show that subjects cannot ignore one of two integral dimensions. For example, if they are instructed to sort color patches into two piles on the basis of hue, they are better when the value of the saturation dimension is correlated with hue and are impaired when the two dimensions are orthogonal. A variety of other criteria have been proposed that provide converging evidence for the distinction between integral and separable dimensions.

Treisman proposes to use Garner's criteria for identifying which dimensions will exist in separate feature maps and then use the predictions of feature-integration theory to either confirm these or propose new candidates for separable features. The predictions are remarkably varied, and I will only

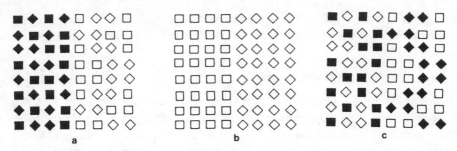

Figure 2.2. Demonstration of texture segregation effects predicted by Treisman's feature integration theory. In (a) the left and right halves of the texture pattern differ in being composed of filled or empty shapes, and this simple feature difference produces rapid segmenting of the pattern into two parts. In (b) the two areas differ in line slopes and, once again, segregation is rapid. In (c) the two halves differ in terms of a conjunction of features. In the left-hand pattern, squares are filled, while diamonds are empty. This relation is reversed for the right half of the pattern. Texture segregation is slow and requires that focal attention be directed to individual members of the texture.

relate a few here. Interested readers can consult Treisman's excellent papers for more.

One prediction is that immediate texture segregation should be possible only if two regions differ in a single feature, for example, two regions containing lines of different slope (Beck, 1981). On the other hand, if two regions have the same features but differ in the way the features are combined, segregation should be slow because of the need for focal attention (see Figure 2.2 for examples). Treisman and Gelade (1980) confirmed this prediction by presenting observers with two areas that differed in the conjunction of shape and color. One area contained red Os and blue Vs; the other area contained blue Os and red Vs. Notice that inspection of either a "color map" or a "shape map" would not allow perception of the boundary between the two areas. Texture segregation would require that information in the two maps be combined, which is presumably the province of the serial, focal attention system.

It might be argued that there is something unique about texture segregation that produces grouping on the basis of single features. Therefore, it would be useful to have converging evidence from other tasks on the validity of feature-integration theory. Treisman and Gelade reasoned that if subjects had to search for a target defined as a conjunction of color and shape, the search would be serial through the display. In contrast, targets that differ from the background in a single feature should be discriminated rapidly without a serial search. They verified this prediction by showing that conjunctive targets produced search times that were a linear function of display size. In addition, the slope of the reaction time versus display size function for target-absent trials was twice as large as that for target-present trials. This pattern of reaction times is consistent with a serial, self-terminating search for the target. Feature search, by comparison, produced nonlinear, display size functions with similar slopes for both target-present and -absent trials.

An interesting issue raised by feature-integration theory is whether large amounts of practice with a pattern composed of separable dimensions can eventually result in a single, unitized "object" (LaBerge, 1975; Shiffrin and Schneider, 1977) that can be perceived as a whole outside of focal attention. Several experiments indicate that practice does not eliminate the need for focal attention in conjoining separable features of a single object. Treisman and Gelade (1980) trained two subjects for 13 sessions in searching for a colored letter. The same serial search pattern that was observed at the beginning of the training was present on the last session. Thus, this amount of training was not sufficient to unitize the separate dimensions of color and shape outside the focus of attention. A similar result was reported by Beck (1981). It remains to be seen whether other dimensions, such as different shape features, can be unitized through training.

Perhaps one of the most intriguing predictions of feature-integration theory concerns the existence of "illusory conjunctions." According to Treisman and Schmidt (1982), if the features in separate maps are not integrated, they may float free and combine with each other to produce the perception of illusory objects. The frequency of illusory conjunctions should increase whenever display durations are too short or attention too overloaded to integrate separate functions correctly. They showed that subjects did, in fact, combine shapes and colors of different objects when attention was diverted. In addition, the likelihood that two objects would exchange a feature was independent of whether they differed in size or spatial separation or whether one object was filled and the other a simple outline.

Illusory conjunctions provide one of the most compelling demonstrations that objects are initially represented as separate dimensions. They have been found in many different situations. For example, Efron and Yund (1974) found that subjects occasionally combined the intensity and pitch of tones in different ears. Lawrence (1971) reported that when subjects were required to identify an occasional uppercase word embedded in a stream of lowercase words, they often reported the word that followed the target. Similarly, Intraub (1985) found that a rectangular frame surrounding one picture in a rapid sequence of pictures was often confidently seen as surrounding a picture that actually preceded or followed the framed picture.

Separability of features is also supported by experiments that show that some kinds of perceptual processes operate on some features and not on others. A striking example is provided by Ramachandran and Gregory (1978). They constructed two random dot patterns that were identical except for a small square area in the center that was shifted horizontally in one pattern with respect to the other. Sequential presentation of these patterns produces apparent movement of the square. Ramachandran and Gregory showed that this apparent motion could be abolished if the dots were defined only as differences in hue, for example, green dots on a red background with the luminances of the two colors matched. Under these conditions, subjects reported that the displays appeared to shimmer without any coherent movement. Apparently, the stage responsible for deriving apparent motion information for these random dot displays does not have access to hue information and can only operate on luminance differences. Similar conclusions hold for binocular stereopsis with random dot stereograms (Lu and Fender, 1972; Gregory, 1977; deWeert, 1979). Thus, there is clear evidence of specialized processing modules that have access to some features of a display and not others.

Evaluation of feature-integration theory

The evidence is fairly persuasive that preattentive vision operates on separate features. In addition to the behavioral evidence reviewed above, there is converging evidence from single-unit recording experiments indicating that there are multiple "cortical maps," each specialized for the analysis of particular features (Zeki, 1976). According to feature-integration theory, selective attention plays a crucial role in integrating these separate features. Two considerations, however, suggest that the situation may be more complicated than this. First, we do not yet know if there will be convergence between the candidates for the basic units of preattentive vision that have been postulated in the different approaches reviewed above. Walters, Biederman, and Weisstein (1983) found that differences in a texture pattern defined by the combination of orientation and spatial frequency were not preattentively discriminable. This is puzzling from the point of view of spatial frequency analysis because the elementary channels in this approach are defined by a conjunction of these two attributes. One possibility is that these two dimensions are conjoined together as a single channel early in the visual system, but are represented as separate features later. Houck and Hoffman (In press) reported evidence consistent with this view. They found that the strength of an aftereffect, which depended on both the color and orientation of lines (McCollough, 1965), was independent of whether or not the lines were attended. Thus, there is evidence that the visual system does have access to conjunctions of color and form at an early level without the need for attention. Access to this conjunction at later levels, however, requires the use of spatial attention. It seems likely the tasks used by cognitive psychologists may be tapping into relatively late representations that are not identical to those being investigated by single-unit recording and selective adaptation techniques.

There is a second reason for being cautious in concluding that attention is the only mechanism mediating feature conjunctions. Preattentive vision may be responsible for computing surface lightness as well as other constancies, and, as the next section shows, these computations often depend on combining several separate features together.

Perception of surface lightness

Todd and Mingolla (1983) point out that

the amount of light that reflects from a surface in any given direction depends on the orientation of the surface and its material composition, the positions and spectral compositions of all luminous bodies that can illuminate the surface directly, and the positions, orientations, and material compositions of all nonluminous bodies that can illuminate the surface indirectly through the process of reflection. (p. 593)

The visual system manages to separate these different factors, allowing the observer to perceive each component separately.

The phenomenon of lightness constancy offers one striking illustration of the visual system's ability to separate these different factors. The perceived lightness of a surface depends on its reflectance and remains constant in the face of large changes in illumination intensity. Explanations of this constancy fall into two groups: "cognitive" explanations that claim that the visual system takes illumination into account (Rock, 1983) and lateral inhibition models. The inhibition model starts with the observation that changes in illumination of a surface and its background will indeed change the luminance of these surfaces. The ratio of light reflected by the two surfaces will, however, remain invariant. Both Wallach (1948) and Heinemann (1955) confirmed that the perceived lightness of a disk depends on the ratio of luminances of the disk and its background. A mechanism that is sensitive to this ratio could thus show lightness constancy.

Cornsweet (1970) showed that simple neural networks are capable of computing the luminance ratio of two adjacent areas. These networks have two properties: a logarithmic response to light intensity and lateral inhibitory connections. The center-surround receptive fields of retinal ganglion cells appear to be ideal candidates for these computing elements. The lateral inhibition model thus claims that lightness depends primarily on "local" luminance ratios in the retinal image.

Although lateral inhibition appears to offer an attractive explanation for constancy, it has serious deficiencies when it is applied to more ecologically valid situations than the typical disk and annulus arrangement. Consider the case where one views a horizontal white card being strongly illuminated from an overhead light source. As the card is rotated away from the horizontal position, it will reflect progressively smaller amounts of light into the observer's

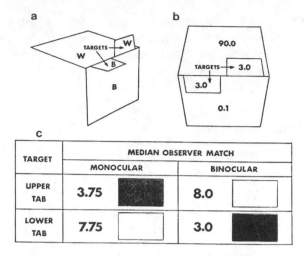

Figure 2.3. Display used by Gilchrist to investigate the role of surface orientation in lightness perception. The display is strongly illuminated from above so that the black tab has the same luminance as the white tab. Top right (b) shows the luminances of the two tabs as well as their backgrounds. When the display is viewed monocularly, the white tab is perceived to be *darker* than the black tab. Binocular viewing allows the observer to perceive the true lightness of each tab correctly. (From Gilchrist, 1980.)

eye, while the amount of light reflected from the background remains the same. According to ratio theory, the card should look progressively darker but, in fact, constancy holds in this situation as well and the card continues to look white.

Gilchrist (1979) confirmed in a clever series of experiments that the three-dimensional geometry of surfaces must be taken into account to perceive lightness correctly. Figure 2.3a shows a display containing four surfaces. The subject is required to judge the lightness of the two small tabs, one white and the other black. The intensity of the illumination falling on the two tabs is adjusted so that their luminance is equal.

When subjects viewed the display monocularly, they were deprived of the depth cues that would have allowed them to see the spatial arrangement of the surfaces correctly. Under these conditions, subjects perceived the black tab as being lighter than the white tab, a powerful visual illusion! This outcome is predicted by the lateral inhibition model because the white tab is surrounded by a bright background and therefore subject to inhibition from the surround. Allowing the subject to view the display with both eyes, however, reversed this outcome and allowed subjects correctly to perceive the relative lightnesses of the two tabs.

Gilchrist and Jacobsen (1983) provided further evidence against the ratio theory of lightness. Their subjects viewed a three-dimensional scene through

Figure 2.4. A computer-generated display depicting an opaque sphere with specular highlights, a transparent sphere with specular highlights for both its front and back surface, and the shadows cast by these objects on a checkerboard background. (From Todd and Mingolla, 1983.)

a window that had a high-intensity "veiling luminance" superimposed on it. Although the presence of this veiling luminance drastically changed the ratios between different parts of the scene, subjects were remarkably accurate at judging the lightnesses of the different surfaces. In contrast, the predictions of the ratio theory were confirmed when the naturalistic scene was replaced by a flat Mondrian-type display consisting of papers having a variety of reflectances. Apparently, the three dimensionality of the naturalistic scene allowed observers to separate the retinal image into two different components: one corresponding to the veiling luminance and the other to the scene. A related finding was reported by Neisser and Becklen (1975), who showed that subjects could selectively attend to one of two event sequences projected to the same area on a video monitor.

Some of the difficulties in interpreting variations in retinal image intensity are shown in Figure 2.4, which is a computer-generated display from a recent report by Todd and Mingolla (1983). They point out that "a fundamental difficulty for analyzing the image of a complex scene is that the contours produced by opaque objects, transparent objects, texture, and shadows are completely indistinguishable within a single local region. To obtain reliable information from these higher-order properties, it would be necessary to adopt a method of analysis that makes use of the global organization of an image in addition to its locally defined gradients" (p. 394). We still do not understand how the visual system achieves such a global organization, but it now seems

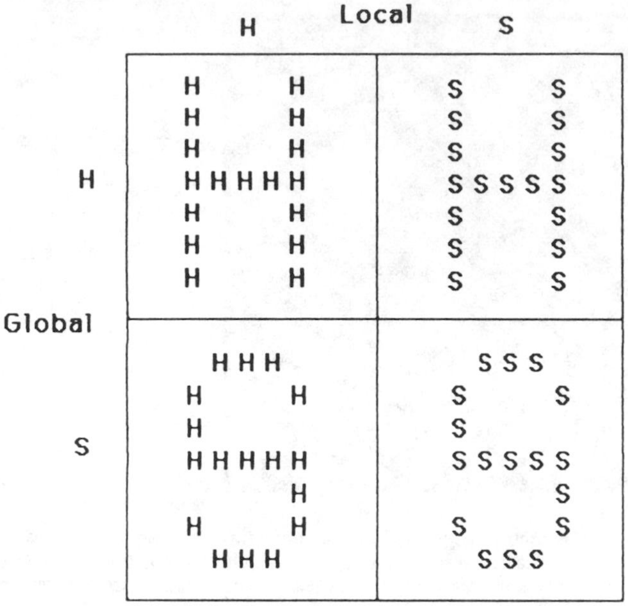

Local

	H	S
H	H H H H H H H H H H H H H H H H H	S S S S S S S S S S S S S S S S

Global

S

HHH
H   H
H
HHHHH
      H
H   H
HHH

S S S
S   S
S
S S S S
      S
S   S
S S S

Figure 2.5. Stimuli of the type used by Navon (1977) to investigate global precedence. Subjects were required to classify either the large or small letter as an "H" or "S." (From Pomerantz, 1983.)

likely that the ability to separate illumination, reflectance, shadows, and surface orientation is dependent on the presence of three-dimensional edge information that is integrated across an entire display and cannot be obtained with strictly "local" computations.

Organizational effects in perception

The inadequacy of elementary feature approaches to perception was, of course, first stressed by the Gestalt psychologists. Consider Figure 2.5, which shows a set of large forms composed of small forms. The large shape is clearly independent of the identity of the small ones and is, therefore, an emergent feature of the whole. This kind of display has a hierarchical structure in which a shape at one level can be analyzed into component shapes at a lower level. A question of interest is whether these different levels are analyzed in a fixed order or independently in parallel.

There is evidence that some combinations of elementary components result in shape discrimination that would not be predicted on the basis of the components themselves. Pomerantz (1981) suggests that these effects result from the creation of "emergent features" that are properties of the whole. Such emergent features may be directly detected as elementary features in their

DISCRIMINATION: POSITIVE vs NEGATIVE DIAGONAL

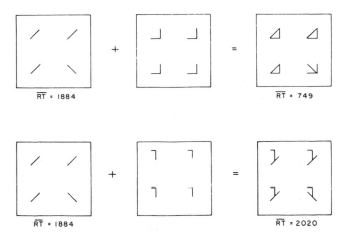

Figure 2.6. Discrimination test: The task is to locate the line segment with a different slope than the rest. In the top row, addition of the context shown in the middle panel produces the display on the right, which actually leads to an improvement in performance. In the bottom row, addition of a context impedes performance relative to the no context case. (From Pomerantz, 1981.)

own right. Indeed, in some cases, they may actually have priority over the perception of their components.

Pomerantz (1981) has introduced several candidates for emergent features as well as useful measurement techniques for deciding which combination of elementary components possesses emergent properties. Figure 2.6 shows some typical examples. The subject's task is to locate the disparate element in the display. In the top left display, this discrimination is relatively difficult, requiring 1884 msec. Adding the context in the middle panel, however, dramatically reduces this time to 749 msec, even though the context itself is uninformative as to the location of the disparate element. The bottom row shows that context is not always advantageous. In this case, an uninformative context actually makes the discrimination more difficult.

What is the difference between the two types of context effects shown in Figure 2.6? Pomerantz (1981) suggests that the addition of the right angle context produces triangles possessing the "emergent feature" of closure. This emergent feature may be directly picked up by the visual system and does not have to be synthesized from the component features. In contrast, when parts are embedded in contexts that do not produce emergent features, they may become more difficult to perceive because of masking, loss of features, and so on.

The concept of emergent features suggests that wholes may be perceived differently than their component parts because of the emergence of new features that occur when parts are combined. A different interpretation would be that some shapes are perceived holistically and are never analyzed into component parts, emergent or otherwise. Treisman and Paterson (1984) have used feature-integration theory to investigate this question. They reasoned that if closure is an emergent feature that acts like other elementary features, such as color and shape, then it should be available to form illusory conjunctions with these features. They confirmed this prediction by showing that subjects were more likely to combine lines and right angles to form illusory triangles when the display also contained circles. They suggest that the closure feature possessed by the circle provides the necessary emergent feature for perceiving a triangle. Consistent with this view is the finding that triangles are not perceived as an indivisible whole. The sides of triangles were capable of combining with an "S" shape to produce an illusory "$" figure suggesting that triangles are represented at some level as a set of component features, one of which is closure (Treisman and Paterson, 1984).

Object superiority effects

There are several other examples in which the discriminability of a component feature is improved by embedding it in an "irrelevant" context. Perhaps the best known is the "object superiority effect" reported by Weisstein and Harris (1974). Using the stimuli shown in Figure 2.7, they had subjects determine which of four possible line segments occurred either alone or embedded in a context that was designed to vary in its approximation to a three-dimensional, coherent object. Like the experiments conducted by Pomerantz, the context itself was uninformative as to the identity of the target. Increases in coherence produced increases in discrimination of the component line segment. Williams and Weisstein (1978) subsequently showed that lines in an object context could even be perceived better than lines presented in isolation.

The ability of an irrelevant context to enhance the perception of component line segments is intriguing, partly because it appears to contradict a hierarchical system that assumes that higher-order features are built from lower-order components. One might explain this effect by postulating that three-dimensional figures are processed as "wholes," independent of constituent features. Alternatively, it may be that coherent figures possess emergent features that aid discrimination. Pomerantz (1981) points out that the target and context lines in the coherent figures of Figure 2.7 form a triangle with context lines and, therefore, have the emergent feature of closure. This suggests that three dimensionality of the figure may not be the necessary condition for producing object superiority. Consistent with this position, McClelland and Miller (1979) report object superiority effects for figures that are consistently judged by subjects to be two-dimensional in appearance.

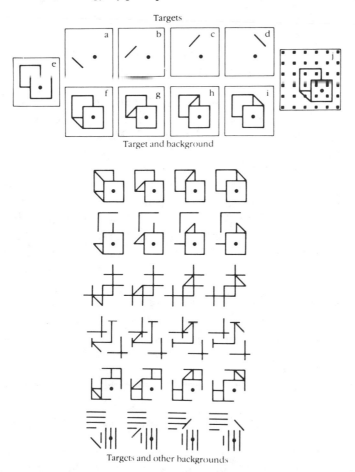

Figure 2.7. Displays used to demonstrate the "object superiority effect" of Weisstein and Harris (1974). The task is to determine which of four line segments (a)–(d) was presented. Addition of the context shown in (e) results in object displays (f)–(i). Discrimination of the target line was superior in these displays relative to the other contexts shown. (From Spoehr and Lehmkuhle, 1982.)

Forests or trees?

The context effects reviewed above raise the general issue of the order in which different levels of information should be extracted. A purely bottom–up strategy would start with elementary features and build higher-level components using suitable combination rules. A top–down strategy would use high-level features to predict the presence of lower-order components. The pervasive use of context, both in visual pattern analysis, as in the preceding examples, and in understanding speech (Miller and Isard, 1963), suggests that a top–down strategy may be an important component in human recognition processes.

Top–down strategies have also been employed in computer scene recognition work. Fu (1974) describes research aimed at developing scene grammars that would start with an overall characterization of a scene or object ("this is a face") and use this information to predict the presence of component features ("this must be an eye"). Apparent confirmation of this global-to-local processing order was provided by Navon (1977) employing stimuli such as those illustrated previously in Figure 2.5. Subjects were instructed to attend to either the large letter or its component small letters and identify the relevant form as an "H" or "S." He found that when subjects were required to attend only to the small letters, they were slow when the large letter was a different form. In contrast, the identity of the small letters was irrelevant when subjects were responding to the global shape. Navon referred to this pattern as "global precedence," suggesting that the processing of the global level inevitably preceded the local level.

The concept of global precedence is an attractive one, not only because it allows the recognition system to take advantage of the redundancy between levels of a shape, but also because it seems to find support in the spatial frequency literature. The global level of a shape could, in some cases, be primarily represented by low spatial frequency channels that could be realized by neurons with large receptive fields. These neurons have been found to have rapid conduction time, enabling them to convey gross shape information to cortical receiving areas prior to the arrival of high-resolution information. Behavioral evidence supporting this view was provided by Breitmeyer (1975), who found that subjects could detect the onset of a low-frequency grating faster than a high-frequency grating. These findings are compatible with the notion that gross shape information has a processing speed advantage over local details.

Navon's original report was followed by a number of papers that call into question the generality of global precedence. The general thrust of these papers is that the relative discriminability of the information at different levels is a more important variable than globality in determining processing dominance. Martin (1979) showed that reducing the number of local forms reduces the global precedence effect, presumably because the shape of the large letter is poorly defined. Kinchla and Wolfe (1979) found that global precedence is critically dependent on visual angle. Global precedence was obtained only when the global level was smaller than about 6° of visual angle. Forms larger than this actually produced a pattern of local precedence. Pomerantz and Sager (1975) also report local precedence. Kinchla and Wolfe (1979) suggest that processing is neither top–down nor bottom–up. Rather, form details at an intermediate size level are processed faster simply because the visual system is maximally sensitive to intermediate spatial frequencies. Hoffman (1980) directly manipulated the visibility of the two levels by distorting the global or local shapes. The undistorted level was dominant regardless of whether it was global or local.

The results show that the different levels of structure in an object are not necessarily processed in a fixed order. Information at several size levels is extracted in parallel, with intermediate sizes having an advantage because of their greater discriminability. In addition, humans have considerable flexibility in attending to different levels (J. Miller, 1981; Pomerantz, 1983; Kinchla, Solis-Macias, and Hoffman, 1983). It still remains to be seen whether these conclusions hold over a variety of other stimuli. The artificiality of the stimuli in Figure 2.5 is readily apparent, and the issue of global precedence needs to be examined with more "ecologically valid" forms. As Pomerantz (1983) has pointed out, the grouping of identical elements into a whole (the trees making up the forest) is only one instance of part–whole relationships that might reveal global precedence. We now turn to a different domain that shows, in a different way, the influence of higher-level knowledge on the extraction of component features.

The role of top–down knowledge in perception

The previous discussion of emergent features and global precedence was concerned with how different kinds of structure inherent in a visual image can be utilized in shape recognition. It is also apparent that recognition is influenced by a different kind of top–down processing: prior knowledge and familiarity. Research in speech perception provides many striking examples of how higher-level knowledge, such as syntax and semantics, is rapidly brought to bear in understanding utterances (Miller and Isard, 1963).

Top–down processing in speech perception

One specific proposal of how bottom–up information about the characteristics of the speech signal can be combined with constraints provided by grammar and semantics is provided by the HEARSAY (Reddy et al., 1973) system. HEARSAY is a computer-based speech recognition system that uses a "blackboard" to combine information provided by several different levels. For example, the bottom–up analysis of the acoustic signal may suggest a set of phonemes as candidates that can then be checked by accessing lexical knowledge about the possible words in the language. Word candidates can, in turn, be compared with syntactic and discourse level knowledge to further constrain the selection of the identity of the message. Thus, HEARSAY is a massively parallel system that uses knowledge from many domains simultaneously to interpret speech.

The specific way in which humans utilize different knowledge sources in decoding the speech signal is suggested by recent experiments on the *phoneme restoration effect*. Warren (1970) replaced the initial "s" sound in the word "legislatures" with the sound of a cough. When subjects listened to a sentence containing this modified word they reported hearing "legislatures" as an intact word with a cough superimposed. In fact, subjects were often quite inaccurate in localizing where in the word the cough had occurred. They apparently restored the missing phoneme based either on their lexical knowledge or their expectation for the word gleaned from the preceding sentence context.

Samuels (1981a, b) has recently introduced a signal detection methodology for studying the phonemic restoration effect and employed it to provide several important insights into the nature of speech recognition. The method involves presenting a subject with two different versions of a word such as "funerals." In the *replacement* version, the medial "n" phoneme is replaced by a burst of white noise. In the *added* version, the white noise is added to the phoneme. The subject's task is to determine, for a particular spoken word, which of these two versions has been presented.

The task is closely based on the perceptual experience provided by phonemic restoration. If the subject restores the missing phoneme, the replaced version should sound like the added version and therefore yield poor discrimination (indexed by low d' scores, the criterion-free measure of sensitivity provided by signal detection theory). Changes in response bias to report a particular word as being the replaced version would produce changes in β, which is a measure of response bias. The introduction of signal detection methodology thus allows one to determine whether manipulations of experimental variables affect perceptual processes or later decision level processes.

In his first experiment, Samuels (1981a) investigated the roles of both bottom–up and top–down factors in the tendency to restore missing phonemes. Bottom–up factors were investigated by examining five different classes of phonemes to be replaced: liquids, stops, nasals, fricatives, and vowels. These phonemes differed in terms of their physical similarity to the replaced sound (white noise). Stops and fricatives, for example, both contain high-frequency components that are similar to the spectral composition of the noise. If this similarity is important in restoration, deletion of these phonemes from words should produce low d' scores.

Top–down effects were represented by three variables: word frequency, location of the deleted phoneme in the word (initial, medial, or final), and word length (two, three, or four syllables). High-frequency words might be expected to offer faster access to the lexicon and, therefore, provide better restoration. Position of the missing phoneme and word length are potentially important variables according to the kind of model proposed by Marslen-Wilson and Tyler (1980). They suggest that subjects use the initial speech sounds to produce a set of candidate words (a cohort) and that later sounds are used to reduce this cohort to a single remaining choice. This model would predict that replacing final position sounds in long words would increase restoration because it is these sounds that would be subject to the greatest amount of contextual knowledge.

The results showed a powerful effect of bottom–up processes, with the degree of restoration directly related to the similarity between the noise and the missing phoneme. Top–down effects were also important. Restoration was greater for long words as well as high-frequency ones. No discriminability effects were observed with changes in phoneme position. The importance of both top–down and bottom–up processes in the phoneme restoration effect suggests that subjects will restore a missing phoneme when the replacement sound confirms an expectation for that sound suggested by the word context itself.

If subjects are using lexical representations of spoken words to restore missing phonemes, then restoration should be greater with words than non-words. Samuels showed that this was the case in a second experiment using a cue paradigm. An intact version of a word or a phonologically legal nonword was presented as a prime followed by the added or replaced version of the prime. The task, as in the first experiment, was to decide whether the sound was intact or was missing a phoneme. This discrimination was considerably poorer for words than for nonwords, suggesting that a substantial component of the phonemic restoration effect is due to the listener's knowledge of words.

An additional experiment investigated the role of higher-level knowledge sources on phonemic restoration. Words were chosen in such a way that restoration of the missing phoneme could produce either one of two different words. For example, the final syllable in "battle" and "batter" was replaced with white noise. Restoration of this missing phoneme could then yield either of the two original words. The question of interest was the effect of context on restoration.

As in the earlier experiment, replaced and added versions of the words were prepared. They were then inserted into sentences that provided either a predictable or unpredictable context. Examples of predictable and unpredictable contexts would be respectively: (1) "The soldier's thoughts of the dangerous battle made him very nervous." (2) "The pitcher's thoughts of the dangerous battle made him very nervous." Subjects were required to make the added/replaced judgment as well as a forced choice judgment between "battle" and "batter."

If semantic level information can affect the processes responsible for phonemic restoration, then one might expect more restoration (and therefore lower d' scores) with predictable sentence contexts. This is reasonable if top–down processes are capable of priming the speech representations of words on the basis of context. This priming might be expected to operate in a way similar to word frequency in making certain lexical items more available than others. In fact, the results showed just the opposite with slightly *less* restoration for predictable sentence contexts. Apparently, higher-level knowledge sources cannot pass their information "down" to the processes responsible for producing phonemic restoration. The slight improvement in d' scores with increasing predictability might reflect an increase in the attentional capacity for making the added/replacement judgment that predictability might afford. Although a predictable context failed to increase the perceptual restoration of missing phonemes, it did increase the likelihood of subjects using the response category of "intact" (an effect indexed by the β measure of signal detection theory). Top–down knowledge does exert an influence in the phonemic restoration effect, primarily through changes in decision criteria.

These findings provide important clues to the roles of top–down and bottom–up processes in the perception of spoken words. First, bottom–up processes play an important role in phonemic restoration. Restoration was much more likely for phonemes that were acoustically similar to the replacement sound. Second, top–down knowledge at the lexical level was rapidly brought to bear in encoding speech sounds as shown by the greater restoration for words than nonwords. Third, top–down knowledge of syntax and semantics is also utilized, but primarily at the decision level and not in the perceptual encoding of the speech input. The picture that emerges from these results suggests that the initial encoding of speech occurs via a lexical module (Fodor, 1983) containing representations of spoken words. The word-level entries in this module can be activated by speech sounds that meet some minimal criterion for similarity. Thus, deletion of a phoneme from a word may still produce activation of an entry in this lexicon as long as the replacement sound is sufficiently similar to the excised phoneme. Those top–down processes that are responsible for computing grammatical and semantic relationships have access only to the output of the module and cannot influence its internal operation.

The claim that there is a lexical module in speech perception is also supported by evidence from a different paradigm. Swinney (1979) showed that an ambiguous spoken word such as "bug" immediately produces semantic priming of associates of both of its two meanings ("insect" and "spy") regardless of context. After a short delay, however, only the meaning consistent with context is still available. This result suggests that the lexical module in speech consists of word representations together with links connecting associated words. Activation of a representation produces automatic spreading activation to "neighboring" items. The total set of activated words is then available to postlexical processes that select the appropriate one on the basis of context. This description is compatible with the modularity suggested by the phonemic restoration effect as well as other priming results (Seidenberg et al., 1982).

It's interesting to compare the preceding description of speech perception with the HEARSAY model. HEARSAY had a poor acoustic representation of the speech signal and tried to finesse this weakness with the heavy use of top–down knowledge at the very earliest stages of analysis. The human speech system, by contrast, can apparently afford to isolate the lexical analysis system from other kinds of knowledge, such as syntax and semantics. These latter sources definitely play a role, but only after lower-level processes have produced candidate words for integration into the ongoing discourse.

The word superiority effect

The rapid use of top–down knowledge in perception is obviously not restricted to the speech domain. A similar phenomenon occurs in reading printed words, as demonstrated by the well-known "word superiority" effect in which a letter is perceived more accurately when it is presented as part of a word than when it is shown alone (Reicher, 1969; Wheeler, 1970). As with the other examples presented of context improving the perception of more elementary components, this result is puzzling because it appears to violate an intuitive processing order in which wholes are constructed out of their constituents.

McClelland and Rumelhart (1981) and Rumelhart and McClelland (1982) have recently presented an "interactive activation" model to account for the word superiority effect. A variant of this model has also been constructed for the speech domain (Elman and McClelland, 1983) and appears capable of explaining the phonemic restoration effect as well as several other related effects. The basic structure of the model is represented in Figure 2.8. Nodes represent the feature, letter, and word levels. These nodes are interconnected by both excitatory and inhibitory links. The key assumption in this model is that activation at one level spreads to subsequent levels before processing on the earlier level is complete (McClelland, 1979; Eriksen and Schultz, 1979). This provides an opportunity for these later stages to, in turn, feed their activation back to earlier stages. This positive feedback is the principal mechanism by which the model is able to account for context effects.

Consider the temporal course of activity in this network upon presentation of the word "trip." The horizontal and vertical line feature detectors are activated, and these nodes both feed into the "t" node. Partial activation of the letter node for "t" spreads to all words in the lexicon that begin with this letter. In addition, all words that do not have this initial letter are inhibited. Activation of the word-level nodes feeds back to the letter level, further

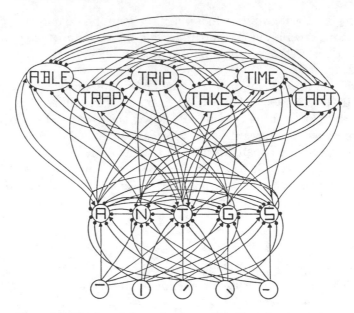

Figure 2.8. The interactive activation model. Three levels are shown corresponding to features, letters, and words. Excitatory connections are represented by arrows, while inhibitory connections are denoted by solid circles. (From McClelland and Rumelhart, 1981.)

reinforcing the activity in the "t" node. Thus nodes on the letter level receive two kinds of activation: a bottom–up component from the features that are present and a top–down component produced by the word-level knowledge. It is this latter source that produces the advantage of words over single letters.

The interactive activation model demonstrates how a parallel and interactive network of relatively simple computing elements can support higher-level cognitive activity. These networks are under active investigation as models of memory (Hinton and Anderson, 1981) and perception (Feldman and Ballard, 1983).

Scene recognition

Our final example of contextual effects in perception arises out of a consideration of what an observer does when viewing natural scenes. The eye makes approximately three to four fixations per second so that each view must be processed in about a quarter of a second. This rapid encoding might depend on the redundancy in the information picked up in each fixation, allowing the observer to use fixation n to create active expectancies about the view on fixation $n + 1$. Recent research, however, suggests that observers do not need this kind of redundancy to encode briefly presented pictures rapidly.

Potter (1975, 1976) presented subjects with a rapid sequence of 16 color photographs and required them to detect the presence of a predefined target picture. The target was defined by either showing the picture to the subject prior to the sequence or by verbally describing it (e.g., "child and butterfly"). Detection accuracy was over 60% even for the fastest rate of presentation (113 msec per picture). In addition, it made little difference whether the target picture was actually seen in advance or merely described. This latter finding suggests that rapid search was not based on the detection of a simple distinguishing feature of the target but instead was based on a "semantic" analysis of each picture. The implication is that subjects can extract the meaning of a picture in considerably less than a sec.

The claim that subjects were not using a visual matching strategy in the verbal description condition of Potter's experiments is supported by a "negative set" study reported by Intraub (1981). Subjects saw a rapid sequence of color photographs of objects drawn from a single semantic category (e.g., "animals"). The task was to detect a single picture that did not belong to that category. The members of the category were diverse in shape, size, and so on; the target did not differ in these characteristics from the rest of the pictures in the sequence. Nonetheless, detection accuracy was 79%, with a presentation rate of 1 picture/258 msec, a rate in the same general neighborhood as the fixation rate. In addition, detection was considerably above chance even at the fastest presentation rate of 1 picture/114 msec.

What properties do pictures possess that allow them to be understood so rapidly? It seems likely that one factor that fosters rapid understanding of visual scenes is that they are "coherent" at several different levels. Biederman (1972, 1981) and Biederman, Mezzanotte, and Rabinowitz (1982) have identified five kinds of relations that hold between the individual objects in a scene (see Table 2.1). Two of these, support and interposition, might be considered to play a "syntactic" role in perception because they have more to do with the physical properties of objects than their semantic relationships. In contrast, probability, position, and size refer to relations that must access semantic level information about scenes. For example, for subjects to know that a fire hydrant is out of place in a picture of a kitchen requires them to access high-level "schemas" for familiar scenes.

Biederman et al. (1982) asked subjects to view briefly presented pictures such as Figure 2.9 and determine whether a predefined target object such as a couch was present. The target object could be violating one or more of the constraints listed in Table 2.1. The question of interest was whether scene violations affected perception of the target and, if so, whether some kinds of violations were more potent than others. One reasonable prediction is that syntactic violations, such as support and interposition, would be more deleterious to object detection than semantic level violations. Interposition, for

Table 2.1 *List of relational violations and examples for a single object*

1. *Support*, e.g., a floating fire hydrant. The object does not appear to be resting on a surface.
2. *Interposition*, e.g., building in the background passing through the hydrant. The background appears to pass through the object.
3. *Probability*, e.g., the hydrant in a kitchen. The object is unlikely to appear in the scene.
4. *Position*, e.g., the fire hydrant on top of a mailbox in a street scene. The object is likely to occur in that scene but it is unlikely to be in that particular position.
5. *Size*, e.g., the fire hydrant appearing larger than a building. The object appears too large or too small relative to the other objects in the scene.

Source: Biederman (1981).

Figure 2.9. A street scene containing an object (the "Goodyear sofa") undergoing a triple violation of probability, size, and support. (From Biederman, Mezzanotte, and Rabinowitz, 1982.)

example, would be expected to impede low-level processes responsible for "segmenting" objects. Several computer programs have been created to perform this segmentation operation for collections of polyhedra (Guzman, 1968; Waltz, 1975), and they are envisioned as precursors to the stage of object recognition. A sequential processing approach would hold that any anomalies encountered at this level would forestall later analyses based on semantic relations.

The results were both clear-cut and surprising. All five relations were found to be important for object detection, but there was no clear ordering of violations on the syntactic–semantic dimension. For example, subjects were actually more likely to miss a target appearing in an inappropriate scene (probability violation) than one that was transparent (interposition violation). In addition, objects became harder to detect if a probability or position violation was added to an interposition violation, a finding that is clearly at odds with the bottom–up first order of processing.

These results show that subjects have something such as a "frame" (Minsky, 1975) or "schema" that stores information about the composition and arrangement of objects in familiar scenes. This knowledge is activated by some aspect of the visual input and serves to speed the identification of objects that are consonant with the schema. It also serves to mark those objects that are novel or unusual so that, if time permits, long-duration fixations can be directed to them (Friedman, 1979). Scene-level knowledge is apparently invoked so quickly that it can interfere with decisions about presumably more primitive relations such as whether objects violate interposition relations (Biederman, 1981).

The ability of scene coherence to improve the detectability of component objects is puzzling in several respects. How is the schema invoked if not by the identification of its constituent objects? One possibility is that a small subset of the objects is identified, and these are sufficient to activate the schema. This activation in turn improves the identifiability of other objects in the scene, including the target. Another possibility is that activation occurs through the presence of "scene-emergent features." Biederman (1981) reports an interesting experiment in which the component objects of a scene are replaced by blocks and cylinders. Even though each object is unrecognizable in isolation, the juxtaposition of these objects produces a compelling impression of the correct scene. Objects placed in these altered scenes also show the usual effects of probability and size violations. Apparently, the long-term memory representation of familiar scenes is highly schematic.

Research on scene perception suggests that the human observer can rapidly "understand" the content of each fixation in viewing normal scenes. This high-speed processing does not depend on redundancy between individual fixations (Potter, 1975, 1976; Intraub, 1981) or on expectancy (Biederman, Teitelbaum, and Mezzanotte, 1983). Instead, partial processing of the scene together with scene-emergent features may be sufficient to activate schematic knowledge about the likely component objects in a scene as well as their interrelations. This top–down knowledge is rapidly brought to bear in the identification of objects in the scene as well as serving to guide the fixation of novel or unusual areas.

Summary and conclusions

Our selective overview of cognitive approaches to perception does not reveal a coherent theory of how the brain achieves such an accurate representation of the identity and nature of visual and auditory inputs. This is hardly surprising given the complexity of the problems being solved by perceptual systems. One can, however, identify several themes that are emerging that should prove important in guiding the construction of vision models.

There appears to be a general consensus that the earliest stages in vision operate by breaking images down into relatively simple, context-independent features. There is considerable debate about the nature of these features; it remains to be seen whether they are better characterized in spatial frequency terms or as lines, angles, and so forth. It should prove especially illuminating to observe whether the physiological and cognitive paradigms converge on the same set of components.

It is also clear that many perceptual functions, such as lightness constancy, are difficult or impossible to explain in terms of the local computations provided by simple feature detectors. Instead, global processes must classify the various edges in a scene as arising from either differences in reflectance or illumination and integrate this information over the entire scene (Gilchrist, 1979). This process must take into account the shape and depth of objects as well as shadows and transparencies (Todd and Mingolla, 1983). Having access to the three-dimensional layout of objects is apparently critical to this process (Gilchrist, Delman, and Jacobsen, 1983).

Shape identification may begin as soon as critical features are detected in the image and may overlap with the actual "parsing" of surfaces into objects (Biederman, 1981). Partial processing of these features may be sufficient to activate memory representations of familiar combinations of features so that lexical knowledge of speech sounds or printed words can be used rapidly enough to enhance the perceptual identification process. Even abstract information regarding what objects are usually encountered in familiar scenes plays an important role in recognition (Biederman, 1981).

These results suggest that the perceptual system is a highly parallel collection of processing modules with interaction of knowledge at several different levels. This system gives the appearance of finding an "intelligent solution" to the problem of determining the identity and layout of the physical world (Rock, 1983). The near future should see increasing progress in specifying the algorithms of perceptual intelligence, the nature of its hardware, and its functional role in behavior.

3 The neurobiology of perception

Richard T. Marrocco

Vision is today, and has been historically, the most intensely studied vertebrate sensory system. Its importance in object recognition and visually guided behavior in people with normal visual systems is critical. Pathology within the visual system has a tremendous impact on individual suffering and the nation's economy. It is indeed a pity that the recent world economic climate has slowed vision research significantly and delayed our understanding of the visual processes.

This chapter on the neurobiology of perception attempts to identify the more important research issues in visual neuroscience of the past decade. I have focused on problems that are amenable to electrophysiological, anatomical, or neurochemical analyses and have discussed neuropsychological studies (i.e., lesion-induced behavioral deficits) only briefly. I have attempted three main objectives: (1) a current description of the anatomy and physiology of the visual pathways, with which the research issues can be understood and evaluated; (2) a contrast of the two sides of the issues and an indication of the strengths and failings of each; and (3) an attempt to indicate what I think are the problems for future research and the methodologies appropriate for their solutions.

I have chosen three highly controversial and one only slightly controversial issue. The issues are introduced here and a detailed description of them is provided later. In the first category is the question of whether visual neurons are best described by their responses to features (e.g., spots or bars of light) or by sine-wave gratings. Second, is information processing accomplished in a serial or parallel fashion? Third, are multiple visual areas in the visual cortex dedicated to the processing of a single stimulus attribute? Finally, and less controversially, is there currently a physiological basis for stimulus constancies?

This discussion completely neglects all other sensory systems. Some may criticize the neglect of other sensory areas, because perception and vision are surely not synonymous. However, the issues chosen are broad and applicable

Preparation of this chapter was supported in part by the National Science Foundation Grant BNS 82–07531.

33

to other sensory modalities. For example, multiple maps of the receptor mosaic can be seen in the cortical regions of the auditory (e.g., Suga, Niwa, and Taniguchi, 1983) and somatosensory cortices (Kaas et al., 1978), and, in both cases, there is evidence that the maps may be dedicated to the coding of different stimulus attributes. Stimulus constancies are present in auditory pitch perception (e.g., Geldard, 1972). In addition, parallel processing is an accepted fact of the tactile system (e.g., Bowsher, 1965), and the auditory system is a Fourier (frequency)-analyzing device. One could also argue for auditory feature detection, in the sense that species-specific calls, frequency-modulated tones, etc., represent groupings of simple stimulus features into biologically important patterns (e.g., Barlow, 1972). In any event, it is clear that the resolution of these issues for other sensory pathways will provide valuable lessons for visual physiologists. In the next sections, I have presented a brief description of the methods used in visual neuroscience and the current understanding of the anatomical and physiological basis for the visual processes.

Methods in visual neuroscience

The visual system can be characterized in terms of pathways originating in ganglion cells in the retina and projecting primarily, though not exclusively, to either the lateral geniculate nucleus (LGN) of the thalamus or to the superior colliculus (SC) or optic tectum of the midbrain. The LGN, in turn, projects to primary visual cortex, which sends connections back to the LGN (corticogeniculate pathway) and forward to other cortical visual areas. In contrast, SC projects to the pulvinar, a second major visual area in the thalamus. The pulvinar, in turn, projects to the visual cortex. Several pathways are omitted in this brief overview, and the detailed structure and function of these pathways is considered below.

Electrophysiological recording

One of the main goals of the neurobiologist is to discover the ways in which single cells or groups of cells in a particular animal species code information about visual stimuli. In order to study the codes it is necessary to place an electrode into the vicinity of a neuron of the central or peripheral nervous system. Animals are anesthetized systemically to prevent painful sensations from reaching consciousness. For central nervous system neurons, a small incision is made in the scalp, and a hole is drilled into the skull. The brain's meningeal covering is also cut, and the electrode is inserted into the brain.

The experimental animal of choice depends, of course, on the question to be answered. Our own work seeks to understand human vision and therefore relies on primates, which have visual systems nearly identical to those of humans.

Responses of nerve cells are recorded with metal or glass microelectrodes, which can sense voltage changes in volumes of tissue about the size of single cells. These changes, or action potentials, are amplified and made useful to the experimenter by displaying them visually on an oscilloscope, making them audible by connections to a loudspeaker, and shaping them for analysis by a computer. Most visual neurons code information through a frequency modulation of action potentials. The neuroscientist attempts to develop a quantitative description that relates changes in stimulus parameters to the frequency of action potentials.

The retina as a whole is a heterogeneous structure, parts of which differ in terms of individual cell properties and in the ways that different cells are connected together. Therefore, the stimuli presented to the eye must be precisely positioned, and the position of the eye must be rigidly fixed. In some experiments, position control may be achieved by training the animal to fixate the visual target. In others, drugs called neuromuscular blockers are used to prevent eye movements. In each case, eye position is checked to assure that the retinas have remained immobile during successive stimulus presentations.

Stimulus focus is also crucial. Whereas the act of fixation by the behaving animal assures that the images of the stimuli are in focus on the retinas, supplementary external lenses are required in the anesthetized animal.

In addition, the composition of the stimulus must be precisely controlled. For example, to study how the visual system codes for stimulus wavelength or color, the wavelength must be varied by small amounts in each presentation, while all other stimulus attributes, such as saturation and brightness, are held constant.

Receptive fields

The earliest study of the mammalian visual system was reported by Kuffler in 1953. He described the responses of single cat retinal ganglion cells to small spots of light. The area of the retina that, when stimulated, was capable of influencing the frequency of action potentials was called the cell's receptive field. Receptive fields consisted of small, circular, central regions and ringlike, or annular, surrounding regions. In about half of all cells, stimulation of the center evoked an excitatory response (i.e., an increase in the action potentials), while stimulation of the surrounding region evoked an inhibitory response (i.e., a decrease in the action potential frequency). In the other half, the reverse pattern was seen. The dimensions of these fields depended on a number of factors, including the ambient level of illumination and the types of stimuli used to stimulate the cells. For the monkey, ganglion cell receptive fields in the foveal region are about 25–40 μm in diameter, and they are slightly larger in cat.

More recent work suggests that some receptive fields may be composed of a series of concentric regions. Some have suggested that the polarity of the zones alternates regularly; others have suggested the presence of a large, inhibitory zone surrounding the receptive field surround. These zones, in most cases, may increase the functional sizes of the receptive fields by a factor of two. In contrast, McIlwain (1964) showed that in some cells of the cat retina, movement of irregularly shaped patterns as far away as 30° of arc from the receptive field could alter the cell's activity. Thus, the sizes of receptive fields can depend greatly on how they are measured.

Methods of visual stimulation

The earliest studies of retinal and central visual system structures used geometric stimuli, such as small light and dark spots. It was soon clear that while the neuron algebraically summed together the responses of the center and surround, the center was usually dominant. Thus, responses to thin lines or to stimuli that extended well beyond the receptive field borders were possible. Colored stimuli were used extensively as well. It was discovered that while the cat possesses only a rudimentary apparatus for distinguishing different wavelengths of light, the monkey visual system has a much more advanced retina that is exquisitely sensitive to color. A more detailed description of cat and monkey ganglion cell responses will be presented below.

For cells of the retina or LGN, the orientation of the stimulus with respect to the receptive field is unimportant. However, some cells of the visual cortex (simple or complex cells) require stimuli of a particular orientation (i.e., responding to a bright horizontal, but not vertical, edge or line). Others (hypercomplex cells) will only respond if the target is of a specific length and width. Still others have been reported to respond only to the picture of a monkey's hand! Thus, the visual system can be explored using stimuli that are simply made and easily presented with slide projectors. In general, the more central the neuron, the more complex its requirements for visual stimuli.

In recent years, many visual neuroscientists have abandoned geometric stimuli for sine-wave grating stimuli, long used by optical engineers to test the quality of lenses. This stimulus resembles a series of defocused light and dark bars. The brightness of the bar at any point varies sinusoidally in a direction perpendicular to the orientation of the bars. The "luminance profile" is thus said to be sinusoidal. Sine waves are specifically chosen because, when combined according to the theorem of Fourier, they can be synthesized easily to approximate the shape of any wave form. In addition, any wave form can be analyzed into its Fourier or frequency components. Synthesis and analysis are the reverse operations, or mathematical transforms, of each other.

Sine-wave stimuli have been used to stimulate neurons visually. The sensitivity of a visual cell to sine-wave gratings of different widths, or spatial frequencies, is called the neuron's contrast sensitivity function. It is constructed by determining the contrast of each of a series of sine-wave stimuli that, when presented to the receptive field, evoke threshold responses. The beauty of this approach becomes clear when it is appreciated that the contrast sensitivity function and luminance sensitivity of the cell's receptive field at a given position are inversely related. Moreover, the contrast sensitivity function completely specifies the spatial properties of the receptive field. A more complete description of this methodology is given below. The advantages and disadvantages of these approaches to the neurobiology of vision are also discussed in detail in a later section.

Neuroanatomy

The second major approach to be heavily exploited in neurobiology in recent years is neuroanatomical. Two major techniques have been developed: (1) the horseradish peroxidase (HRP) method, which is used primarily for identifying the locations of the cell bodies, or somata, of neurons that project to a known area of interest; and (2) the autoradiographic (AR) approach in which radioactively labeled amino acids are used primarily to discover the axonal terminations of cells, the somata of which reside in known locations. The methods depend on intra-axonal transport of the HRP enzyme toward amino acids and other organic molecules away from the cell body.

The labeling material is injected into tissue a few days before the animal is perfused. The HRP tissue is then reacted with a chromagen, making the enzyme visible. In AR, the labeled proteins or enzymes within cells are invisible. Thin sections of tissue must be placed onto photographic emulsion-coated slides. The radioactivity then exposes the emulsion and produces an image of the labeled cellular processes.

Both of these techniques have virtually unlimited application in neuronal tissue and are considerably more sensitive than previous methods of tracing neuronal connections. The HRP technique, in addition, may allow the size and shape of the soma to be seen. Descriptions of results obtained with these methods will appear in several subsequent sections.

Anatomical and physiological substrates of vision

Retina

This overview will focus primarily on neural structures. It will not discuss such nonneural operations as the refraction or absorption of light by the intraocular media, and only a brief description is given of retinal photochemistry. The interested reader is referred to Rodieck (1973) for a comprehensive summary of these topics.

The transducers for light lie at the back of the eye behind a vascular network and all other retinal elements. All retinal cells interposed between the light and the receptors are almost transparent. The first substantially absorptive cells, the rod and cone photoreceptors, contain highly sensitive photopigments. Mammalian cones have three varieties of pigments differing in the spectral sensitivities: peaks at 420, 540, and 556 nm have been reported for macaque cones by Bowmaker, Dartnall, and Mollon (1980). Rods have a single photopigment absorbing at 502 nm.

Absorption of light by the rod pigment results in a complex chain of photochemical events, which in turn cause electrical changes in the cell membrane. These changes in the membrane voltage are communicated to secondary neurons by modulation of the release of chemicals known as neurotransmitters. There is good evidence that, in the dark, some vertebrate photoreceptors continuously release a substance known as dopamine onto bipolar and horizontal cells, causing the electrical charge in these cells to be changed as well (e.g., see Kaneko, 1979 for a review). Changes in the membrane voltage caused by an increase in the electrical negativity of the cell's membrane are called hyperpolarization. The changes caused by a decrease in negativity are called depolarization.

The electrical responses of receptors and horizontal cells to light are hyperpolarizing only. Bipolar cells, however, can be hyperpolarized or depolarized, depending on where light stimuli fall on the retina. The bipolar cell's receptive field has center and surround regions as those described earlier for ganglion cells. Direct receptor to bipolar connections mediate center responses, while receptor connections with horizontal cell and then to bipolar cell mediate surround responses. In fact, the receptive fields for ganglion cells are simply elaborations of the bipolar cell receptive fields.

A bipolar cell that hyperpolarizes to light flashed in its field center is said to be an "off" center type. A second type of bipolar cell (the "on" center) shows the reverse response.

Horizontal cells are said to be laterally transmitting because they mediate interactions between light stimuli in adjacent retinal regions. The other laterally transmitting neuron, the amacrine cell, is actually a diverse group of cells, all of which have very large receptive fields. There are morphologically and physiologically distinct varieties of amacrine cells in some species of fish (Kaneko, 1973; Murakami and Shimoda, 1977). In addition, there are a half-dozen types of amacrine cells that are distinguishable by the types of neuropeptides they contain (Brecha, 1983), and there is yet another amacrine cell that uses dopamine to affect horizontal cell activity of the outer retinal layer (Dowling, 1979). At least five transmitter substances have now been localized in amacrine cells as well (Brecha, 1983).

Amacrine cells make frequent lateral contacts with other amacrine and ganglion cells, forming a large spatial network of neurons (e.g., Dowling, 1979). This may be the basis for the large size of the ganglion cell surround. The inclusion of amacrine cells in the inputs to ganglion cells (Werblin, 1970) may also provide the latter with Y-like properties (e.g., transient responses, responding at stimulus onset and offset). However, not every ganglion cell receives inputs from amacrine cells. Those ganglion cells that do not contact amacrine cells probably belong to the X class (see section on projections of the retina). Ganglion cells are also of the "on" and "off" types corresponding to the type of direct bipolar cell input.

Using light and electron microscopy, ganglion cells can be grouped into a number of morphological types (e.g., Ramon y Cajal, 1909; Dowling and Boycott, 1966), but the extent to which these represent functionally important groupings is unclear. More recently, however, the question of types has been reexamined by injecting horseradish peroxidase (HRP) into known central visual nuclei (Bunt et al., 1975; Leventhal, Rodieck, and Dreher, 1981). Axons in these nuclei transport the enzyme retrogradely to the somata and dendrites of ganglion cells, where it can be visualized. The question is whether the location of the projection determines the structure of the cell body. Six morphological types of neurons have been identified in this work. β Cells have small cell bodies, very narrow dendritic fields, and project to the lateral geniculate nucleus (LGN). α Cells have large cell bodies, wide dendritic fields, and project to the LGN and superior colliculus (SC). γ Cells have small somata, large dendritic fields, and project to the SC and pretectum. ε Cells look like γ cells but project to the inferior pulvinar. Two other varieties did not conform to any of the above patterns.

Using extracellular recording techniques, de Monasterio (1978a) has identified six functional classes of ganglion cells: (1) Type I cells have concentric fields and are color opponent. (2) Type II Cells have coextensive fields and are also color opponent. (3) Types I and II are probably the β cells described above. (4) Type III cells are spatially concentric, with broad band spectral sensitivities and are probably a mixture of α and β cells. (5) Type IV cells, intermediate in character between types I and III, are probably α types. These have a rudimentary color sensitivity produced by an imbalance in the spectral mechanisms between center and surround. (6) Types V and VI are broad band cells that require moving visual stimuli and are likely to be γ cell types.

The functional classes of retinal cells can be grouped further in several different fashions. Briefly, there are three major schemes based on (1) tests of spatial summation, (2) response vigor, and (3) a battery of tests including conduction velocity, receptive field dimensions, response time course, etc.

Spatial summation tests show whether a neuron's receptive field sums light signals algebraically (X type) or nonalgebraically (Y type). Response vigor tests are based on whether a neuron's response rate to stimuli is high or low, somewhat of a subjective judgment. In the battery of tests, response time course refers to whether a cell responds during the entire stimulus duration (sustained response) or adapts rapidly (transient response). Of the three, the most foolproof is that of spatial summation. Linear summation is so radically different in design that nonlinear summation has to represent a fundamental difference in function. In contrast, only quantitative differences along the continua separate cells in the other schema. A detailed description of these is presented in the section on parallel versus serial processing.

Projections of the retina

The projections of the retina are numerous, and all of the types of cells projecting to each central location have yet to be discovered (e.g., see Rodieck, 1979). All vertebrates have the following central projections: (1) hypothalamus, (2) accessory optic system, (3) pretectum, (4) tectum (superior colliculus), (5) ventral thalamus, and (6) dorsal thalamus. Since the pathways through both the tectum and dorsal thalamus eventually reach the visual cortex, they are considered to be in parallel. However, cells within either of these parallel pathways are connected to each other in a serial fashion. The importance of parallel and serial forms of organization is discussed more fully below.

According to Rodieck (1979), X cells (types I and II color-opponent variety) innervate the hypothalamus, pretectum, and dorsal thalamus (LGN). Y Cells (types III and IV) innervate the accessory optic system, pretectum, tectum, and dorsal LGN. W Cells (types V and VI) innervate the pretectum, accessory optic system, tectum, and ventral and dorsal LGN. In primates, Leventhal et al. (1981) have recently identified two other neurons, one of which apparently projects directly to the pulvinar (dorsal thalamus), and the other has unknown projections. This chapter will focus only on the two major projections to the LGN and tectum.

As mentioned previously, there are approximately six varieties of retinal ganglion cells and an equal number of morphological types identified in HRP studies. There is some evidence for biochemically distinct varieties of ganglion cells as well. For example, some may contain the neuropeptides somatostatin and substance P (Brecha, 1983). The exact biological role of retinal neuropeptides and the identification of the cellular machinery associated with the production of classic transmitters are yet to be determined.

Lateral geniculate nucleus

The monkey LGN offers the clearest view of parallel processing in the visual thalamus. Although comprised of six major layers [four small or parvocellular (P) layers and two large or magnocellular (M) layers], some cells lie in between these layers (interlaminar cells) and others cluster into small groups beneath the ventral most layers [S laminae (Kaas, Guillery, and Allman, 1972)]. Table 3.1 summarizes the properties of P and M cells in the monkey LGN. The most dramatic differences between the P and M laminae are spectral tuning and contrast sensitivity (Marrocco, McClurkin, and Young, 1982a; Kaplan and Shapley, 1982). Recently, the distinction between P and M layers has been made histochemically as well. Butyrylcholinesterase, a potential precursor of acetylcholinesterase (AChE), is found almost exclusively in the P layers, while AChE is found preferentially in M layers (Graybiel and Ragsdale, 1982). In general, however, the differences between P and M laminae listed in the table appear gradually as an electrode descends through the LGN from dorsal to ventral (Marrocco et al., 1982a). Surprisingly, there are no morphological differences between cells in the P and M layers other than soma size and dendritic orientation (Wilson and Hendrickson, 1981). There are, however, four times more interneurons in M than P layers (Hamori, Pasik, and Pasik, 1983).

The LGN not only receives retinal afferents but also inputs from the tectum, pontine, mesencephalic and thalamic reticular formations, the dorsal raphe nucleus, locus coeruleus, and the striate and prestriate cortices. The input from the striate cortex is massive, and one estimate places the number of cortical inputs equal to the number of retinal inputs (Beresford, 1962). Several studies (Karlsen and Fonnum, 1978; Baughman and Gilbert, 1979) have indicated that the cortical cells that provide the input preferentially accumulate glutamate and aspartate, suggesting that these may be the neurotransmitters that affect LGN cells.

Several laboratories have recently examined the physiological action of the corticogeniculate pathway, but with somewhat conflicting results (e.g., Kalil and Chase, 1970; Richard et al., 1975; Baker and Malpeli, 1977; Geisert, Langsetmo, and Spear, 1981). Our recent results have shown that the cortical modulation of LGN relay cells can be excitatory or inhibitory, depending on the manner in which the cortex is stimulated. There seems, in fact, to be an antagonistic spatial structure in the cortical outflow resembling the shape of a large receptive field. Our results suggest that several cortical neurons influence a single LGN cell. Thus, the receptive field analogy is only approximate.

We, and others, have further shown that the cortical influence alters the responsiveness of the LGN cell center, or surround, or both (Geisert et al., 1981; McClurkin and Marrocco, 1984). As a result, the effective sizes of the

Table 3.1. *Properties of monkey LGN cells*

	Parvocellular (P) laminae		Magnocellular (M) laminae	
	X-like	Y-like	X-like	Y-like
Proportion of total	97%	3%	60%	40%
Conduction latency	Long	Short	Short	Short
Soma size	Small	Small	Large	Large
Dendritic orientation	Parallel to laminae	Parallel to laminae	Perpendicular to laminae	Perpendicular to laminae
Spectral tuning	Color opponent	Broad band	Color opponent and broad band	Broad band
Spatial tuning	Low pass to bandpass	Low pass to bandpass	Low pass to bandpass	Low pass to bandpass
Summation mode	Linear	Nonlinear	Linear	Nonlinear
Temporal tuning	Low frequency	High frequency	High frequency	High frequency
Movement sensitivity	Low	?	High	High
Contrast sensitivity	Low	Low	High	High

receptive field response mechanisms change, thereby altering the spatial selectivity of the neuron (McClurkin and Marrocco, 1984). In many cells, the
temporal selectivity and, in a few cells, the chromatic selectivity are also
changed. These results suggest a reevaluation of the function of the LGN. It
can no longer be considered simply as a relay nucleus, doing little for information processing beyond that already accomplished by the retinal input.
Rather, LGN responses reflect a fixed retinal and a variable cortical input,
the latter determined by the spatial pattern of visual stimulation.

Pulvinar

The second major pathway through the thalamus originates in the retino-
pulvinar projection. The vast majority of ganglion cells in this system synapse
first in the superficial layers of the tectum (Marrocco and Li, 1977). Tectal
cells in turn project to the inferior and lateral pulvinar. While both W and
Y ganglion cells project to the tectum, only W cells are relayed to the inferior
pulvinar (Marrocco, McClurkin, and Young, 1981). In addition, some ϵ-type
cells having small somata project directly to the inferior pulvinar (Leventhal
et al., 1981).

The pulvinar complex is extensively and reciprocally interconnected in a
visuotopic manner with many visual cortical areas (e.g., Spatz, Tigges, and
Tigges, 1970; Benevento and Rezak, 1976; Ogren and Hendrickson, 1979;
Berson and Graybiel, 1983). Moreover, the number of fibers in the cortical,
in contrast to the subcortical, input suggests that the response properties of
pulvinar neurons may more mimic the former than the latter. Recent electrophysiological studies of pulvinar neurons tend to support this (Gattass,
Oswaldo-Cruz, and Sousa, 1979; Bender, 1982; McClurkin, 1984). Many cells
are directionally and orientationally selective and have multiple discharge
centers that are characteristic of prestriate but not tectal neurons.

The most dramatic differences between pulvinar and cortical neurons are
receptive field size and breadth of orientation and directional tuning. Receptive field sizes of striate neurons are approximately 1° arc in width; those of
pulvinar cells average about 10° and may be as large as 60°. Half-peak bandwidths for orientation tuning in striate neurons is about 20° (De Valois,
Albrecht, and Thorell, 1982), while pulvinar cells average about 120°
(McClurkin, 1984). It is indeed curious that many prestriate neurons have
the degree of specificity characteristic of striate neurons, when subjected to
such dense innervation by pulvinar efferents. However, few studies have
specifically tested for visual responses from the receptive field periphery in
striate or prestriate neurons. It may well be that the suppressive fields reported
for V4 neurons (Schein, Desimone, and de Monasterio, 1983) in fact represent
influences from very large receptive fields of pulvinar cells.

At present the function of the pulvinar is uncertain. The coarse structure of the receptive fields suggests that it does not analyze stimulus form, at least in the "epicritic" fashion characteristic of striate neurons. Neuropsychological studies in lesioned primates suggest that the nucleus may be involved in moving attention to different parts of visual space (Ungerleider and Christensen, 1977). The pulvinar appears not to be involved in the initiation of eye movements, as the cells respond after saccades and are silent during movements in the dark (see Keys and Robinson, 1979). Therefore, they are not likely to be involved in the disengagement of fixation or in the programming of the saccade itself. However, they may be involved in refixation, which necessitates the termination of the activity evoked by stimulation of the previous site of fixation. Pulvinar cells could inhibit cortical neurons to prevent them from responding to the new peripheral stimuli brought about by the saccade (i.e., saccadic suppression).

The idea of a saccadic suppression is not new, but its application has usually been limited to the geniculostriate pathway (e.g., Noda and Adey, 1974). It is difficult to imagine how the punctate receptive fields found in this pathway could provide the widespread inhibition apparently required by the spatial character of the suppression (see Brooks and Fuchs, 1975). It might be possible to test the saccadic suppression idea by simultaneous recordings in alert monkeys from pulvinar and cortical neurons whose fields are located in the same part of visual space. Clearly, more data are needed.

Striate cortex

Both the dorsal LGN and pulvinar project to the visual cortex. The visual cortex occupies almost the posterior one-third of the cerebrum and is composed of the striate and prestriate areas. The striate area receiving primarily the afferents from the LGN is situated in the posterior pole of the occipital lobe. The prestriate cortex lies anterior to the striate area and receives primarily pulvinar and striate afferents. The visual cortex contains six laminae that lie parallel to the cortical surface. Some of these laminae contain sublaminae. The uppermost lamina is made up mostly of fibers, while the lower five layers are each defined by clusters of cells. The input and output to the cortex tend to be located in different laminae.

Few areas in the visual system have been as vigorously studied in recent years as the striate cortex. If forced to choose two papers of seminal influence on our current understanding of this area, I would choose Hubel and Wiesel's 1962 work on functional architecture and the anatomical studies of Lund et al. (1975) of the cortical laminar organization.

Hubel and Wiesel discovered that the shape requirements for cortical neurons were very different than those of more peripheral cells. They defined three main types (simple, complex, and hypercomplex) and showed that these types were roughly clustered in laminae parallel to the cortical surface. Simple cell receptive fields can be divided into zones of excitation and inhibition. They require oriented stimuli of the appropriate width, are insensitive to length, and are found in the middle laminae. Complex cell receptive fields lack spatially separate zones, require oriented stimuli, are less length- and width-sensitive than simple cells, and are found in the more superficial and deep laminae. Hypercomplex cells are length-, width-, and orientation-sensitive and reside in the same laminae as complex cells. In addition, Hubel and Wiesel developed the notion of ocular dominance and orientation columns, which lie perpendicular to the cortical surface, contain cells driven predominantly by one eye, and have similar orientation preferences. A complete set of orientation ocular dominance columns is termed a hypercolumn.

The columnar structure can be demonstrated with electrophysiological and neuroanatomical techniques [e.g., autoradiographic labeling of ocular dominance columns (Hubel, Wiesel, and LeVay, 1977), or metabolic activity labeling of orientation columns by 2-deoxyglucose (Hubel, Wiesel, and Stryker, 1977)]. In addition, others have presented evidence that a columnar structure also exists for the dimensions of spatial frequency (Tootell, Silverman, and De Valois, 1981, but see Maffei and Fiorentini, 1977), binocular disparity (Blakemore, 1970), and color (Michael, 1981; Livingstone and Hubel, 1984). None of these attributes has been as well documented as the original. Assuming, however, that they are indeed present, the hypercolumn becomes a functional module containing nearly all of the relevant stimulus attributes necessary for an analysis of a small patch of visual space.

The other major recent work has clarified the laminar nature of the cortical input/output pathways (Lund et al., 1975, 1979). Using radioactive amino acids and HRP tracers, Lund and her colleagues have shown that LGN terminals end in cortical layers 1, 2, 3, 4A, 4Cα, 4Cβ, and 6. By far the largest input is to layers 4A and 4C. Afferents from the pulvinar terminate in layers 1 and 2. The output from the cortex to the LGN and pulvinar superior colliculus is from layers 6 and 5, respectively. Intracortical projections originate from layers 3, 4B, and 6. There seems to be a tendency for the input and output to be localized in different laminae.

Lund and others have also shown that the middle cortical laminae are populated by stellate neurons whose axons project superficially as well as deeply to the remaining cortical laminae, except layer 1. Is there a relationship between cell morphology and receptive field type? The answer is a qualified "yes." Many cells in the middle layers tend to have "simpler" receptive field types (Kelly and van Essen, 1974; Gilbert and Wiesel, 1979). That is, the majority of stellate neurons are simple cells, but some are complex as well. Most pyramidal cells, located in all the remaining cortical laminae except layer 1, have complex receptive fields, but some are simple cells as well. Since both simple and complex cells receive geniculate afferents (see section on serial versus parallel processing), the early stages of cortical processing appear to be done in association with stellate neurons, while the output to the other cortical and subcortical cells is carried out by pyramidal neurons.

Much of the recent microelectrode research on receptive field organization is relevant to the issue of serial versus parallel processing, and a discussion is deferred to that section.

Prestriate cortex

The prestriate (or extrastriate) cortex occupies the cortical surface area anterior, lateral, and superior to the striate cortex. Once believed to be cytoarchitectonically homogeneous, this region is actually a collection of about a dozen different areas (Figure 3.1). Much of this region was previously called the visual association area (e.g., see Polyak, 1957) because it did not receive direct inputs from the sensory thalamic relay nuclei. Rather, it communicated with the "primary sensory areas" via short chain "associational" (Meynert) neurons. The prime function of these areas was thought to be multimodal, that is, they showed signs of visual, tactile, and auditory convergence. We now know that there are (1) direct afferents from thalamic sensory relay as well as nonsensory thalamic nuclei to the prestriate areas and (2) that the information processed within many of these regions is primarily visual.

Recent studies have focused on three areas: (1) the physiological differences that accompany the cytoarchitectonic distinctions between areas; (2) the interconnectivities between the functionally distinct areas; and (3) the topography of the visual maps within each area. This chapter will concentrate on the macaque monkey, since the greatest number of similarities are likely to exist between the gyrencephalic cortices of Old World monkeys and humans.

In general, the amount of prestriate cortex represents a larger fraction of the visual areas than the striate in higher mammals, while the reverse is true for lower species. This suggests that the more complex visual functions characteristic of primates may be associated with the prestriate areas.

Van Essen (1979) and van Essen and Maunsell (1983) have recently proposed a cortical hierarchical organization based primarily on the contributions of geniculate and striate inputs to each prestriate area (see Figure 3.2). While there are some inconsistencies between papers with respect to the locations of areas V3 and MT, the models are otherwise in agreement. The first level structure is V1, which receives thalamic input; the second is V2, whose primary input is from V1. V2 projects to V1 and other prestriate areas. The third level is composed of areas receiving V1 input and substantial inputs from other cortical visual areas. This level includes V3 and ventral posterior

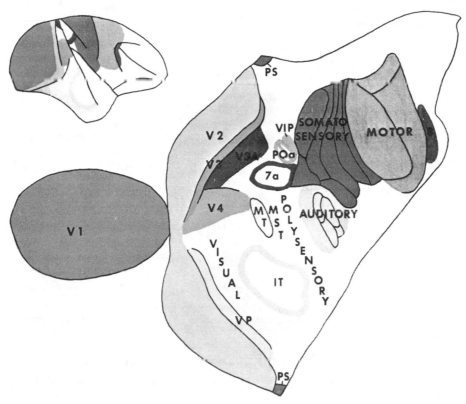

Figure 3.1. Cortical visual areas in the macaque monkey, shown in a lateral view of the right hemisphere. The picture has been drawn in such a way as to eliminate the complex foldings of the cortical surface, allowing the areas within the sulci to be visualized. Abbreviations for the cortical areas: MT, middle temporal; MST, medial superior temporal; VIP, ventral intraparietal; VP, ventral posterior; IT, inferotemporal; POa, Brodmann's parietal area; PS, prostriate area. (From van Essen and Maunsell, 1983, with permission of the authors and Elsevier Biomedical Press.)

(VP) cortical area. The fourth level includes areas V4 and middle temporal area (MT); the fifth areas are 7a, medial superior temporal (MST), ventral intraparietal (VIP), Brodmann's parietal area (POa), and inferotemporal (IT); the sixth and last area at present includes only area 8, the frontal eye fields. Thus far, all connections seem to be reciprocal.

There have been a large number of studies concerned with cellular function in prestriate cortex beginning with those of Hubel and Wiesel's (1965) paper on areas V2, V3, and the Clare–Bishop area in cat. I shall postpone the description of these papers until later sections, since they form the core of the major issues on the properties of individual areas in prestriate cortex.

Topographic maps have been determined for several prestriate areas and are something of a topological delight to study. Neighboring areas may be

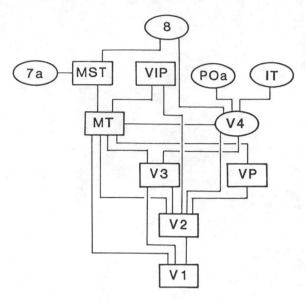

Figure 3.2. Hierarchical organization of the striate and prestriate cortices. For abbreviations see Figure 3.1. (From van Essen and Maunsell, 1983, with permission of the authors and Elsevier Biomedical Press.)

related to each other in simple or complex ways. For example, area V1 shows a first-order transformation of the visual field, while V2 has field discontinuities, making it a second-order transformation (e.g., Allman and Kaas, 1974; van Essen, Maunsell, and Bixby, 1981). Recently, MT has been shown to have a "mosaic representation" (van Essen et al., 1981). A more complete description of the physiology of the prestriate cortices is reserved for the section on individual areas in prestriate cortex multiple visual maps. The following sections will contain the discussion of the major theoretical issues in the neurobiology of perception.

Features versus frequency?

The issue of feature versus frequency can be divided into two separate issues: First, which is the best stimulus for visual receptive fields? More formally stated, what visual stimulus provides a luminance profile that most closely matches the sensitivity profile of the receptive field? Second, how does the visual brain combine neuronal responses to these stimuli to construct an image of the visual world? It is hard to imagine a topic more fundamental to pattern recognition than this, and it is unlikely that any single area in visual neuroscience has generated as much research as the opposing camps in the feature versus spatial frequency issue.

Feature theory suggests that the best stimulus for some visual neurons is one that evokes a vigorous response and is related to the shape of objects in the organism's environment. Pattern recognition occurs when a central set of cells receives inputs from a group of "object" neurons in a pattern that matches stored perceptual templates.

In contrast, frequency (spatial frequency, Fourier) theory affirms that the best stimulus evokes vigorous responses because its luminance profile best matches the sensitivity profile of the receptive field (e.g., see Watson, Barlow, and Robson, 1983). No similarity with environmental shapes is necessary. Neurons in the brain analyze visual stimuli into their underlying spatial frequencies. Sine waves have been used because of their mathematical simplicity and ease of generation, but other functions (e.g., cosine, Gaussian) could in theory be used as well. The unique combination of amplitudes, frequencies, and phase signals from each neuron at each point in visual space is then synthesized into recognizable patterns, although how this is done remains to be theorized.

In discussing this topic, it is of utmost importance to keep in mind several points. First, the theories differ dramatically in their formality. Feature theory, as originally posited, is informal and loosely organized. Frequency theory, in contrast, is formal and quantitative. Second, feature theory was restricted in scope and originally meant to describe only the first stage of processing within the striate cortex. In contrast, frequency theory, while originally used only to describe psychophysical experiments on sine-wave grating visibility (e.g., Campbell and Robson, 1966), has been extensively applied to all types of neurons from the retina (e.g., Enroth-Cugell and Robson, 1966) to area V4 in prestriate cortex (Schein et al., 1983). In fact, the use of sine-wave gratings has become as frequent as the use of bars and edges in visual testing of receptive field organization.

Feature theory

Historically, some of the most influential studies on feature detection were reported by Lettvin and colleagues in 1959. These investigators concluded that the retina of the frog contained neurons ideally suited for the detection of small, moving, dark objects, most likely bugs. Other later studies were less "adaptive" in their interpretations of optimal receptive field stimuli. No one seriously proposed that evolutionary pressures were responsible for an LGN cell's responsiveness to light and dark spots of light. In contrast, Hubel and Wiesel (1962) showed that geometric objects, such as lines, edges, and angles, were more efficient stimulus features than circles for cortical neurons. The adaptive value of detectors for such shapes seemed more straightforward.

The feature model was epitomized in interpretations of the work of Gross,

Rocha-Miranda, and Bender (1972). These investigators reported that some neurons in monkey inferotemporal (IT) cortex responded best to pictures of an appropriately oriented monkey hand. The same report, however, showed that the "hand" cells responded (less vigorously) to other stimuli as well.

Evaluation. Visual sensory neurons respond to many changes in stimulus attributes, for example, wavelength, intensity, saturation, orientation. Any given discharge frequency could be associated with an infinite number of different combinations of these attributes. Thus, there is no unique meaning to a particular discharge rate. Patterns within the discharge train might be useful, but none have been conclusively demonstrated in the visual system. To my knowledge, no violation of these facts has ever been substantiated for any sensory neuron. In principle, it seems that sensory neurons operate according to a "law of multiple response determinants." This law suggests that the breadth of tuning and feature specificity are incompatible.

Recently, Ewert (1980) has concluded from his extensive investigations of toad vision that no specific detectors exist in the retina or optic tectum. Some neurons had small receptive fields and were likely to respond to preylike objects. Others have large fields and orientation preferences that made them suitable for detecting non-prey-like objects. Finally, still others had very large receptive fields most suitable for the detection of predators. There was a strong correlation between activity in the small and large receptive field cells and approach behavior, and between activity of the very large field cells and avoidance behavior. In no case, however, did the response rate of any neuron uniquely specify a particular stimulus shape. This last finding, in the very species that gave birth to the feature model, would seem to be anathema to the feature model. Finally, Schwartz, Desimone, Albright, and Gross (1985) have argued cogently that IT neuron responses can be quantitatively analyzed without any reference to features.

Frequency theory

Frequency theory has arisen from the application of the theorem of Fourier to human vision (e.g., see Campbell and Robson, 1966). According to this view, the visual world produces retinal images whose intensity profiles can be viewed in Fourier terms (i.e., as a sum of sine waves of the appropriate frequency, amplitude, and phase). The visual system is thought to consist of a number of narrow, independent filters, each sensitive to a different range (i.e., bandwidth) of spatial frequencies or sizes. A receptive field is actually a spatial filter that responds to the image in proportion to the amplitudes of the sine-wave components within its bandwidth. A number of neurons with similar spatial properties constitute a channel. The process of decomposing the image by a channel is called Fourier analysis. Shape perception is the result of the recombination of the outputs of the filters at some later stage.

Unlike feature theory, there are three assumptions that must be made before frequency theory can be applied to any system. These are homogeneity, linearity, and isotropy. Homogeneity requires that the human eye be spatially uniform at all retinal positions. Linearity requires that signals interact algebraically. Isotropy requires that the visual system characteristics be the same at along all meridians. The reader is referred to Cornsweet (1970) for a full discussion of these assumptions. It is probably the case that the assumptions are violated in our normal visual environments. In most experimental settings, conditions are restricted to satisfy the assumptions.

There are two variants of spatial frequency theory that appear in the literature. They share the same view that sine waves provide the best fit to receptive field profiles, but differ in the extent to which spectral analyses are thought to occur. In the first, the visual cortex performs a *global*, two-dimensional, Fourier analysis on the entire retinal image (Pollen, Lee, and Taylor, 1971). In this formulation, a cell tuned to a very narrow band of spatial frequencies would respond to the local stimulus in proportion to the amplitude coefficient at every frequency. The second (and more conservative) view suggests that cortical neurons with somewhat wider bandwidths may simply perform a *local* spectral analysis, with response amplitudes approximately equal to the power at the observed frequency (e.g., Pollen and Ronner, 1982).

The application of frequency theory to visual physiology first appeared in the classic paper of Enroth-Cugell and Robson (1966). In this study, cat retinal ganglion cells were classified on the basis of their responses to the introduction and withdrawal of a sine-wave grating on their receptive fields. X Cells were sensitive to the phase of the grating with respect to the receptive field center. A phase relationship was always found that ensured that the grating was silently substituted for a homogeneous field of the same mean luminance (Figure 3.3). At the other phases, responses usually more characteristic of the center or surround were evoked. This is to say that the X ganglion cell linearly summed the light increments and decrements. Y Cells did not show linear summation. Silent substitution was not achieved for any phase angle. This suggests then that the linearity assumption is at least satisfied for X cells.

With respect to which stimulus provides the best match to receptive fields, the evidence from the frequency camp is clear. Careful mapping of receptive fields of retinal ganglion cells with narrow stimuli (small spots or thin lines) produce profiles that are best described by the difference between two Gaussian functions having different amplitudes and space constants (e.g., Enroth-Cugell and Robson, 1966). This is the familiar "Mexican hat" profile (e.g., see Fig. 3.4A), first described for cat retinal ganglion cells by Rodieck and

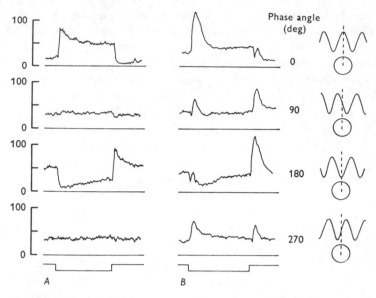

Figure 3.3. Averaged responses of cat retinal ganglion cells to the introduction and withdrawal of a stationary sinusoidal grating. Left-hand side: off-center X cell; middle: off-center Y cell; right-hand side: phase relationship of the stimulus with respect to the receptive field. Note the lack of response in the second trace for the X cell and the responses at all phases for the Y cell. (From Enroth-Cugell and Robson, 1966, with permission from the authors and the *Journal of Physiology, London*.)

Stone (1965), and for monkey by deMonasterio (1978b). The same fundamental shape occurs in neurons across a wide variety of species, including fish, frog, birds, and mammals. Elongated receptive fields, such as the cortical simple cell, resemble the Gaussian profile integrated along one axis to produce a field with finite length (see Figure 3.4B).

Albrecht, De Valois, and Thorell, (1980) have recently shown that the cortical simple cell was much more selective to sine waves than to square waves (i.e., bars) of equivalent contrast and nominal width. They conclude the sinusoid was a much better fit to the receptive field profile of a simple cell than the bar. Watson et al. (1983) have recently made psychophysical measurements of the detection of Gaussian luminance profiles and have argued that the ideal detector resembled a receptive field (presumably cortical) with multiple sidebands.

Other receptive fields appear to be less easily characterized by Gaussian functions, as, for example, directionally selective or suppressed by contrast neurons, but careful, quantitative techniques have not been applied to these classes of cells. Indeed, it seems reasonable to assume that "bug detectors" originally described by Lettvin et al. (1959) will be shown to have elongated, movement-sensitive fields that respond more efficiently to a moving Gaussian

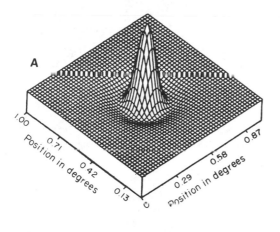

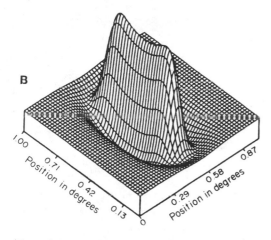

Figure 3.4. Two-dimensional profiles of the sensitivity of cells at the level of the lateral geniculate nucleus (A) and striate cortex (B) to stimuli presented to different parts of their receptive fields. (From Daugman, 1980, with permission of the author and Pergamon Press.)

edge than an octoped. The twofold advantage to "debugging the retina" and "debarring the cortex" in favor of sine waves is that (1) teleological arguments are avoided and (2) quantitative descriptions permit new rigor to receptive field analyses that may allow us to make predictions about responses in novel stimulus situations.

Predictions of responses to novel stimuli in the Fourier domain are also possible once the contrast sensitivity function is known (see Figure 3.5A). The sensitivity profile of the receptive field itself can be constructed by knowing the contrast sensitivity function (see Figure 3.5B). The cell's responses

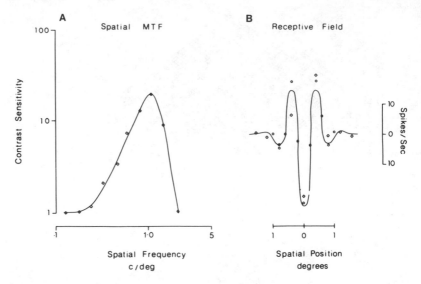

Figure 3.5. Frequency and space representations of a cortical simple cell. (A) Contrast sensitivity function determined with sine-wave gratings. (B) The receptive field sensitivity profile predicted by inverse Fourier transformation. Data points represent actual sensitivity measurements made using thin, light bars as stimuli. (From De Valois, Albrecht, and Thorell, 1978, with permission from the authors and Springer-Verlag.)

to any waveform is then predicted by convolving the contrast sensitivity function with the stimulus' spatial luminance profile.

De Valois, Albrecht, and Thorell (1978) have recently utilized this approach for striate cortical neurons in cat and monkey. They obtained spatial tuning curves for simple cells with drifting sine-wave gratings and were able to satisfactorily predict the cell's responses to narrow and wide and light and dark bars (see Figure 3.6).

De Valois, De Valois, and Yund (1979) have also extended the frequency analysis to two spatial dimensions. They examined the responses of simple and complex cells to checkerboard patterns. Checkerboards offer a unique opportunity to test the Fourier approach because the power at different spatial frequencies is oriented at an angle of 45° to the perceived rows or columns (hence the edges) of the checks. Thus, if the cells are edge detectors, the maximum response ought to be along the horizontal or vertical orientation; if the cells are frequency sensitive, they should have their maximum responses where the power is highest (i.e., at an angle of 45° to normal). The results of the study clearly supported the latter interpretation. Thus, these investigations lend considerable support to the idea that these neurons behave as two-dimensional linear spatial-frequency filters. These types of investigations also suggest that the assumption of isotropy may be satisfied for small retinal areas.

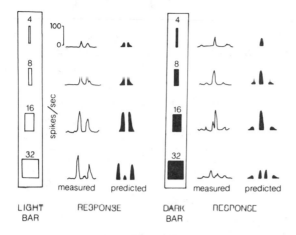

LIGHT BAR   RESPONSE   DARK BAR   RESPONSE

Figure 3.6. Predicted and measured averaged responses of cortical cells to black and white bars of varying widths. The predictions were made from contrast sensitivity functions such as those shown in Figure 3.5. (From De Valois, Albrecht, and Thorell, 1978, with permission of the authors and Springer-Verlag.)

A similar approach has recently been taken by Linsenmeier et al. (1982), who were able to generate receptive field center and surround sizes and their sensitivities from responses to drifting sine-wave gratings. Their results were generally in close agreement with the values determined by standard mapping with small spots of light.

It should be kept in mind that the predictions of responses to bars or the transformation between the space and frequency domains are a consequence of linearity of spatial summation (i.e., superposition). The introduction of significant nonlinearities (such as those found in Y and some X cells) can, but does not necessarily, complicate the picture. As shown by Linsenmeier et al. (1982) and Marrocco et al. (1982a), the Y cell's first-harmonic response component to drifting gratings is only slightly less linear than that of the X cell. In contrast, the use of the average elevated discharge (DC) component of Y cell responses results in significant deviations from linearity and predicts receptive field dimensions unlike those obtained with conventional stimuli. Thus, carefully chosen stimulus parameters and response indices produce linear behavior in most neurons. The perceptual consequences of forcing linear neurons into nonlinear ranges, however, are not clear.

Although a great deal of data suggest that frequency and amplitude are encoded in cell responses, few physiological studies have systematically examined the parameter of phase. The most conservative view of local Fourier processing requires that the phase relationships between the parts of a complex stimulus be encoded. Robson (1975) has theorized that properly aligned pairs of receptive fields, having "mirror image" spatial organization, could signal relative stimulus phase. In cat striate cortex, Pollen and Ronner (1981) have recorded from adjacent pairs of simple cells having the same organization, but a 90° difference in response phases. They hypothesized that if a second pair with receptive field properties of mirror image organization were also present in the same orientation column, then the ensemble could specify all the amplitude and phase information for a grating of a single orientation, spatial frequency, and direction of movement. To complete the local processing, pair relationships would be needed for cells having widely different peak spatial frequency preferences and directional selectivities. If such quartets are indeed present, this could constitute dramatic support for the frequency model.

Evaluation. At present, a global Fourier analysis is considered untenable by most theorists for two reasons. First, the spatial bandwidths of striate neurons is not sufficiently narrow. Second, phase information must be accurately coded over large areas of visual space, a task for which most cortical cells are ill-equipped. However, most physiological studies show overwhelming support for a local frequency model of visual processing.

The frequency theory deals somewhat more successfully with the problem fatal to the feature model, that is, the law of multiple response determinants. For example, it can be shown that while changes in orientation or spatial frequency may change response rates in a particular cell, these changes alter neighboring cellular activity in compensatory ways. Thus, the Fourier model can locally circumvent the law of multiple response determinants for some stimulus attributes. However, changes in absolute luminance, or color, or motion may not affect neighboring neurons in frequency compensatory ways and may thus hamper the integration of spatial information with other stimulus attributes.

There may also be residual difficulties encoding phase in an unambiguous manner. Pollen and Ronner (1981) have shown that response phases of neuronal pairs to a single grating stimulus may be separated by 90°. In a linear system, the encoding of the relative stimulus phase of a two-component grating ought to be a transform of the response phases of the neuron pair to a single stimulus. However, De Valois and Tootell (1983) have demonstrated some marked phase-dependent inhibitory effects of compound gratings on single cortical neurons. Therefore, the transformation from response phase to stimulus phase might not be straightforward. In addition, Pollen and Ronner (1981) show that for some pairs, the relative response phases may vary substantially from 90° for shifts in stimulus spatial frequency less than 1 octave from the preferred frequency. This would, of course, have adverse perceptual effects on the position of object edges, as the visual system is highly phase sensitive (e.g., see De Valois and De Valois, 1980, for a review).

Many questions about frequency theory remain for future research. For one, how well will prestriate neurons beyond V2 be characterized by sine-wave stimuli? A nontrivial problem is that the large receptive field size of these cells encompasses retinal regions that are not homogeneous or isotropic. Thus, it is not clear a priori whether Fourier analysis can be justifiably applied to such neurons.

Perhaps the most puzzling aspect of Fourier theory is the process of synthesis. How is patchwise frequency, amplitude, and phase information combined into global pattern perception? No clear answers have been suggested. One possibility would be to converge information from earlier onto later visual neurons. This convergence, or synthesis, would represent frequency and amplitude values from a group of cells simultaneously stimulated by the corners, edges, and so forth of a retinal image (i.e., having fixed phase relationships with other). The convergence model proposed here, while a compromise between the theories, may not, unfortunately, account for the rotational and translational invariances so nicely predicted by the untenable global Fourier model.

Parallel versus serial processing?

The serial/parallel issue revolves around the question of whether visual processing is best described in terms of multiple, independent channels or in terms of a single, sequential channel. In parallel processing, channels are conceived as groups of quasi-independent neurons, each tuned to different ranges of values along distinct physiological and anatomical continua. The parallel model has been developed primarily from studies of retinal ganglion cells, but has been generalized successfully to lateral geniculate and striate cortex neurons. More recently, parallel processing has also been hypothesized on a system level to describe connections between nuclei.

The current view of serial processing is somewhat more difficult to define, since there are fewer vocal proponents of the model, at least in its original form. It would be reasonable, I believe, to consider two noteworthy and one trivial possibility. Trivially, all processing is serial to the extent that sequential filtering and recoding occurs at each synapse. Nontrivially, Hubel and Wiesel's (1962) neuronal, serial model states that a specific sequence of convergence occurs between LGN cells and visual cortical neurons. Also nontrivially, Ungerleider and Mishkin (1982) suggest a specific sequence of serial connections between cytoarchitectonically defined prestriate areas.

Is parallel processing incompatible with serial processing? The answer is clearly no for any level of serial processing. The view to be taken in this chapter is that parallel and serial processing are coexistent, naturally complementary properties of the central visual system. Are they the only useful organizing principles? Probably not. For example, it is possible that functional systems may be defined by the presence of common neuromodulating peptide hormones that may exist across parallel or serial systems. What follows is an analysis of the strengths and weaknesses of both models.

Parallel processing

The first tenet of the parallel model is that the neural "channels" result from distinctive clusters of physiological–anatomical properties. The first site at which functionally distinct clusters appear is at the retinal ganglion cell level. Clear anatomical (e.g., see Figure 3.7), pharmacological, and physiological differences exist between cat retinal ganglion cells (e.g., Kirby, 1979; Wässle, Boycott, and Illing, 1981; Peichl and Wässle, 1983; Saito, 1983a, b; Ikeda and Sheardown, 1983).

In general, Y (nonlinear) cells have large somata, thick axons, wide dendritic fields, high conduction velocities, tend to be color-blind (broadband), are located more densely in the peripheral retina, and may use acetylcholine as a neurotransmitter (Ikeda and Sheardown, 1983). The X (linear summators) cells have medium-sized somata, thinner axons, narrow dendritic fields, slower conduction velocities, are generally color-opponent (at least in monkeys), are located more densely in the fovea (Stone and Fukuda, 1974; de Monasterio, 1978b), and may use aspartate as a neurotransmitter (Ikeda and Sheardown, 1983). W Cells (some of which have concentric receptive fields and behave like X cells and some of which have nonconcentric fields and behave more like Y cells) also have the smallest somata, thinnest axons, wide dendritic fields, the slowest conduction velocities, and in the cat are evenly spread throughout the retina (e.g., see Wilson, Rowe, and Stone, 1976). The neurotransmitter for these cells is not known. The question is, how are these diverse properties to be meaningfully grouped to represent functionally relevant sets of cells?

Three major grouping schemes have been proposed: (1) taxonomic, (2) spatial summation, and (3) discharge rate. The taxonomists (e.g., Rowe and Stone, 1977) argue that cell function can only be understood when tested under many different situations. They have suggested three fundamental cell groups and have adopted the noncommittal group names W-like, X-like, and Y-like. They criticize the more descriptive terms, such as sustained and transient, as being "essentialist," i.e., neglecting other important visual operations that these cells perform.

The taxonomic argument is somewhat weakened by important species differences. For example, while cat W cells conduct slowly, the cells with equivalent receptive field types in monkey (rarely encountered cells) have conduction latencies in the X and Y cell range. Similarly, most monkey color-opponent

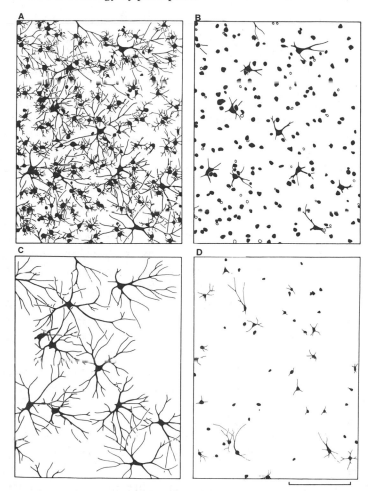

Figure 3.7. Ink drawing of ganglion cells in the cat retina filled with horseradish peroxidase (HRP). (A) All cells. (B) α Cells (with dendrites) and β cells (dark blobs). (C) α Cells only. (D) γ Cells only. (From Wässle, Boycott, and Illing, 1981, with permission of the authors and the Royal Society.)

cells are X, while all are W in cat. If, as seems reasonable, the W cells are performing the same operations in both species, they ought to be conveying their information at similar speeds to the central nervous system. The absence of this suggests that the same group of W cells may instead have different functional "niches" between species.

The spatial summation approach, based on the work of Enroth-Cugell and Robson (1966) (see the section on frequency versus feature), identifies only two major groups, X (linear summators) and Y (nonlinear summators). The Y-cell class is essentially just the absence of linear summative behavior and thus is negatively defined. The same is true for W cells, which are neither X nor Y. This scheme has no room for cells that are neither X nor Y. At the time this work was published, however, the W-cell group was unknown.

The third scheme identifies a cell as either brisk or sluggish, based on the response discharge rate to flashing stimuli (Cleland and Levick 1974a,b). Properties of the brisk cells (i.e., those having a high discharge rate) of the transient variety and the brisk sustained cells correlate highly with Y and X cells, respectively. All other cells are grouped into the sluggish category, including suppressed by contrast, color-opponent, etc. The discharge rate at which a sluggish cell becomes brisk is never specified, however. In this writer's view, the advantages of this scheme over the others are not obvious, and it will not be considered further.

For the sake of parsimony, it is worthwhile asking to what extent these schemes overlap. Does a cell identified as X-like from the taxonomic view correspond to a linear summating X cell? Linear summation was first tested in the cat retina. In later primate experiments, different tests were used and the cell groups changed to the X-like/Y-like/W-like (Dreher et al., 1976; Sherman et al., 1976). It soon became clear, however, that the X and X-like categories were not equivalent, in other words contained different populations of cells (Marrocco et al., 1982a). Many, but not all, of X-like cells are linear summators. Similarly, many but not all Y-like cells are nonlinear summators. In addition, about half of all W cells (usually the concentric variety) are linear summators, while the remainder are nonlinear (e.g., Sur and Sherman, 1982). Thus the classification schemes are not identical, and the opposing camps have attempted to stress the biological utility of each's approaches. Perhaps the more important measure of a scheme's adequacy is its ability to incorporate new findings. In this regard, the taxonomic approach appears the strongest.

The second tenet of the parallel model is that functionally distinct groups project to different locations in the brain (Figure 3.8). X Cells project predominantly to LGN in all species; in the cat a few terminate in the midbrain as well (Illing and Wässle, 1978). Y Cells project to the LGN and also send a branch to the superior colliculus. W Cells, in the cat, project to the LGN and superior colliculus. In primates, the vast majority of W cells project only to the colliculus, something of a problem for a taxonomic approach.

In cat dorsal LGN, X and Y cells are intermingled in all layers, but dominate the A layers (e.g., Stanford, Friedlander, and Sherman, 1983). W Cells are found exclusively in the C laminae. In monkey, X Cells represent 97% of the P laminae and about half of the M laminae (Marrocco et al., 1982a; Kaplan and Shapley, 1982). In contrast, 3% of Y cells are in the P layers and about 97% in the M laminae. Among M neurons, however, there are about equal numbers of X and Y cells. No W cells (with the exception of the rare non-

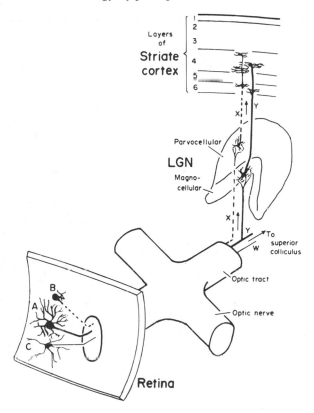

Figure 3.8. Schematic representation of the projections of the monkey retina to the lateral geniculate nucleus and superior colliculus. The A, B, and C cells probably correspond respectively to the α, β, and γ cells shown in Figure 3.7 for the cat retina. (From Stone and Dreher, 1982, with permission of the authors and Elsevier Biomedical Press.)

concentric variety of color-opponent cells) are known to project to the monkey LGN.

It appears as though the cortical projections in primates are segregated as well. Virtually all cells from the monkey LGN terminate in area V1. However, many LGN Y cells in cat also project to and provide a major input to area V2. Likewise, W cells from the C laminae also project to and provide a major input to areas V2 and V3. Within V1, afferents from the parvocellular laminae terminate in sublaminae 4A, 4C β, and layer 6 (see Figure 3.8). Those from the magnocellular laminae terminate in sublamina 4C α and layer 6. It has recently been shown that the small LGN laminae called "intercalated" and "S" project to cortical layers 1 and 3 (Weber et al., 1983). It seems as though, therefore, there is more divergence of LGN output in the cat as compared to the monkey visual system.

Beyond the first cortical synapse, however, the channel segregation is much less clear. Unfortunately, parallel models have relied strongly on conduction velocity measurements in retina and LGN. In the cortex, however, it is technically difficult to obtain conduction latency measurements for neurons beyond the first synaptic stage. Moreover, the high degree of convergence of influences from other cortical areas lessens the importance of a single latency value. Some researchers have assigned cells to channels on the basis of receptive field properties such as stimulus speed sensitivity (e.g., Movshon, 1975; Orban, Kennedy, and Maes, 1981) or linear/nonlinear spatial summation (Movshon, Thompson, and Tolhurst, 1978a,b). Indeed, Movshon et al. (1978a,b) have elegantly shown that virtually all complex cells are nonlinear, while the simple group is a mixture of linear and nonlinear. Hypercomplex cells were not tested. Can we then assume that complex cells are rapidly conducting, and at least some simple cells are slowly conducting? Unfortunately not, for all types of neurons appear to receive fast monosynaptic inputs from the optic radiations. Thus, the high correlation between receptive field type and conduction speed so obvious in the retina is apparently deemphasized in the striate cortex.

In primates, it appears safe to assume that most, if not all, V2 neurons receive primarily striate inputs. Few studies have systematically examined areas V2 and beyond with the notion of X/Y/W classifications in mind. The difficulties mentioned above make it unlikely that this will be accomplished in the future.

Strictly speaking, parallel processing implies independence of channels. In the most extreme case, if environmental pressures have indeed favored the development of channels for one kind of information, crosstalk between channels might be perceptually maladaptive. In some mammalian systems, crosstalk between X and Y cells does occur (e.g., Singer and Bedworth, 1973; Fukuda and Stone, 1976). It is not clear where this occurs, but it is first evident in the LGN. In monkey, the most likely site is the magnocellular laminae. Cells in these laminae are both X and Y (Marrocco et al., 1982a) and have dendrites that invade adjacent laminae (Wilson and Hendrickson, 1981). In addition, the corticogeniculate pathway relays X and Y inputs to some individual X and Y LGN neurons (McClurkin and Marrocco, 1984). Thus, segregation within mammalian parallel systems is not strict. What we do not yet know is whether the visual system with, for example, crosstalk between 10% of neurons performs differently from one with no interactions.

More recently, cortical parallel processing has been hypothesized for the recognition of visual objects and their localization in space. Ungerleider and Mishkin (1982) have proposed that the visual output from V2 traces a pathway to V4 and IT and is used for object recognition. In contrast, V2 also projects to MT, which in turn projects to the intraparietal and superior temporal sulci. This pathway is thought to mediate spatial vision (i.e., the ability to localize objects in space independently of their form). According to this model, the output from the IT and IP regions is reintegrated in the frontal lobe and limbic systems. The properties of neurons within these regions appear to be compatible with a parallel "wiring" arrangement, but this point has not yet been seriously studied.

Whereas the parallel anatomical structure in this formulation is evident, so is the serial. At the single cell level, however, it is quite difficult to conceive of a V4 or IT neuron as the endpoint of a serial sequence beginning in V1. For example, the organization of V1 color-opponent neurons has been shown to be double-opponent and half or more cells with foveal receptive fields are color-sensitive (Michael, 1978a; Livingstone and Hubel, 1984). In contrast, only about 8% of neurons tested appear to be double-opponent in V2 (Baizer, Robinson, and Dow, 1977). Receptive field sizes in both areas are comparable, and it is therefore unlikely that significant convergence occurs between V1 and V2. Neurons in V4 have large receptive fields and there is good evidence for converging inputs from V2. The proportion of all color cells varies from 16% to 100% in different reports. All reports agree, however, that double-opponent color organization is absent (Zeki, 1978; Schein, Marrocco, and de Monasterio, 1982; Schein, Desimone, and de Monasterio, 1983; van Essen and Maunsell, 1983).

Similarly, at this time it appears impossible to model neurons of the inferior parietal lobe as the serial endpoint of intermediate neurons. For one, IP cells have inputs from oculomotor centers, which are not seen in striate neurons (e.g., Wurtz, 1969) and have not been tested in V2 neuronal responses. For another, there are multiple inputs from other cortical and thalamic regions, which themselves are not well characterized, and whose contributions to IP neurons are difficult to assess.

Serial model

The serial or hierarchical model of visual processing (Hubel and Weisel, 1962) was originally intended to explain the organization of receptive field types found in the cat striate cortex. The crux of the model is convergence. Lateral geniculate terminals converge on neurons in the middle cortical layers to form simple receptive fields. Those in turn converge on cells in the more superficial and deep layers to form cells with complex properties. These in turn converge on cells in still more superficial and deeper layers to form hypercomplex receptive fields (Figure 3.9). In a later study, Hubel and Wiesel (1965) also noted that parastriate cortex (V2) had, on the average, fields of greater

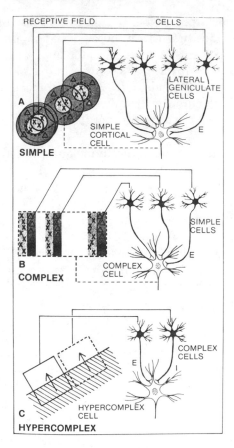

Figure 3.9. Synthesis of receptive fields in visual cortex according to the serial model. (A) simple cell, (B) complex cell, (C) hypercomplex cell. E, excitatory input; I, inhibitory input. (From Kuffler and Nichols, 1976, after D.H. Hubel and T.N. Wiesel, with permission of the authors and Sinauer Associates.)

complexity than those in V1. The hierarchy thus included cytoarchitectonic areas, as well as cells within these areas. The issues, therefore, are whether the evidence supports a receptive field hierarchy and shows laminar and areal differences in receptive field types.

Michael (1978a,b,c,1979) has extensively described the color properties of macaque striate cortex neurons. All neurons described were double color opponent, and the spatial properties could be modeled by convergence from concentric to simple to complex to hypercomplex. These studies lend clear support to the serial processing of spatial and chromatic information. Several unusual features of these papers merit comment. First, no blue/yellow opponent cells were found in spite of their presence in large numbers in the LGN (e.g., Wiesel and Hubel, 1966; De Valois et al., 1966; Marrocco et al.,

1982a). Second, concentric double opponent were most frequently encountered in layer 4C, the sublamina in which other investigators (e.g., Livingstone and Hubel, 1984) fail to find color-opponent neurons. In contrast, subsequent reports find most double-opponent cells in layers 2 and 3, where Michael finds few if any color-selective cells. The reasons for these discrepancies are not clear because the methods appeared to be quite similar. What is clear is that the double-opponent organization is the sole form of color sensitivity in the striate cortex.

Recent work on the properties of cortical neurons has shown that synaptic interactions are vastly more complex, and the divisions between cell types are less clear than originally conceived. I will discuss four areas of research that are particularly problematic for the serial model: (1) response latencies, (2) endstopping, (3) spatial summation, and (4) intracortical inhibition. The interested reader should consult Stone, Dreher, and Leventhal (1979) for a fuller treatment of this topic.

1. Response latencies. The conduction latencies of the different classes of cortical cells have been studied in some detail. Stimulating electrodes placed in the optic radiations can be used to activate cortical neurons electrically (e.g., Stone and Dreher, 1973; Tretter, Cynader, and Singer, 1975; Bullier and Henry, 1979a). While this method cannot identify all the inputs to a cortical neuron, it can identify the fastest. There is, however, no guarantee that the terminals providing the input are from the LGN. To circumvent this problem, many studies also compare responses evoked from the optic tract with those from the radiations to get a measure of conduction velocity.

In theory, however, it may still be possible to excite neural pathways through the tectum and/or pulvinar rather than the LGN. Assuming for the sake or argument that this is not the case, the following results have been obtained by Bullier and Henry (1979c). All classes of cortical neurons receive monosynaptic input from the optic radiations, some neurons at short latencies and others at longer latencies. Most of the shorter latency cells were located in layers 4–6; those with longer latencies were found in layers 2, 3, 5, and 6, but not 4. Unlike the results typically found for cells in more peripheral locations, a single class of receptive field structure was not correlated with the latency of input. Rather, complex and B (cells with both simple and complex properties) cells received Y-type input; nonoriented cells received X-type input. All of the nonoriented cells and half of simple cells received X-type inputs, and the remainder received Y-type inputs.

In general, simple cells are found most frequently in layer 4 and 6; complex cells in layers 5 and 6; B cells in 3 and 5; nonoriented cells in 4B and 4C. The conclusion to be drawn from these latency experiments is that there are slow and fast "streams" of information through the striate cortex: The former passes through simple, and nonoriented cells; the latter through complex, B, and simple cells. The main difference between this and the original work is that simple (or nonoriented) cells appear not to be the gateway to the cortex. Rather, each type of cell receives short latency input from the LGN and can send fast and slow streams of data in parallel to postsynaptic neurons in the upper and lower layers.

2. Endstopping or endstopped inhibition. These terms refer to the property of some cortical neurons that makes them respond to lines of a particular length as well as width. This feature in the serial model was originally thought to be the sole property of hypercomplex cells (Hubel and Wiesel, 1962). More recent work has suggested that a subclass of simple cells (Sh) also has this trait (Palmer and Rosenquist, 1974) and, more generally, that endstopping is present in varying amounts in nearly all cell classes (e.g., Schiller et al. 1976a; Rose, 1977; Gilbert, 1977). In any case, the presence of endstopping can no longer be considered as de facto evidence for a single cell class.

3. Mode of spatial summation. Movshon, Thompson, and Tolhurst (1978a,b) and De Valois et al. (1978) have demonstrated that simple cells in cat and monkey are usually linear summators (i.e., X cells) and complex cells are nonlinear (Y cells). A small fraction of simple cells did show significant non-linearities, and these usually had unusual receptive field structures (e.g., see Robson, 1975). The behavior of these cells was phase dependent, however, unlike complex cell responses.

These studies weaken the serial model because of the failure to explain how a complex cell's receptive field, which is so grossly nonlinear in its summation, could be formed from the direct convergence of linear, simple cell axons. One could hypothesize rectifying interneurons, but this feature would be a serious modification of the serial model (Stone et al., 1979). Moreover, it is hard to imagine why the visual system would go to such trouble to keep separate the subcortical X and Y pathways only to mix them in the cortex. There is also reasonably sound psychophysical evidence suggesting that the two systems are separable behaviorally (e.g., Tolhurst, 1973; but see Lennie, 1980). Therefore, the serial model needs to be revamped to account for newer data.

4. Intracortical inhibition. Several different types of experiments have shown that the orientation and direction specificity of complex cells is derived not from simple cells, but from LGN cells (Sillito, 1975; Daniels and Pettigrew, 1975). Similarly, Sillito and Versiani (1977) have demonstrated that the orientation specificity but not the length selectivity (by endstopped inhibition) of hypercomplex cells is determined by inhibition from simple, rather than from complex, cells. Finally a paper by Morrone, Burr, and Maffei (1982) has suggested that the lack of simple cell response to visual noise is a consequence of inhibition from complex cells. Thus, in summary, the evidence does not support a strict serial model. Simple cells provide input to complex cells, but so may other classes of neurons. Likewise, hypercomplex cells may receive inputs from complex cells, but some may also get simple cell and LGN W cell inputs (e.g., see Dreher et al., 1980) instead.

Evaluation

Most observations support the parallel model of organization at the neuronal level (e.g., see Stone et al., 1979). Its descriptive usefulness seems to be greatest at the retinal and geniculate levels, but it is tempered somewhat by crosstalk between channels. The neuronal distinctions are less clear-cut in the tectum, pulvinar, and striate cortex, and there is substantially more complexity of organization than had been originally conceived.

In evaluating the serial model, it is worth emphasizing that Hubel and Wiesel did a remarkably accurate job of identifying and classifying the neuronal cell types with relatively qualitative measurements and crude techniques. Any objective observer who has repeated their experiments is struck by the accuracy of their observations. Subsequent experimenters have had some difficulty confirming all their observations because they used different classifying criteria and more sophisticated stimuli.

The finding that more "simple" receptive field organization is found in proximity to the geniculate input and more "complex" organizations at greater synaptic distances has been generally confirmed by many researchers. Do the prestriate areas contain cells with more complex fields? It seems clear that cat prestriate cortex is not well described by serial relationships. This is due in part to the high incidence of simple cells found in V2 and V3. What Hubel and Wiesel did not know was that the LGN also projects to areas V2 and V3, and provides the major inputs to these areas as well as V1. In contrast, the monkey system appears to be nicely serial. Important species differences can also be seen in studies of the effects of cryogenic blockade of V1 on the responses of V2 neurons (Schiller and Malpeli, 1977; Sherk, 1978). Most neurons in the cat show little change in responses, underscoring their importance of the LGN projections. In monkeys, V2 is virtually silenced by blocking V1.

Perhaps the more important issue is a model's usefulness as a predictor. The bulk of the data from the prestriate cortices suggest that completely new and less easily categorized varieties of cells are formed by inputs from many different structures. It appears likely that this level of complexity may be problematic for both the serial and parallel models. It should be remembered that there are conceivable models of cortical organization that are neither serial or parallel, and failure of one is not necessarily evidence for the other. Finally, while the serial model currently seems to be of limited usefulness at the neuronal level, it should attain immortality as one of the most potent stimulators of research ever seen in this field.

Are individual areas in prestriate cortex devoted to the analysis of individual stimulus attributes?

One of the most fascinating developments in recent neuroscience research is the discovery of multiple neural maps of visual space, each residing in a distinct region of prestriate cortex. While cytoarchitectonic subdivisions within the prestriate cortex have been known for some time (e.g., see Polyak, 1957 for a review), more recent work has revised the locations of many of the classical area boundaries. The revisions are probably due in part to increases in the sensitivities of new techniques, but are also due to the use of more rigorous criteria for the identification of new areas. Typically, these criteria include the connections with cortical and subcortical structures, the nature of the topographic maps, and, most recently, the receptive field properties of the resident neurons (e.g., see van Essen and Maunsell, 1983).

The question that arises immediately is why our visual systems need multiple visual areas in addition to the striate cortex. One hypothesis that has been suggested is that each visual area processes a different stimulus attribute (Zeki, 1978). The purpose of this section is to evaluate the issue of whether different cortical visual areas are attribute specific. I shall first describe why certain attributes are considered important. I will then present some criteria for what should constitute an attribute-specific area (ASA) and evaluate how well the physiological evidence provides support for these criteria. I shall conclude by suggesting how to structure future research to clarify the functions of the multiple vision areas.

What are the important, or fundamental, stimulus attributes? The "fundamental" attributes are fundamental primarily because they are easy to produce in the laboratory. Few if any theoretical considerations guided the choices of fundamentals, although color may be an exception to this. What, for example, is fundamental about the choice of a light disk on a dark background beyond the pragmatic considerations of stimulus construction?

A conceptually bold approach has recently been taken by Marr (1982). In his theory, the fundamentals of form, for example, are primitive symbolic blobs, edges, and bars attached to the edges of objects. These convey information about object position, orientation, contrast, length, etc. One should not confuse symbolic blobs with "bloblike detectors" (i.e., concentric receptive fields). Indeed, Marr specifically avoids discussing cells or even levels at which these representations may be implemented in vision. While vagueness regarding implementation frees the model from certain constraints, Marr's theory may leave the reader puzzled about much of the function of the peripheral visual system. This chapter will deal with more traditional models of visual processing.

There are no guidelines in the literature for what should constitute the requirements for an ASA. I offer the following suggestions and will evaluate the neuronal data with reference to them:

1. Cells in this area should be narrowly tuned to one attribute and broadly tuned to all others.
2. The area should be characterized by substantial convergent inputs.
3. Input areas should not be tuned as in requirement 1; output areas should be more like requirement 1 than input.
4. Reversible lesions of this area should block acquisition of tasks using this attribute. Given requirement 2, lesions of the input area should not have the same effect.
5. This area should be metabolically active during tasks demanding the use of the attribute.
6. The most dramatic effects of rearing an organism in an environment lacking a specific attribute should be seen in the prestriate cortex, not in the LGN or striate areas.

To date, the coding of three attributes has been assigned to different prestriate areas: (1) color to V4, (2) stimulus disparity to V2, and (3) motion to MT. In what follows each assertion will be examined in turn.

1. Is V4 a color-specific area? Before dealing with the central issue of this section, it is worth determining whether color can be coded adequately in the striate cortex. The answer is probably no. While striate cortical neurons show wavelength sensitivity (e.g., Michael, 1978b), it is relatively broad. Broad spectral tuning does little more than partition the visible spectrum into two regions (excitatory or inhibitory). Since humans can recognize scores of unique colors, narrower neural tuning would appear necessary.

A more pressing problem is the multivariate nature of frequency codes, described earlier. Ignoring patterning within the discharge train, even the most narrowly tuned color cell can be aroused to the same frequency by manipulating several stimulus variables. In order to circumvent this, evidence must be presented showing that cells sensitive to one dimension are insensitive to all others. For example, color-specific cells should be broadly tuned for orientation, brightness, width, and velocity.

Three lines of evidence have been presented in support of a color area (Zeki, 1977, 1980). First, the proportion of color-coded cells is much higher than in other prestriate areas. Second, the spectral bandwidths of color-opponent neurons is substantially narrower than those in other cortical cells.

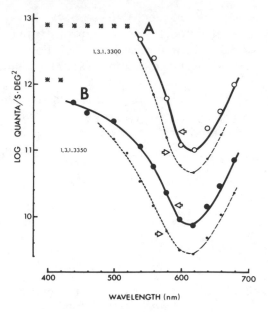

Figure 3.10. Comparison of the spectral bandwidths of color-opponent cells in prestriate area V4 (solid lines) with color-opponent neurons in the retina (dashed lines)
of the macaque monkey. (A) Color opponent. (B) Color biased. (From Schein, Marrocco, and de Monasterio, 1982, with permission of the American Physiological Society.)

Third, cells in this area respond as if they discount the background illumination, that is, show the property of color constancy.

The first two claims have not, in general, been supported by subsequent
research. First, several laboratories (Kruger and Gouras, 1980; Schein, Marrocco, and de Monasterio, 1982; van Essen and Maunsell, 1983) found that
the percentage of color-opponent and color-biased (a reduced form of color
coding) neurons is about the same as found in other visual areas. Second,
the half-peak spectral bandwidths of color-opponent prestriate cells are similar
to those of cells found at the retinal geniculate and striate levels (Schein et
al., 1982; DeMonasterio and Schein, 1982; van Essen and Maunsell, 1983)
(see Figure 3.10). There have as yet been no attempts to replicate the report
that V4 cells are involved in color constancy.

Several methodological differences are evident between the more recent
and the original papers, and these may be the sources of the discrepancies.
For example, stimulus configuration appears to be critical. A special property
of V4 neurons only recently described is the silent surround (Schein et al.,
1983). These surrounds allow color interactions across large areas of the visual
field and dramatically affect spectral sensitivity.

The extent to which color-sensitive neurons are selective for other stimulus
attributes is also controversial. Most investigators in this area agree that V4

cells are not motion sensitive. However, it has been claimed that color-selective cells are not orientation, shape, or disparity sensitive (Zeki, 1980). The shape insensitivity could be due, in part, to the apparent convergent input to V4 neurons from V2, satisfying requirement 2 above. However, many of the color-opponent cells are orientation sensitive (Desimone and Schein, 1983; Schein et al., 1983), and some of them are reportedly static disparity sensitive as well (Burkhalter and van Essen, 1982). Thus, it would appear that the responses of V4 neurons do not represent a significant departure from the behavior dictated by the law of multiple response determinants.

While there are some important procedural differences between Zeki's work and that of others, several differences remained to be explained. The claim (Zeki, 1983) that the color-specific effects are restricted to the anterior bank of the lunate sulcus, rather than the posterior bank of the superior temporal sulcus, may be important, but Schein et al. (1982) failed to note any differences between these areas.

Is there evidence to support the remaining requirements? For example, if V4 is devoted to the analysis of color, then its destruction ought to impair color vision. Studies of brain-damaged primates are always difficult to evaluate, since the inherent serial connectivities between regions may be responsible for deficits distant from the lesion site. Nevertheless, is it reasonable to expect that lesions of V4 would have consequences for color discrimination? It seems that an answer depends critically on the task used. It is probably not involved in color naming, which may require narrower spectral bandwidths than have been shown to exist. However, V4 could be involved in color contrast discriminations. A recent behavioral study (Dean, 1979) examined the ability of monkeys to make successive hue discriminations. Lesions of V4 and V4A (anterior lunate sulcus and gyrus) were ineffective in preventing the retention of learned discriminations or the acquisition of new ones. While this study implies that mechanisms for hue discrimination are not located in V4, it does not eliminate a role for V4 cells in other types of color discrimination.

Does V4 become metabolically active when the animal sees color? It would appear to be straightforward to stimulate visually the retina of an anesthetized monkey with monochromatic light and measure the activity of cells in V4 with the 2-DG technique (Sokoloff et al., 1977). If this area is intimately involved in the processing of color, it should become more metabolically active, as prescribed by requirement 5 above. Unfortunately, this is a technically difficult experiment, requiring the removal of all brightness cues from the stimulus, and thus the empirical support for requirement 5 will require technically demanding experiments.

Is V4 affected by environmental conditions during development? It is, unfortunately, technically difficult to eliminate a narrow wavelength band of visual stimulation from the visual world of a young monkey. The overlap, for example, between the middle and long wavelength pigment absorption curves would assure that at least two cone types will be stimulated. However, attempts to restrict stimulation to narrow bands have been reported for the ground squirrel, which has fewer cone pigments than the monkey. Although there are fewer color-opponent mechanisms in animals raised in red-only light compared to those raised in white light, there are no differences in their spectral responses (McCourt and Jacobs, 1983). Neither behavioral data nor metabolic activity labeling has been reported in these animals.

Finally, an as yet unexplored implication of ASAs is that it represents an endpoint in visual processing. For example, all color processing has been hypothesized to occur in V4. What happens to the information subsequently? Surely the V4 recipient zone of IT should also be color specific. If we assume that V4 is indeed the "end of the line" and that color constancy is the highest form of color vision, then all other forms of color discrimination must be accomplished prior to V4 and be of a more primitive nature. This suggestion receives some support from physiological studies of wavelength and saturation discrimination (De Valois, Abramov, and Mead, 1967; De Valois and Marrocco, 1973). Unfortunately, developing a rank ordering of color discrimination tasks on the basis of complexity is difficult without theoretical linking hypotheses. Formulations such as Marr's may provide valuable guidance in these matters.

Evaluation. The weight of the evidence supports the conclusion that while some V4 cells may be involved in color analyses, they may be doing other analyses as well, and many neighboring cells are not color selective. Thus, the area cannot be dedicated to the analysis of one attribute.

2. Is V2 a disparity-specific area? Hubel and Wiesel (1970) reported that little measurable disparity sensitivity could be found in binocular neurons of striate cortex in paralyzed macaques, but that a range of sensitivities was found for cells in area V2. At the time, this finding was difficult to reconcile with results from other species (i.e., cat and sheep) whose striate cortices had neurons with a broad range of optimal disparities. The conclusion that depth sensitivity resided in area V2 in monkeys has since been weakened by the work of Poggio and Fischer (1977). Using alert, trained monkeys, these investigators found many striate neurons with sensitivities for a range of either crossed or uncrossed disparities.

The second visual area (V2) has also been suggested as an area that may be specialized for detecting motion in depth, that is, dynamic disparity (Cynader and Regan, 1978). There are special requirements for neurons in such an area. First, the cell must be able to differentiate between both different interocular target velocities and directions of movement. Second, the cell should be sensitive to different interocular rates of change in the size of the retinal image. About 25% of cells in cat V2 (Cynader and Regan, 1978) and in monkey V2 (van Essen and Maunsell, 1983) have such properties.

Do V1 neurons also display dynamic disparity sensitivity? Unfortunately, virtually all studies have tested V1 cells with stimuli moving only in the frontoparallel plane, that is, disparity is always static during stimulus motion. Those striate neurons tested with changing disparities and showing sensitivity to dynamic disparity are extremely rare.

Evaluation. Partial support for the hypothesis that V2 is an ASA for disparity has been presented. What is not clear, at this point, is whether or not V2 neurons may respond to different static (stationary) values of disparity or whether continuously changing disparity is the essential condition. This is difficult to determine since many cells do not respond robustly to stationary stimuli. Many others do, however, and it would be possible to test these alternatives. Also lacking is the extent to which V2 neurons are tuned to other stimulus attributes. In any case, the major outputs of V2 are to V3, V4, and MT, and we should expect to see some dynamic disparity-sensitive neurons in these areas also. Some cells with these properties have been reported (Sakata et al., 1983) in the posterior parietal association area (POa) (Fig. 3.1).

Most of the requirements listed above have not been adequately tested for the disparity area. It would appear premature to think of it as an ASA at the present time.

3. The motion area (MT,V5). MT was originally described by Zeki (1974) and currently appears to be much less controversial than the color area. The motion area is characterized by a high proportion of directionally selective, movement-sensitive cells. However, some neurons also appear to be binocular disparity sensitive, Maunsell and van Essen (1983b), and still others are orientation selective (Maunsell and van Essen, 1983a). While most of the latter neurons appear to be tuned to static disparity, there are also dynamic disparity-sensitive units. None of the cells appears to be color or form sensitive.

A property of MT neurons not seen at earlier levels is the motion-sensitive surround (Miezen, McGuiness, and Allman, 1982), which may be motion sensitive in a direction that is diametrically opposite to the sensitivity of the center. It is this feature that makes the cell sensitive to relative movement and different from V1 or V2 cells. These neurons may be involved in visual tracking. It will be interesting to see whether the receptive fields of both eyes have similar or dissimilar characteristics. If similar, the tracking function is probably only effective for fronto-parallel motion. If dissimilar, tracking in depth could be signaled by these neurons. The receptive fields of MT units are also much larger than striate fields, which may account for their ability to signal rapid stimulus motion. While this capacity may be due in part to convergence of information from V1 or V2, it could also reflect processing of high stimulus velocities present in the tectum (Marrocco and Li, 1977) and relayed to the prestriate area via the pulvinar (Marrocco et al., 1981).

The output of MT is the intraparietal (IP) sulcus and the processing of motion information therein appear more complex than those of MT. Does this implicate IP as a movement area? The literature (e.g., see Lynch, 1980) suggests not. Neurons in IP but not in MT do, however, differentiate between stimulus motion on the retina and motion produced by eye movement across a stationary target. This behavior is not unique to IP, as some cells in the superior colliculus also differentiate between stimulus and self-produced motion (Robinson and Wurtz, 1976).

Motion perception also occurs under circumstances in which there is no retinal image motion (i.e., during tracking), but the percept is slightly different than that experienced with a moving retinal image (Dichgans and Jung, 1969). One could argue that a true movement area is one in which cells respond to the two types of motion in qualitatively similar but quantitatively different ways, as needed to explain the differences in percepts. Cells responding during visual tracking have been reported in IP, and they probably receive extra-retinal input, are subject to attentional effects, or are a combination of both. Neurons in the thalamus and striate cortex, while not tested during tracking eye movements, fail to discriminate between real and self-induced rapid image motion (Wurtz, 1969). It would seem unlikely that they would discriminate between the different types of slow (tracking) motion.

Zihl and colleagues (1983) have recently examined the capacity of brain-damaged humans to discriminate direction of motion. Computerized tomography scans indicated bilateral damage to the temporoparietal and occipital areas, and unilateral damage to the medial and superior gyri of the visual cortex. The subject could not see motion in depth in her central visual fields. Motion perception along horizontal or vertical axes was somewhat impaired, while motion in all planes was intact in her peripheral visual fields. According to the subject, moving objects appeared frozen and would later appear displaced in her field of view. Motion aftereffects and stroboscopic movement were also absent. However, movement perception elicited by acoustic and tactile inputs was normal.

No studies to date have looked at changes in the short- or long-term metabolic activity as a result of visual movement stimulation. As with the case of color, it is not at all clear what the appropriate stimulus should be. Perhaps the best stimulus should be a montage of patches, each moving in a different direction. Since the neurons in MT also seem to be unselective for shape, a random motion dot or bar pattern could be used.

Evaluation. Middle temporal areas appear to be highly specific for, but not exclusively devoted to, motion detection. While not all of the above requirements have been tested, the majority (e.g., numbers 1–4) support this interpretation. Conclusions about V2 cannot be made safely at present with the current data. Finally, area V4 appears not to fulfill most of the above requirements and thus appears not to be specific for color.

Directions for the future

Understanding the visual functions of the different prestriate areas will necessitate careful selection of visual stimuli. For example, demonstrating that animals with prestriate lesions fail to discriminate between a circle and a square sheds no light on function because of the complexity of the cues in the geometric shapes. Is stimulus size, number of convexities, the presence of angles, or the orientation of contours crucial for prestriate neurons?

In this writer's view, attention to the local versus global cues is required. The most successful use of such cues to examine perception was done by the Gestaltists. Using the principles of proximity, good continuation, closure, and so forth, it is possible to develop stimulus displays that force visual analysis by examination of individual elements or by the overall pattern. This approach has been successfully used by Julesz for understanding the mechanisms of stereopsis and is interwoven into Marr's (1982) theory as well. It has recently been applied to animal research by Sprague, Hughes, and Berlucchi (1981).

Analysis of the prestriate movement area, therefore, might be accomplished by devising stimuli that otherwise have little global structure, but have a local structure on the basis of common fate grouping, that is, a structure is detected by neurons because selected individual elements are moving together in the same direction. The success of this approach awaits future analyses.

Physiological bases for stimulus constancies

It is a remarkable fact that the responses of the visual system are resistant to (1) minor changes in stimulus conditions and (2) alterations in the ambient level of stimulation. In the first case, changes in stimulus size, shape, and orientation can occur with little change in perceptual response. In the second instance, overall alterations in brightness and color are generally ignored by observers. Recent work has established that several lower species respond in certain situations as though controlled by constancy mechanisms.

What physiological explanations can be used to account for these visual phenomena? In the following section, I will examine some of the requirements needed by neurons that are alleged to support constancies. I will also discuss neuronal data that indicate possible mechanisms for these phenomena.

Size constancy

Size constancy reflects the tendency of human and primate observers to ignore the retinal image size of familiar objects and judge them to be larger or smaller based on previous experience with these objects. This occurs under conditions in which the stimulus' visual context is rich with information about depth. When the contextual cues are absent, observers respond in agreement with the physical size of the retinal image. Therefore, a neuron that mediated size constancy must have a large receptive field or a smaller one influenced by stimulation of distant retinal areas.

Second, size constancy has been tested psychophysically under restricted conditions of viewing, in which the visual angle of the target is held constant. Physiologically, we should like to know over what limits of retinal image size constancy can occur. The classic paper by Holway and Boring (1941) suggests that size constancy holds for distances between 10 and 120 ft (3 and 36 m), although greater and lesser distances were not tested. Thus if object size were allowed to vary as it would under real conditions, its smallest target (53.2 mm), which subtended 1° at 3 m, would subtend only 4.5' of arc at 36 m. Thus the constancy neuron should give equivalent responses to targets in this range of visual angles. The equivalence should only be seen when peripheral stimulation is present and should disappear when only the target is on the receptive field.

Response equivalence could be accomplished in at least two ways. First, neurons with large receptive fields could show response equivalence for many different stimulus sizes that fall within equally sensitive areas. Such mechanisms have been suggested as the physiological basis for stimulus equivalence across retinal translation (Gross and Mishkin, 1977). It should be remembered, however, that because of the multivariate nature of responses, equal response rates to stimuli of different size probably would be attained at the expense of loss of information along some other attribute (e.g., luminance). Parenthetically, there may be an accompanying decrement in the perception of luminance not previously appreciated. It may be possible for the psychophysicist to use a stimulus probe to test luminance during conditions giving rise to size constancy.

Alternatively, the process of shifting the place of focus may affect the size of the receptive field itself, that is, cortical input from intra- or extraocular muscles (e.g., Buiserret and Maffei, 1977) could alter the effective center size by depressing or enhancing the inhibitory flanks. Whatever the mechanism, contextual effects must influence the receptive field sizes as well.

Several reports of neurons in prestriate cortex indicated possible candidates for mediation of size constancy effects. Marg and Adams (1970) have suggested that receptive fields in human visual cortex plotted at several distances remained about the same size. Gross et al. (1972) have found neurons in IT cortex that appear to show size equivalence, i.e., neurons unconcerned with stimulus areas. In fact, this is not an uncommon attribute for deep collicular and lateral pulvinar neurons (Marrocco and Li, 1977; McClurkin, 1984). Blakemore, Garner, and Sweet (1972), however, presented data that apparently obviate subcortical mechanisms. They demonstrated that size constancy, as shown in spatial frequency adaptation effects, occurs beyond the first point of binocular interactions.

Unfortunately, in none of these studies was the effect of context (i.e., background stimulation) tested. In addition, the Marg and Adams (1970) experiments were done with less than adequate fixation stability and control of accommodation. For another, the physiological experiments were performed under systemic anesthesia and neuromuscular paralysis, which may not be optimal conditions for demonstrating contextual effects of peripheral stimulation. It is difficult to imagine, then, how any of the studies cited above could reflect the true mechanisms of size constancy. It would seem that the alert, behaving primate would be the subject of choice for these studies, since this species almost assuredly behaves according to size constancy rules (e.g., see Walk, 1965).

By using dichoptic viewing, it has been possible to suggest pre- or poststriate neuron centers as possible loci for some psychophysical phenomena. This method must be carefully applied and interpreted (see Long, 1979). It may be possible in future studies to localize them further by making use of other anatomical constraints. To illustrate, the corpus callosum joins the two halves of the visual field, and its influence is seen neuronally in prestriate visual cortices. It may be possible to ask which conditions allow, for example, spatial interactions across the vertical meridian. As a case in point, Land et al. (1983) recently demonstrated that when the ambient illumination is presented across the midline from a test stimulus, color constancy effects are present in normal observers, but fail to occur in patients with callosal transsections. These results are consistent with the localization of color constancy mechanisms in area V4 of prestriate cortex.

Color constancy

Color constancy is the tendency of the visual system to ignore conditions of ambient illumination in judgments of hue. Not only do hue relationships not change between daylight and fluorescent lighting, but colored displays are resistant to hue shifts despite wide-band chromatic filtering. There is, of course, a limit to constancy. Narrow-band filtering allows only the hue of the filter to be perceived.

Is it possible that subcortical, chromatic-opponent neurons could subserve color constancy? It is simple to show that broad-band changes in the color temperature of a light shone on the receptive field do not dramatically or differentially affect any class of color-opponent cells, as each cell class will have its spectral neutral point shifted by about the same amount and toward the same end of the spectrum. Most would agree that differential effects on color-opponent cell classes are probably necessary to change color appearance or color names (e.g., Boynton and Gordon, 1965). However, the responses of color-opponent cells to colored stimuli are dramatically affected by narrow-band filtering (Nothdurft and Lee, 1982). Monkey LGN cells respond under these conditions only to the brightness of the stimuli. All categories of opponent cells behaved similarly. It is thus unlikely that mechanisms for color constancy appear subcortically.

Zeki (1980) has described cortical cells that appear color selective in spite of varying conditions of illumination and seem to respond to colored papers in the same qualitative manner as the experimenters. In these studies, Zeki recorded from single neurons in V4 (anterior bank of the lunate sulcus). The stimulus was a "Mondrian-like" collage of color patches that varied in brightness and dominant hue. Zeki found that cells that normally responded only to a red patch in neutral illumination did not do so when the patch was illuminated with narrow-band colored light. Seen with narrow-band light, the patch appeared to differ from its neighbors only in brightness, not in color. Moreover, the cell responded in a similar way despite changes in the energy passed through the narrow band filters. The response constancy is probably due to the silent surround (see section on individual areas in prestriate cortex). Thus, the behavior of these neurons corresponds quite closely to the conscious percept and may represent the functional basis of color constancy.

Brightness constancy

In brightness constancy, the overall intensity of illumination of objects is ignored and the relationship of brightness between object and background or surround is sensed, that is, the visual system measures the ratio of the brightness of an object and its surround. If the ratio remains constant despite changes in ambient illumination, brightness constancy is said to hold. Many organisms respond in ways suggesting that brightness constancy controls visual responses (see Locke cited in Woodworth and Schlosberg, 1954, for primate experiments).

Cells in the peripheral parts of the visual pathway (e.g., retina or LGN) are predictably affected by shifts in ambient illumination. Increasing the ambient illumination decreases the responses to increments and decrements of light in roughly a ratio fashion (Jacobs, 1965). In a concentric neuron, the higher the luminance falling on the antagonistic surround, the smaller the response of the center (e.g., Singer and Creutzfeldt, 1970). The sizes of the receptive fields appear inadequate, however, to account for constancy effects over very large retinal areas typically tested in psychophysical experiments. In addition, constancy appears to break down when the ambient illumination is visible to the viewer, a result not easily reconcilable in terms of peripheral receptive field mechanisms.

There have been no published reports in which brightness constancy has been studied in cortical neurons. However, it is likely that prestriate neurons with large fields are the most likely candidates for those signaling constancy. It is possible, for example, to train a monkey to respond to two, centrally viewed stimuli only when they are of the same luminance. If constancy is present, it should be possible to vary systematically the surround luminance of one of the targets and cause little change in behavioral responding. Simultaneously, the activity of a prestriate neuron could be monitored to learn under what conditions the cell's activity mimicked that of the monkey. It should also be possible to record from different hierarchical levels (e.g., see van Essen and Maunsell, 1983) to find when the cellular behavior first appeared in the visual pathway.

In conclusion, there is little firm evidence for physiological mechanisms that can account for constancy phenomena. Prestriate neurons with large fields and/or silent surrounds would seem to be the cells to investigate in the alert monkey. The law of multiple response determinants suggests that the perceptual constancy of one attribute may be accomplished at the expense of information about other attributes. This chapter has provided some guidelines to help in the search for the neural bases of these unique phenomena.

4   A neurobiological view of the psychology of perception

Richard T. Marrocco

The aim of this chapter is to evaluate potential areas of interaction between sensory neuroscience research and the cognitive approaches to perception lucidly presented by Dr. Hoffman in Chapter 2. The interaction would seem to be a natural consequence of the growth within each field and the interdisciplinary nature of neuroscience research. If feasible, this interaction would have positive consequences for both disciplines. While I have based my comments on areas within vision research only, the principles discussed have general validity within other sensory systems as well. In what follows, the terms neuroscience and sensory physiology will be used synonymously.

Is there a basis for interaction between cognitive psychologists and neuroscientists? There are at least two levels to an answer to this question. On one level, there is certainly common intellectual ground that could, with sufficient planning, be explored jointly. If this is true, why has there been relatively little interaction thus far? The answer to this question brings up a second level, motivation. It seems as common in the literature for the neuroscientist to ignore the important behavioral considerations in the design and interpretations of experiments as it is for the cognitive psychologist to lose track of the known neural mechanisms that underlie perceptual behaviors.

The lack of motivation is reflected in attitudes expressed by individual researchers. Neuroscientists criticize cognitive psychologists for failing to become involved in the fundamental operations of the nervous system, that is, the roles of ions, molecules, and small assemblies of neuronal elements in behavior. In general, the behaviors of interest to neuroscientists are inherently simpler than those considered by cognitive psychologists. Cognitive psychologists, in turn, accuse neuroscientists of being unwilling to consider more global approaches to a field. At the heart of the issue, then is an unwillingness to consider alternative scopes to relevant problems. To this must certainly be added the major differences in methodologies. Nonetheless, more interaction would take place if seen to be mutually beneficial to both camps.

This is not to say that there is no interaction at present. There are clear examples of advances in the understanding of perceptual processing that have

80

arisen from physiological experiments. For example, key papers on the mechanisms of visual masking (Judge et al., 1980), the neural basis of hue and saturation discrimination (De Valois et al., 1966; De Valois and Marrocco, 1973), and several elegant investigations in nonvisual areas, e.g., mechanisms of tactile and thermal discriminations (e.g., Knibestol and Vallbo, 1980; Johnson and Lamb, 1981) have recently been published.

However, the interactions appear asymmetrical, that is, the impact of cognitive psychology on the field of sensory physiology appears much less than the reverse. The neuroscientist, especially the electrophysiologist, monitors neuronal activity with microelectrodes at a single synaptic level. It is currently difficult for the cognitive psychologist to design perceptual experiments that shed light on a particular brain area, i.e., synaptic level. It is likely that improved communication would follow the development of behavioral tests that will reliably discriminate between neural functions at different synaptic levels.

The neuroscientist also benefits to a lesser extent from interchanges than the cognitive psychologist for what can be called anthropomorphic reasons. The neuroscientist is frequently less interested in *human* perceptual processes and more frequently interested in the behaviors of lower organisms than the cognitive psychologist. Indeed, some neuroscientists assert that significant advances in understanding the neural substrates of behavior in the reasonable future will only be found using invertebrate simple systems. This rather myopic view, however, disregards recent dramatic achievements in understanding the neural mechanisms of vision in the mammalian retina, the functional anatomy of the mammalian hippocampus and vertebrate olfactory bulb, and the neural correlates of learning in the mammalian cerebellum, to mention but a few.

What are the key points of interaction between cognitive psychology and sensory physiology? To begin with, there are several assumptions held in common. Both would agree that the perceptual experience is the combined influence of afferent feature-detecting (or frequency-selective) channels and efferent modification of these channels by past experience and the like. Indeed, there is substantial agreement on the nature of the afferent mechanisms which underlie bottom–up processing. There is less agreement on the role of top–down processing, but this is primarily due to the neuroscientists' inability to manipulate higher cognitive processes in physiological paradigms. Both groups assume that there are no emergent properties of the perceptual system that will not eventually be comprehensible through an understanding of the constituent parts.

Are the important questions the same in the two fields? Naturally, both groups are ultimately interested in explaining the mechanisms of the perceptual process. More specifically, spatial frequency analysis and, to a lesser extent, feature detection and constancies are common areas of interest. Global information processing is also a common area of interest but, at present, has been studied by the neuroscientist only indirectly by assessing context effects in receptive field organization (see section on stimulus constancies).

The first context effects in cortical neurons were reported by Blakemore and Tobin (1972). These investigators found that the orientation specificity of striate neurons was affected by the presence of oriented contours in the periphery of the cell's receptive field. Maffei and Fiorentini (1976) found that most cortical neurons had zones that did not influence the cell's discharge when directly stimulated, but which did when jointly presented with bars in the cell's receptive field proper.

More recently, John McClurkin and I have shown that most simple and complex cells in the monkey's striate cortex may be influenced by linear (X) and nonlinear (Y) inputs from parts of visual space quite distant from the receptive field surround. Hypercomplex cells showed these influences infrequently. Thus, it would appear that the definition of a receptive field may soon be modified to include more distant regions of visual space. Although the "context" manipulated in these experiments may not be identical to that found in perceptual experiments, the principle appears to be the same. Yet to be answered is whether a single mechanism accounts for most contextual effects and, if not, where the multiple mechanisms are organized.

In truth, the key areas of overlap represent only a small fraction of the material presented in the chapters on perception and neuroscience. A quick comparison of recent texts in perception and sensory physiology gives the same impression. This suggests that, in general, there are more dissimilarities than similarities in the questions of interest. However, it should be remembered that these are entirely *post hoc* comparisons. Clearly, more overlap will occur when there is better agreement on which questions ought to be answered. Following this, the joint planning of experiments from the inception ought to be significantly better than attempting to discover common grounds after the fact.

Do the two disciplines use the same language? As a neuroscientist, my task in preparing this chapter is a little like that of a traveler in a foreign country. With an incomplete knowledge of grammar and syntax, the traveler attempts to understand the fine points of the country's culture. The problem is both idiomatic (familiar terms, such as features and attention, are used in unique ways) and neologistic (for example, textons and perceptrons). In addition, there are words, such as transparency, emergent features, primal sketches, and integral dimensions, not used in the traveler's native language.

The task, then, is to develop something akin to a Berlitz dictionary, perhaps having the title of "Perceptual Psychology for the Traveling Neuroscientist." To illustrate the translation, consider the feature integration theory of Treisman and Gelade (1980). This theory suggests the existence of feature maps devoted to a single sensory attribute and postulates that a stimulus acquires multiple features when attention is directed to the stimulus' spatial location.

The first job of the neuroscientist is to translate the terminology into physiological terms and assign the postulated operations to known brain locations. There is some evidence (see Chapter 3, section on areas in prestriate cortex) that different cortical maps may exist for some stimulus attributes. It is also known that the different cortical areas are interconnnected in a retinotopic fashion, so that all maps presumably lie in register with one another. The anatomical site of attention (currently unknown) presumably has retinotopic connections with each area and probably acts as a gate to facilitate information processing between the relevant map areas.

Since the status of attention in the anesthetized animal is unknown but probably impaired, feature integration ought to be impossible. However, the use of alert, behaving animals in physiological paradigms should allow parts of this theory to be tested. One ought to be able to train a monkey to respond differentially to a color/form discrimination in a Wurtz-type paradigm (e.g., see Wurtz and Goldberg, 1972). For example, a monkey could be taught to choose between an equiluminant red square and green triangle, either on the basis of hue or form. It should be possible to test whether reversible blockade of V4 or IT with the drug muscimol (which temporarily causes a cessation of neural function) disrupts responding based on one type of cue or the other. Naturally, the crucial experiments are the controls, and several convergent lines of evidence would be necessary before the results could be accepted.

Alternatively, it might be possible to monitor neuronal activity in all the areas corresponding to the attributes contained in a simple stimulus. One could predict that when the stimulus was the target of a saccade (presumably requiring attention to it), cells in the relevant attribute areas ought to be more active than cells in the nonrelevant attribute areas (same part of the visual field) or during periods when the stimulus was not the target of the saccade.

A second example of translation can be illustrated for the experiments by Gilchrist (1979), who showed the influence of perceived depth on brightness judgments. Under monocular viewing conditions, brightness appears to be governed by lateral inhibitorylike mechanisms, resulting in illusory brightness percepts. This suggests that monocular brightness detectors are inadequate. Under binocular viewing, veridical brightness percepts were reported despite changes in ambient luminance. Since stereoscopic cues are present in the binocular display, it would appear that constancy must be computed at or beyond the level of the striate cortex, where stimulus disparity is first computed and complex periphery effects are commonly recorded. Perhaps the silent surrounds of prestriate cells (see section on individual areas in prestriate cortex maps) play an important role under these conditions.

Are there common constraints between the fields? The constraints are more evident in the perceptual domain than the neuroscientific domain. In spatial frequency analysis, for example, four channels have been postulated (Wilson and Bergen, 1979), and constraints on the channel bandwidths have been dictated by physiological and psychophysical experiments. It should be mentioned that the actual number of channels is testable and, in fact, is incorrect. All spatial frequencies appear to be represented in a continuous fashion among neurons in the dLGN and visual cortex (Schiller et al., 1976b; Tolhurst and Thompson, 1981; Marrocco et al., 1982a; Kaplan and Shapley, 1982; De Valois et al., 1982). In addition, texton and feature integration theories use building blocks whose properties are constrained by known properties of central visual neurons.

Other areas, however, appear to be less constrained. The most obvious is the field of computer pattern recognition, where virtually none of the models appear to be constrained by current physiological principles. Producing a device that allows an object to be recognized depends on algorithms that work under a limited range of conditions. The models do not as yet have the flexibility of a biological system, perhaps because they have ignored both the properties of visual cells and the laws of neuronal connections. There is one notable exception to the above among computational models. Marr (1982) has designed a model of the first stage of filtering in the visual system that closely resembles the receptive fields of retinal ganglion cells.

Are there unrecognized benefits that might accrue from an interaction between cognitive and neuroscientists? The blending of disciplines nearly always assures that new ideas will be generated. I would expect the cognitive psychologists to learn more of the machinery of perception, which in turn would constrain experiments and theories. A likely benefit for the neuroscientist is that theories of perception based on known biological principles would have a beneficial effect in guiding physiological experiments. This would be particularly fortunate for the visual neuroscientist for, in the present view, one major element of the scientific process that is generally lacking is the presence of coherent, testable theories of vision. This point has recently been made in an forceful way by Marr (1982).

It is also likely that new techniques will be developed with which to approach a range of novel problems. For example, development of a high-resolution, real-time brain scanner for noninvasive monitoring of electrical activity appears to be some years away, but may be developed more rapidly through an intensive interdisciplinary effort.

Finally, both types of scientists have moral and ethical responsibilities to humanity. At some point, what the researcher does should have some impact on our understanding of the normal or pathological functions of humans. To this extent, it would appear beneficial for the neurobiologist to be more aware of problems in cognition. Perhaps experimental questions might be posed in ways that would have human implications, even if small.

What are the differences between the disciplines? While some are real, others are merely semantic. The following paragraphs will describe some of the real difficulties encountered by the neuroscientist in dealing with the topics of memory, global processing, texton theory, and global precedence effects. I will then describe semantic difficulties and illustrate these with the topics of feature detection and attention.

The phenomenon of memory poses a real barrier. Memory plays a large role in perceptual research and a very small role in sensory physiology. Familiarity, context, or "sets" are concepts that imply a stored representation of objects and their relationships with other objects. For the neuroscientist, the way in which representations are stored (and learning and memory accomplished) physiologically is currently unclear. For simple nonassociative behaviors, there is excellent evidence pointing to presynaptic calcium modulation. For more complex behaviors, there is some evidence for large, consolidating neural circuits, but no evidence for holographic models of perception or memory. Regardless of the mechanism, it is unlikely that the electrophysiologist would be able to recognize a memory representation among sensory signals (but there are exceptions to this in the invertebrate literature). Even if recognizable, the code in a single cell is likely to be only a fragment of the complete trace. Thus, the neuroscientist is ill-equipped to deal with the topic of memory.

Perhaps it could be possible to monitor memory-related activity of a few cells for small visual objects, for example, the representation of characters or others that subtended small visual angles. For these objects, the volume of active tissue would be small and monitorable by extensions of current electrophysiological techniques (i.e., arrays of electrodes). On the other hand, there is no guarantee that the retinotopic relationships between the objects and their stored traces are maintained, that is, representations may be stored in sizes that are independent of actual object size. It may be possible to identify neurons participating in stored representations by other means. For example, the anatomical localization of cells in a network may soon be visualizable by the development of high-resolution autoradiographic or computerized tomographic techniques. It does appear that current single-cell recording techniques may be of minimal use in discovering cell assembly properties.

Presumably the sensory physiologist would face similar difficulties when faced with the analysis of global processing, as distinct from context effects. Some writers, nonetheless, have attempted to equate global recognition systems with low spatial frequency detectors having high conduction velocities. This equation would appear premature, for none of the cells with fields large enough for global processing has ever been tested for either spatial frequency selectivity or conduction velocity. Conduction velocities may be meaningless at this level because of the high degree of convergence and divergence of connections from other areas. Moreover, the high/low selectivity in X/Y cells is questionable physiologically and in some cases psychophysically as well.

In contrast to single-cell approaches, the neuropsychological (ablation) approach has recently produced an extremely successful analysis of the neural basis of global versus local information processing (Sprague, Hughes, and Berlucchi, 1981). These investigators found that visual behavior in cats guided by global cues was unimpaired following lesions of V1 and V2. In contrast, when behavior was controlled by local stimulus "features," the same lesions dramatically disrupted visual discriminations. These results are consistent with the idea that the large receptive fields in the prestriate cortices may be more suited to the analysis of form than cells at earlier levels. Because there are always problems with the interpretation of lesion studies, future work should attempt to block cortical activity reversibly and more discretely. At present, then, surgical ablation is one of the few techniques available to the neuroscientist for the study of global form perception.

Attempts to translate texton theory into testable physiological hypotheses meet with some difficulty. In this instance, the logical deductions from the translation appear to be incorrect. This can be illustrated with the concept of the preattentive system. Let us assume for the sake or argument that the behavioral phenomenon of serial attention is equivalent to the physiological concept of stimulus selection (see Wurtz et al., 1980). Since stimulus-selection phenomena can be seen in the superior colliculus and pulvinar, this would appear to place the preattentive textons in the brainstem, thalamus, or retina. Does this suggest that preattention (textons) first appear subcortically? None of the possibilities is likely because none of the cells would appear capable of responding selectively to the postulated texton shapes. Selectivity to form, closure, and orientation are minimal requirements for the discrimination shown in Hoffman's Figure 2.1 in Chapter 2, but few researchers would entertain the possibility that these capacities were subcortically mediated. The more likely possibility is that serial attention is a phenomenon quite distinct from stimulus selection.

Other differences between the disciplines are evident in concepts that are simply not translatable into the language of the other discipline. For instance, "global precedence effects" are currently not translatable from cognitive to neuroscientific thought. The key problem for the translation is that most cellular processes, as we currently know them, are tuned to either high or low spatial frequencies, but not both, that is, the spatial frequency bandwidths are too narrow to permit the cells to have the capacity to process both fine and coarse stimuli. Thus, certain paradigms appear to be less amenable to the physiological approach than others.

Some differences are merely semantic. Consider, for example, the difference in the use of the term "feature." For the cognitive psychologist, a feature refers to a set of abstractions independent of retinal location, brightness, size, or contrast. For the neuroscientist, features are considered by some to represent stimuli best equipped to stimulate a particular receptive field and are therefore retinotopically tied.

There would also appear to be some disagreement as to whether, in top–down processing, hard-wired, feature-detecting modules can be modified by attention. Central to the problem is the definition, and anatomical location of a module. To the cognitive psychologist, the definition is largely operational and anatomically noncommittal, whereas to the neuroscientist it is usually linked to a specific type of neuronal element. Because the location of the attentive mechanism must reside more centrally than the modules, feedback pathways to more peripheral levels must mediate the modifications. Feedback circuits exist between the prestriate and striate visual cortices and the thalamus or midbrain. It is thus possible that feature detectors at striate cortical and subcortical levels are modifiable (e.g., see McClurkin and Marrocco, 1984).

Most cognitive psychologists, however, would probably place the modules at higher levels in the visual pathway. Almost the entire prestriate visual system appears to be reciprocally interconnected, that is, feedback is as frequent as feedforward in higher visual centers. The most common loop is between areas that are separated by a single synapse. However, examples of more distant loops are also known, for example, between V4 and V1, perhaps providing a prestriate mechanism for attentional effects on hard-wired modules. Cognitive experiments designed to restrict anatomically the locus of cells controlling behavior would represent a significant advance in localizing the feedback mechanisms.

While most of these points of disagreement are minor, others are thought to be more fundamental. According to Rock (1975), the cognitive psychologist assumes that by manipulating stimulus and task he will discover the nature of representation in the process of perception. The most fruitful approach is to identify commonalities between perceptual experiences, thus establishing the generality of a phenomenon. According to this view, the understanding of perception can be advanced without dealing directly with the activities of single cells. In contrast, the neuroscientist is more likely to investigate the underlying mechanisms of a single type of experience, which, while helping understand the neural machinery of perception, may fail to develop a general theory of perceptual experiences. Thus, Rock believes that the central tenets of cognitive psychology and neuroscience are distinctly different.

In the opinion of the present author, the foregoing differences in scope are bridgeable. There is nothing to prevent the neuroscientist from looking for neural processes common to several instances of similar perceptual experiences. Indeed, it would seem that the failure to obtain converging lines of evidence would reflect simply on the lack of thoroughness of the individual investigator rather than the field as a whole.

I feel there are distinct advantages to reconciling differences between the disciplines. The reconciliation could open some of the traditionally neglected cognitive areas to neuroscientific research. For instance, the concepts of prior knowledge and familiarity, common terms in perceptual research, have no counterparts in neuroscience. Studies of visual development, which show how cortical neuronal properties in adult animals depend critically on the nature of the stimuli present in the early environment, would seem to be the closest conceptual counterpart to familiarity. Unfortunately, the principles derived from the physiological results are not easily translatable into cognitive research because the precise role of single-cell receptive fields of the striate cortex is unclear.

As an improvement, it may be more fruitful to examine experiential effects on structures whose relevance to form perception may be clearer. Such a structure is the inferotemporal cortex, which contains cells apparently selective to complex stimuli such as faces and hands (Gross et al., 1972; Desimone et al., 1984). Are the properties of these cells hard wired, or do they depend on visual experience and familiarity? Assuming that specific deprivation of the sight of faces led to a reduction in the number of face-responsive cells and an otherwise normal visual pathway, one could determine whether the absence of these cells prevented the development of familiarity to faces.

In conclusion, this discussion has focused on some of the difficulties that have impeded scientific interchange between cognitive psychology and neuroscience. Many, but not all, of the problems are surmountable. The key to improvement lies in the acceptance of perceived mutual benefits for both areas, leading to more frequent dialogues and joint participation in the process of experimental design.

A psychologist's reply

James E. Hoffman

This dialogue has identified concerns that are both common and unique to the fields of neuroscience and cognitive psychology. The methods of the neuroscientist seem best suited to studying the structure of "early" or preattentive processes. In contrast, cognitive psychologists are primarily interested in how knowledge and attention can be used to integrate the information provided by early visual processes into stable representations of recognizable objects. Are these two areas fundamentally different or are there useful interactions that can be pursued?

One reason for pursuing a dialogue is that each field suffers shortcomings that might benefit from insights and constraints offered by the other. The physiological approach cannot, by itself, provide information on the functional significance of receptive fields. Cognitive psychology, in turn, needs converging evidence for inferences based on behavioral measures, such as accuracy and latency of responses. Consider first an example in which behavioral data proved useful in guiding the search for receptive fields with known functional significance. Regan, Beverley, and Cynader (1979) pointed out that when an object moves toward an observer, he/she can determine whether the object is on a collision course with the head by noting the relative direction and velocity of movement of the retinal images in the two eyes. An object directed at the head will produce movement of the retinal images in opposite directions. A noncollision path will result in movement of the two retinal images in the same direction. The speed with which this information can be extracted suggests that there may be a specialized channel or module dedicated to this function. They confirmed this conjecture in two ways. First, they found that subjects who were adapted to constant movement of objects along a collision path showed lower sensitivity to detecting threshold movement along the same path. The neural basis for these channels was revealed by single-unit recording of cells in the cortex of cats. Single cells were found to be selectively sensitive to movement along different paths relative to the head.

This research project serves as an elegant example of the power of a combined behavioral and neurophysiological approach to perception. Marrocco's comments on my chapter provide several additional examples of areas in which this kind of interaction might be fruitful. Visual attention, in particular, is an area with a long history of this combined approach, and recent developments suggest that impressive progress is being made. Evidence from behavioral experiments, event-related brain potentials, single-unit recording, and neuropsychology is being brought to bear on determining the site of attentional mechanisms in the nervous system.

89

This is the kind of approach that Marrocco has in mind when he suggests that the physiological evidence may be in conflict with Julesz's (Julesz and Bergen, 1983) "texton theory," which suggests that reasonably abstract properties such as closure may be computed by preattentive processes. If attention operates at the level of the superior colliculus, for example, then it would be difficult to claim that these kinds of properties are extracted at earlier levels of the nervous system because their receptive fields appear to be incapable of the required analysis. The available evidence, however, appear to implicate posterior parietal cortex as one of the major sites of visual attention. Wurtz and Goldberg (1972) found that cells in the superior colliculus showed enhanced responses to light falling in their receptive fields when these lights were the targets of saccadic eye movements. Thus these cells were involved with the oculomotor system but were not "attention cells" because movement seemed to be a necessary determinant of the enhancement. This is not true of spatial attention. Only cells in posterior parietal cortex were found to show an enhanced response to attended lights independent of the requirements to make motor movements (Bushnell, Goldberg, and Robinson, 1981).

It is interesting to note that there is now considerable converging evidence for the claim that the posterior parietal cortex plays a key role in visual attention. Patients with damage to the parietal area show difficulty in directing attention to the contralateral direction (Posner, 1984). In addition, such patients show reductions in attention-related components of the event-related potential (Knight et al., 1980).

More recent evidence (Moran and Desimone, 1985) indicated that two earlier sites in the visual system are affected by voluntary attention. Cells located in areas V4 and IT in the monkey cortex show modulation of activity within individual receptive fields as a function of where the monkey is attending in visual space. These results, together with work on the parietal area, indicate that selection is occurring relatively "early" in the visual system and at several levels.

The ability to record single-unit activity in awake, behaving monkeys makes it possible to explore the cellular basis of a variety of cognitive processes. This kind of approach will be of continuing interest to cognitive psychologists and fosters a collaborative approach. Marrocco's suggestion that such an approach could be applied to exploring feature-integration theory is a good one, and I hope that this kind of interdisciplinary approach will be of interest to the community of scientists in the fields of cognitive psychology and neuroscience.

5 A psychological view of the neurobiology of perception

James E. Hoffman

In Chapter 3, Marrocco reviews a large body of evidence concerning the structure and function of individual cells in the visual nervous system. One can only be impressed at the ingenious new methods that have been developed to trace the origin and destination of synaptic connections between nerve cells. In addition, new kinds of stimuli, based on sine-wave gratings, have been employed with increases in the quantitative specification of cell responsiveness. The principle questions that this evidence poses for a cognitive psychologist are the following: Do these findings provide useful constraints on theories of perception? To what extent should cognitive psychologists make use of findings based on the activity of single cells? Finally, can any useful areas of convergence be identified between the two fields?

In attempting to answer these questions, it will be useful to make explicit the different concerns of investigators in the areas of cognitive psychology and neurophysiology. Cognitive psychology is concerned with an information-processing approach to vision that seeks to discover the nature of internal representations and processes that are the basis of vision, imagery, and other spatial skills. A good example of this approach was provided by the late David Marr and is eloquently presented in his book *Vision* (Marr, 1982). He defines a representation as a "formal system for making explicit certain entities or types of information, together with a specification of how the system does this" (Marr, 1982, p. 20). The nature of a representation is closely tied to the kind of algorithms that are applied to the representation for the purpose of solving a particular problem. As an example, Marr points out that although Roman numerals provide an adequate representation for quantity, they do not easily lend themselves to arithmetic operations such as multiplication. Thus, the choice of representation will have important consequences for what kinds of information can be easily and quickly extracted.

Marr (1982) further points out that questions about the nature of algorithms and representations are independent of questions about the nature of the hardware implementation. Clearly, this latter level of analysis is the domain of neurophysiology. To the extent that these two domains are separate and

91

independent, the disciplines of cognition and neurophysiology will remain separate with little in the way of fruitful interaction. An extreme version of this view is the claim, often repeated in arguments about levels of analysis, that one could not understand the chess-playing behavior of the local campus mainframe by examining successive states of the computer memory over time. There seems to be two reasons for the apparent plausibility of this argument. First, it is clear that the sheer amount of detail regarding the behavior of individual memory elements would be difficult to relate to the complex behavior of chess playing. Second, computer memory consists of "general purpose" components that are completely unrelated to the kind of information that they can contain. Their contents must be interpreted by a general purpose processing unit.

The situation may be considerably different in the human brain. Instead of a single processor accessing a large, undifferentiated memory, the brain is more like a set of relatively independent and specialized modules. The instruction set of this computer is directly encoded in the structure of synaptic connections between neurons. These massively parallel, distributed processors are being actively explored as plausible models for reading (McClelland and Rumelhart, 1981), speech perception (Elman and McClelland, 1983), shape recognition (Hinton, 1981), and strereopsis (Marr and Poggio, 1976). There are two reasons why these models make the chess-playing scenario lose some of its force. First, there should now be a relation between location and function so that any given brain area computes only certain kinds of relations. Second, knowledge of structure may provide valuable clues about function.

The preceding arguments are meant to suggest that although cognition and neurophysiology are directed at understanding different levels of the vision problem, there are likely to be areas of fruitful interchange as well as areas that remain the unique province of each discipline. I would like to hazard a guess, first, at what aspects of perception are likely to benefit from an exchange of ideas and, second, what problems in the domain of cognition currently remain beyond the reach of neurophysiology.

A key tenet in Marr's (1982) theory of vision is that there are several distinct representations constructed by the visual system in perceiving a scene. One broad distinction may be made between "early" and "late" processes. Early processes result in what Marr calls the "2½-D" sketch, which represents information about the orientation, distance from the observer, and presence of discontinuities between local surface patches. The representation is in a "viewer-centered" coordinate system, that is, it represents the appearance of the visual world as seen from the particular vantage point of the observer. The 2½-D sketch is constructed without reference to memory for objects, and thus is the result of bottom–up or data-driven processes.

Later stages are responsible for constructing a description of the scene that is suitable for matching to representations stored in memory. Thus, it is only after early processes have provided detailed information about the local geometry of visible surfaces that factors, such as expectancy, context, and attention, can play a role. Early vision forms a "module" that takes as input the retinal image and delivers an output consisting of the 2½-D sketch. Marr argues that the algorithms used in the construction of the 2½-D sketch can be investigated by carefully examining the nature of the physical world to see what kinds of constraints it affords and then building a computer implementation that exploits these constraints. When possible, the algorithm should account for known psychophysical data and be consistent with single-cell data. Marr and Paggio (1976), for example, constructed an algorithm for finding shapes in the random dot stereograms invented by Julesz (1971). This algorithm worked by assuming that, first, any given point in one eye should match a single point in the other and, second, that changes in disparity between neighboring points in the image should be small. These assumptions, based on the properties of surfaces in the three-dimensional world, provided the necessary constraints to arrive at a unique solution. In addition, the algorithm accounts for several phenomena known to occur in human stereovision.

It seems likely that single-unit recording data will be most useful in understanding processes at the level of the 2½-D sketch. Such data will be especially useful when they are collected with an eye for the kinds of problems that early visual processes must solve. Finally, it seems likely that later visual processes, with their dependence on attentional and memory processes, may be difficult to study with single-cell techniques. Nonetheless, other kinds of physiological measures may prove useful in understanding these phenomena as well.

Areas of convergence

Spatial frequency analysis

There are several examples of useful exchanges between neuroscientists and psychologists interested in vision. Marrocco (Chapter 3) provides a good review of the utility of spatial frequency concepts in characterizing the response of single units. This approach was initiated by Campbell and Robson (1968) when they measured the contrast sensitivity function for human observers using psychophysical techniques. A large number of subsequent studies (reviewed by De Valois and De Valois, 1980; Graham, 1980) have firmly established that the visual image is initially analyzed in separate channels, each one most sensitive to a different spatial frequency. Wilson and Bergen (1979) estimate that there are four such channels, a conjecture that was recently supported by Sekuler, Wilson, and Owsley (1984), who showed that the pattern of deficits in spatial vision that occur with increasing age is consistent with loss of sensitivity in various subsets of four channels.

Marrocco contrasts the spatial frequency approach with the older theory that cortical receptive fields are responsible for extracting "features," such as lines, angles, etc. The feature approach is nonquantitative. In addition, it is clear that the receptive fields described initially by Hubel and Wiessel (1959) are incapable of unambiguously signaling the presence of features. For example, simple cells are not only activated by lines but also by spots and a variety of other stimuli. Nonetheless, it is still possible to maintain that early vision is designed to extract features. Frisby (1980) points out that although any single receptive field has multiple response determinants, edges and lines may be detected by examining sets of receptive fields. A spot, for example, might stimulate both a simple cell and a circular-surround receptive field. Thus, the unique "signature" of a particular feature might lie in the across-fiber pattern in a set of detectors.

Marrocco also points out that sine-wave gratings have proved to be more effective stimuli for cortical cells than lines or edges, consistent with the claim that these cells are performing a Fourier analysis of the image. The spatial frequency approach might yield similar benefits in evaluating the perception of complex forms such as letters for humans. Gervais, Harvey, and Roberts (1984) recently showed that the confusions that subjects make in identifying briefly presented letters are better explained by similarity in the spatial frequency domain than by overlap in features. Thus, considering the spatial frequency content of a stimulus together with the contrast-sensitivity function of the visual system allows for relatively accurate predictions of perceptual errors.

It is clear that the spatial frequency approach offers a compact, quantitative description of some of the spatial characteristics of early vision. Nonetheless, there are some definite limits to this approach that suggest caution in accepting the proposition that the visual system computes a Fourier analysis on the retinal image. For example, it is reasonable to ask whether there are any advantages for the visual system to transform the input in this way. Pinker (1984) notes that recognition might proceed by comparing the frequency spectrum for an unknown shape to a set of spectra in long-term memory corresponding to known objects. One advantage of this approach is that the matching process will be invariant under certain kinds of transformations of the visual image. Changes in the position of an object would primarily change the phase spectrum of the image so that matching based on the frequency spectrum would be unaffected. In addition, because information about gross shape is represented primarily by low spatial frequency channels, while sharp edges are contained in high frequencies, the matching process could still be successful even when the edges in the shape are slightly distorted or embedded in noise. This separation of information at different size scales is one of the most attractive aspects of the Fourier approach.

Thus, a transformation of the retinal image to the spatial frequency domain does offer some advantages in implementing a shape-recognition process. Nonetheless, Pinker (1984) points out that this approach faces difficulty when faced with the kind of cluttered scenes that are found in the "real world." The Fourier analysis of a scene will produce a spectrum that is a complicated combination of spectra for individual objects. In other words, this approach does not address the problem of how objects are separated or parsed from their backgrounds. It would appear that this separation is necessary for object recognition to be successful.

The alternative of trying to match the representation for the entire scene to memory is unattractive because objects can be arranged in an essentially infinite number of ways that would produce corresponding changes in the frequency spectrum. There certainly are "scene level" effects on object perception as shown by Biederman (1981), but these effects cannot be conceptualized as trying to match a literal view of a scene with a copy in memory. Scenes must be represented in a much more abstract way than this. Once a scene has been transformed to the frequency domain, figure-ground separation would appear to be even more difficult than it would be if something like features were employed. The only theories of the figure ground process that we presently have, those due to Guzman (1969) and Waltz (1975), are based on interpreting the nature of junctions between lines representing edges of intersecting objects.

It may be that the visual system is employing channels sensitive to different spatial frequencies, not to perform a Fourier analysis of the image, but to filter it into representations having different scales. Phenomenologically, we can attend to a shape at several different levels of detail. For example, in viewing a face, we can either attend to the whole configuration or switch our attention to a component part such as an eye. This "object" can in turn be decomposed into pupil and iris. This hierarchical decomposition can be continued until limits imposed by acuity are reached. This phenomenological description is supported by experiments that show that when subjects are required to recognize shapes at different levels of a form, they appear to match each level to memory in a serial fashion (Hoffman, 1980; Kinchla, Solis-Macias, and Hoffman, 1983).

X and Y cells

A second area reviewed by Marrocco, in Chapter 3, that has affected theorizing in perception is the distinction between X and Y cells. X Cells are described as having slow conduction velocities, sustained response, and small receptive fields, making them sensitive to high spatial frequencies. In contrast, Y cells have fast conduction times, a transient response, and large receptive fields. These properties suggest the presence of two separate analysis systems, perhaps operating in parallel. The different properties of these two channels have been the basis for explanations of two different perceptual phenomena: visual masking and global precedence.

Breitmeyer and Ganz (1976) and Breitmeyer (1980) have suggested that visual masking can be understood as an interaction between the sustained and transient channels. Consider the case of nonmonotonic masking functions in which the identity of a target form is impaired by flanking forms presented at some delay after the onset of the target (metacontrast). Breitmeyer and Ganz suggest that the sustained activity initiated by the target is interfered with by transient activity produced by the mask. This interference may result from inhibition between the two systems. The finding that maximum interference is obtained only when the mask is delayed relative to the target appears to be a natural consequence of the faster transmission times enjoyed by the transient system.

The Breitmeyer and Ganz (1976) model is a good example of the strategy of using data derived from single-unit recording experiments to explain a perceptual phenomenon. On the positive side, it seems natural to use information about the properties of early visual processes as heuristics for constructing models on a psychological level. As Uttal (1981) points out, however, these models are vague in specifying the link between neural activity and perceptual experience. Target identification requires at least some of the shape-recognition processes described earlier and is thus a fairly complex process. The model does not specify precisely how activity in sustained channels is used to recognize the target nor what the role of various decision and memory stages would be in this task. In fairness, however, it should be pointed out that this level of specificity in psychological models is more the exception than the rule.

Very similar considerations can be applied to evaluating the hypothetical role of these channels in explaining why large shapes are sometimes recognized more quickly than small component shapes. This global precedence could result from low-frequency, transient channels transmitting the global shape information prior to the slower, sustained channels, which would be responsible for the fine details of the shape. This kind of "explanation" is, however, only a very loose analogy and may, in important respects, be misleading. First, if the shapes used by Navon (1977) in his original investigation of "global precedence" are high-pass filtered so that the low spatial frequencies are removed, the global shape is still clearly visible. Thus, in these forms, global shape is not specified by low spatial frequency information.

Second, the global precedence effect has turned out to be a complex one, and the claim that the visual system invariably accesses large shapes before small ones is clearly not the case. The order of access to levels is at least partially under control of the subject and appears to be a manifestation of a spatial attention process. In addition, as I pointed out in Chapter 2, the visibility and arrangement of the parts making up the whole play an important role in determining which level achieves precedence.

Cortical maps

A final area reviewed by Marrocco, in Chapter 4, is the intriguing evidence that the cortex contains several spatiotopic maps, each specialized for different dimensions, such as color, movement, binocular disparity, and so forth. These studies provide at least a rough confirmation of the psychological evidence reviewed in Chapter 2 that the image is initially processed into separable features. There, is, at present, no single-unit evidence on the second claim of Treisman's theory (Treisman and Gelade, 1980) that it is spatial attention that integrates features together into unitary "objects."

Marrocco, in Chapter 3, suggests that once again, it is wise to be cautious about claiming too close a correspondence between the nucrophysiological results and inferences drawn from performance measures gathered in cognitive psychology experiments. He lists several useful criteria for deciding whether a cortical map is really devoted to processing one dimension of an input. One of these states that the cells should be activated by only one feature. The available evidence suggests that cortical cells do not possess the required level of specificity. For example, many of the cells found in V4, the so-called color area, are also sensitive to shape features such as orientation. Evidence from the cognitive domain, however, indicates that color and shape are separable dimensions. Texture segregation cannot be accomplished, for example, if two areas differ in a combination of color and shape (Treisman and Gelade, 1980).

This conflict between results from single-unit recording and inferences regarding the nature of feature detectors derived from cognitive psychology experiments may indicate that these methods are tapping into visual processes at different levels. In Chapter 2, I pointed out that spatial frequency and orientation act as integral features in psychophysical procedures, such as selective adaptation, but appear to be separable in performance situations, such as texture segregation.

A similar dissociation appears to hold for color and shape. According to Treisman's taxonomy, these dimensions are separable and should require attention to be integrated. There are, however, aftereffects, such as the one discovered by McCollough (1965) that depends on a combination of color and shape. The occurrence of such contingent aftereffects suggests that the visual system may treat these dimensions as a composite. Houck and Hoffman (1984) confirmed this conjecture by showing that the strength of McCollough aftereffects was independent of whether or not the adaptation gratings were attended. Thus, both the single-unit data and psychophysical adaptation data suggest that color and shape are treated as integral dimensions in early vision only to act like separable dimensions in texture and search experiments. Resolving these discrepancies should provide important information about the interface between early vision and later processes that depend on memory and attention.

In summary, these different areas show that single-unit data provided by neurophysiologists continues to be used to guide the construction of psychological theories. The correspondence between levels, however, is a loose one. It is doubtful, for example, that any psychological theory will stand or fall on the basis of new findings from the single-unit area.

Beyond the 2½-D sketch

The correspondence between physiological data and behavioral outcomes is likely to be even fuzzier as one leaves the data-driven processes used to construct the 2½-D sketch and considers processes involved in shape recognition. Marr (1982) claims that the 2½-D sketch does not contain representations of objects, only surfaces. Attention and memory must play a role in the linking of parts into objects and the subsequent match of this representation to memory for recognition. Little is known at the single-cell level about the nature of object recognition. Gross, Rocha-Miranda, and Bender (1972) report a cell in the inferotemporal cortex of monkey that appears to be responsive to the shape of a monkey hand. However, as Marrocco points out, this cell gives some response to the objects that clearly are not hands, so it seems unlikely that this cell is the terminal point in a pattern recognition system for hands. In addition, it is not clear that the discovery of such cells will necessarily reveal much about the shape recognition algorithm. One wonders whether the activity of such cells will be invariant with changes in viewpoint, occlusion, etc.

Other methods besides the single-unit approach may prove useful in understanding higher-level processes in perception. For example, theories of reading developed in standard cognitive paradigms have proved useful in understanding deficits resulting from various kinds of brain injuries (Coltheart et al., 1983). The pattern of reading deficits that are observed in patients can, in turn, add converging evidence for reading theories as well. Its interesting to note that certain kinds of brain injuries result in the inability to experience certain specific perceptual experiences. Treisman (1985) summarizes a number of such cases in which patients are unable to perceive smooth, continuous motion or are unable to integrate separate features into a unified object. Although there are well-known difficulties in interpreting the extent and nature of injuries in patients, these cases may provide converging evidence for well-developed theories of different perceptual phenomena.

Event-related potentials

Another technique that may prove useful to cognitive psychologists is the event-related brain potential (ERP); ERPs have proved to be a useful clinical tool in assessing visual function in adults and babies (Hillyard, Picton, and Regan, 1978). Recent research indicates that they may provide some insight into higher-level processes as well. For example, the claim that subjects may process multidimensional stimuli as separate features has received support from a number of recent ERP experiments (Previc and Harter, 1982). Hillyard and Munte (1984) recently showed that when subjects were instructed to attend to bars of a certain color in a particular location, two separate ERP components could be observed corresponding to these two dimensions. Stimuli that were in the attended location produced an enhancement in the amplitude of early components (100–200 msec after stimulus onset). Stimuli that also has the attended color produced additional negativity somewhat later in time. These results suggest that subjects first test a stimulus for the presence of the location feature, and only if this test is positive do they go on to test for the presence of the color feature. Thus the ERP data are consistent with a serial, self-terminating feature testing model. They also add converging evidence for the claim that separate features belonging to a single object are processed separately at some point in the brain.

Event-related potentials are presently used primarily as a chronometric tool to study the timing of different mental processes. They are not fundamentally different than a reaction time. They do offer a few advantages, however, that may justify the difficulties involved in measuring them. First, the order of different ERP components (and thus the order of processing stages) can be directly observed. Although there are reaction time (RT) methods that allow one to infer *which* stages may be present (Sternberg, 1969), their order is a matter of guesswork.

The weakness of ERP methods is that, at present, it is difficult or impossible to localize the source of particular ERP components in the brain. New techniques based on magnetic fields should improve spatial resolution. In addition, animal analogues of cognitive experiments should allow depth and ERP recordings to be simultaneously collected and compared. These improved techniques should enhance the usefulness of ERP measures, which are, at present, virtually the only way to gather physiological data about human brain processes that occur in millisecond time periods.

Summary

In summary, single-unit data gathered at different levels of the visual system are likely to be of continuing interest to cognitive psychologists. This level of analysis should provide insights into the nature of early visual processes. Higher-level processes, such as attention and the use of top–down knowledge, may be more difficult to study with current single-cell techniques. Studies of brain-damaged patients and the use of event-related brain potentials may be more suitable measures at this level.

A neurobiologist's reply

Richard T. Marrocco

Since the purpose of this treatise was to contrast different approaches to the study of single areas of behavior, it is worth assessing what we have learned from our comparative survey of approaches to the study of perception. One theme was apparent in both the discussions of Hoffman and Marrocco. The potential for an enhanced interchange of ideas between cognitive and neural science exists in several areas and is highest for studies of the early stages of visual processing. Areas of high potential include the role of X and Y cells in determining different size spatial frequency channels, the specificity of cells in different cortical areas for stimulus dimensions, and the ability of cortical neurons (e.g., the middle temporal area) to discriminate between figure and ground. The least interaction will likely occur over the later stages of processing, and these areas will probably remain the exclusive property of cognitive science for the near future. Areas that are likely to remain separate include the areas of imagery, expectancy, and reasoning. At the risk of being overly optimistic, some of the apparently separate areas may be brought closer by advances in technology. Others appear to be genuinely beyond reach of the most intimate collaboration.

An examination of the foregoing chapters suggests that an interchange has taken place, and it is thus worth asking what new communication has occurred. In assessing the novelty of the dialogue, let us consider the discussions of the specificity of cortical areas for stimulus dimensions. Current theories of feature integration (e.g., Treisman and Gelade) are based in part on earlier findings, suggesting a strict attribute specificity within different cortical areas. Because specificity appears in more recent findings to be more relative than absolute, we may conclude that attention to a particular attribute involves the gating of connections to several prestrite areas. In addition, it now seems reasonable to consider the possibility that attention must be able to couple or decouple attributes. How this might be accomplished is unclear, but it might help circumvent the problems raised by the physiological law of multiple response determinants. In any case, the dialogue has laid the foundation for testable congitive and physiological hypotheses.

One of the more promising approaches for the future is to teach the alert primate to perform in ways that will reveal its cognitive abilities. I am not going to claim that this preparation will allow us to understand all the many of the major issues facing cognitive science. However, I do believe that significant progress will be made if this approach is carefully planned.

There are at least two problems that must be met. First, one must train the monkey to perform in a paradigm that explores top–down problems. While the training task will surely be near-Herculean, the rewards will justify the effort. In fact, several promising starts have already been made toward this end (e.g., the studies of Wurtz et al., 1980; Mountcastle et al., 1981) on mechanisms of attention, the studies of transient memory and expectancy by cells in the prefrontal lobes by Fuster (1980) and Komatsu (1982) and the studies of short-term visual storage in inferotemporal neurons by Gross et al. (1979). Indeed, it would seem that many current psychophysical studies of human vision are currently adaptable to monkey subjects.

The second, and more difficult problem is the choice of response measure. While some top–down tasks may benefit from single-cell analysis, many others will not. For those areas that may benefit, the convergence of the cognitive problem within the physiological preparation offers a unique setting for explicit theory testing, that is, here is an opportunity for cognitive theorists to make specific, immediately testable, physiological predictions. Although it is probably true, as Hoffman (Chapter 5) points out, that a cognitive theory's viability is not dependent on any single physiological outcome, several unfulfilled physiological predictions should be a sufficient impetus to revise the theory.

Finally, the dialogues concerned with spatial relationships (e.g., global recognition effects) were the most speculative. I have argued that single-cell approaches may be unproductive, and I believe Hoffman's suggestion to combine single-cell and event-related potentials in primates is well worth exploration. Nonetheless, the dialogue potential between cognitive and neural scientists for this new field appears to be good. However, the discussions of several areas (e.g., speech recognition, word superiority, etc.) were more like monologues than dialogues, and an understanding of the neural bases for these phenomena appears to lie well beyond the capabilities of current electrophysiological methods.

Part II

Attention

6   The psychology of attention

William Hirst

People use the concept of attention as if they know what the word means, complaining about their inability to "pay attention" or their habit of letting their "attention wander." As William James (1890) noted,

Everyone knows what attention is. It is the taking possession by the mind, in clear and vivid form, of one out of what seems several simultaneously possible objects or trains of thought. Focalization, concentration, of consciousness are of its essence. It implies withdrawal from some things in order to deal effectively with others. (p. 403)

The present chapter outlines the attempts of psychologists to give scientific muscle to James's description. This chapter does not cover everything that might be discussed in a fuller treatise on attention. It is concerned solely with the cognitive psychology of attention. Any fully articulate theory of attention will have to concern itself with both mind and brain, yet, to a great extent, the cognitive psychology of attention has proceeded with little regard to neuroscience. The narrow focus of this historical account reflects, in the main, the narrow theorizing of most research in attention.

The study of attention usually falls into two separate areas: divided attention and selective attention. To attend to an object is, no doubt, to process selectively that object over others also stimulating the system. The study of the mechanism that allows an organism to select one object or message over others constitutes the field of selective attention. The study of divided attention is concerned with limits on the amount that an organism can select. People can easily selectively attend to one object, but may find it harder to attend to two or three objects at the same time. Students of divided attention determine the number and kinds of things a person can attend to at the same time. Inasmuch as researchers of attention have treated the study of selective attention and divided attention as separate areas of research, the present chapter does as well. The research on selective attention will be reviewed initially; a review of the literature on divided attention then follows. The

I want to thank Edward Levine, Ron Cordova, Catherine Hanson, and Liz Hatch for reading and commenting on earlier drafts. Support from NIH Grant 17778 is gratefully acknowledged.

105

reader, however, should be aware that the theorizing in one field inevitably shaped the theorizing in the other, and these connections are noted at the appropriate points.

Selective attention

For the first half of this century, the study of attention was not considered a legitimate area for scientific study. According to the dogma of the time, attention was without a doubt a mentalistic concept, and mentalism and science did not mix. The decline of behaviorism and the birth of cognitive psychology, however, changed the scientific scene. Not only was attention a legitimate area of study, it proved one of the most amenable to scientific study.

When cognitive psychologists, particularly Cherry (1957) and Broadbent (1958), began to study attention in earnest, they approached their chosen subject with a clear idea of how the human mind was configured. They compared it to a communication channel. Information impinges upon this channel, enters, is processed, and, in the course of this processing, is transformed into an output. The telephone, for instance, takes as input a human voice, transforms it into electrical impulses, sends them through wires, often boosting the electrical impulses along the way, then translates these impulses back into sound waves that resemble a slightly degraded human voice, the output of the process. Computers are another example of a chain of processors mediating input and output. You type numbers into the computer, these numbers are transformed, rearranged, and worked upon by the processing mechanisms internal to the computer, and finally, an appropriate output emerges on the computer screen.

The analogy to the human organism is clear. Sound waves, for instance, enter the system, are processed for meaning, and are often transformed into a response, realized by moving the mouth. This "information-processing" metaphor has guided much of the research in cognitive psychology since the early 1960s. As Neisser (1967) showed, the job of the cognitive psychologist is (or was) to study the flow of information through this system, monitoring the processing and transformations that occur between input and output.

One assumption is central when the information-processing metaphor is applied to the study of attention. Just as telephone wires can transmit only so many messages and computers can process only so much information at a time, so is the capacity of the human organism limited. The human organism will become overwhelmed unless it selects from the multitude of information impinging on it the one message that it wants to process accurately and meaningfully. A cocktail party, for instance, is a hubbub of conversation.

People could not function in such a noisy setting if they were unable to select one conversation over other competing conversations.

Much effort has gone into understanding the character of selective attention. As the present chapter will make clear, the mechanism can be studied without relying on introspection. More importantly, the central questions can be phrased without introducing terms such as "consciousness." Nevertheless, whereas the study of attention may have proceeded on solid scientific grounds, the exact nature of the phenomenon has proved difficult to articulate.

Determining the location of the bottleneck

The metaphor of the human mind as an information-processing channel presents a straightforward means of conceptualizing selective attention. Processing can be either serial – where only one process can go on at a time – or parallel – where more than one process can occur concurrently. Clearly, selection must occur at some point within the processing channel. As information enters the system, it must, at least initially, be processed in parallel. Selection could occur quite early on in the chain of processing, in which case the system would have to convert the parallel processing of new information quickly into serial processing. The point of conversion could be thought of as a bottleneck. At this bottleneck, the desired message is selected and transferred to a serial processor. Because the incoming information has been barely processed, selection at this stage must be based on the stimulus characteristics of the message.

On the other hand, the bottleneck may occur late in the processing channel. Parallel processing could occur until the the message is more deeply processed, even processed for semantic content. Selection, then, would occur right before or synonymous with the mechanism responsible for decision making or response selection. As a consequence, selective attention could be guided by semantic characteristics.

Kahneman (1973) had this scenario in mind when he argued that the study of selective attention could be viewed as a search for the bottleneck in the human information-processing system. At the time Kahneman made this claim, it appeared to capture the extant research. According to Kahneman, the issue was whether the bottleneck occurred early on in the chain of processing or toward the end, or, to put it another way, was selection based on stimulus characteristics alone or on semantic characteristics as well?

Although this question appears straightfoward enough, the brief history to follow will indicate that there is no fixed bottleneck at a fixed location. The first psychologists investigating the whereabouts of the bottleneck placed it quite early in the processing chain; a second generation of psychologists placed it toward the end; and a third generation sought a compromise position and allowed the bottleneck to move around the information-processing system to meet task demands, thereby weakening the vigor of the original metaphor.

Early bottlenecks

The dichotic listening experiment. The search for a bottleneck involved an ingenious experimental paradigm first suggested by Cherry (1957) and fully developed by Broadbent (1958). Called dichotic listening, its origins are found in the cocktail party in which several messages impinge upon the sensorium simultaneously, and the organism must select one to attend to. In the dichotic listening paradigm, subjects listen to two messages presented through earphones. They are told to attend to one of the messages and ignore the other. In order to make sure that subjects are indeed attending to the designated channel, the experimenter instructs the subjects to repeat everything that they heard on the attended message, a procedure called shadowing. Shadowing errors are recorded.

The dichotic listening experiment proved a remarkable tool for discovering the basis on which the selection of one message over another is made. It was easier to shadow the attended message when the two were read by people of different gender than when they were read by the same person (Treisman, 1960), when the two messages were spatially separate than when they are located in the same space [for instance, piped into different ears or piped into the same ear (Broadbent, 1954)], when the messages were in different languages than when they were in the same language (Treisman, 1964), or, most generally, when the two messages could be easily distinguished on the basis of some stimulus characteristic than when the two messages could only be vaguely distinguished. These findings led Broadbent to posit a stimulus filter.

Broadbent's early filter model. Broadbent (1958) located the bottleneck early on in the chain of internal processing. Specifically, he posited a filter that blocked out all incoming messages different from those of the attended message and allowed only the "selected" message to be further processed. Figure 6.1 contains a schematic representation of his model. Information enters the system through the senses and rests temporarily in a short-term store. Up to this point, everything that entered the system is processed; it is at this point that the processing is whittled down. The system must select the message that will be entered into the limited-capacity p system. The message up to now has not been processed much beyond the stimulus level. In the p system, it will be processed fully and yield either a memory or a response. Broadbent called the selection mechanism a stimulus filter because it based its selection on the stimulus features of the competing messages.

According to this model, messages that do not share the same stimulus

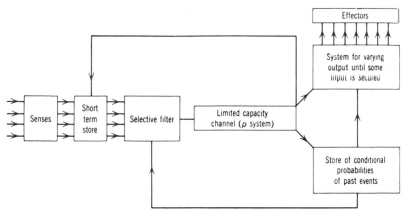

Figure 6.1. Broadbent's filter model. (From Broadbent, 1958.)

characteristics as the attended message should not get through the filter. But a person at a cocktail party can hear his name mentioned by a women even though he is attending to a man's voice. Moray (1959) nicely incorporated this observation into a dichotic listening experiment. The unattended channel in his experiment consisted of commands. Subjects were required to shadow the attended channel, but also to follow the commands of the unattended channel if they heard them. Half of the commands were preceded by strongly affective words, such as curse words. On the average, subjects obeyed 51% of the commands preceded by a high affect word and 11% of the commands preceded by a neutral word. Subjects heard the affective word and were drawn to the unattended channel. They did not hear the neutral word and consequently continued to ignore the unattended channel. The findings suggested that selection could occur on the basis of things other than stimulus characteristics.

Treisman (1960) reinforced this finding with an ingenious variation of the dichotic listening experiment. She asked subjects to shadow the message sent to the right ear while ignoring the message to the left ear. Location, then, was the stimulus feature on which the selection was putatively based. The messages on the two ears, however, switched input channels at various points in the experiment. For example, subjects might hear:

> Attended channel: leaving on her passage an impression of grace and/ singing in
> Unattended channel: singing men and then it was flying in the tree/ charm and a

Treisman found that subjects had a marked tendency to continue shadowing the message for at least a second or two even though it had switched to the channel that they were supposed to ignore. Thus, for the example above, they might shadow "leaving on her passage an impression of grace and charm and" before realizing their mistake and only then switch back to the message on the left ear. The experiment demonstrated that selection can be based on more than the stimulus features of the attended message. In Treisman's experiments, subjects must have based their selection in part on the semantic characteristics of the message. Otherwise, they would not have shadowed the "wrong" input channel at the points that they did.

Treisman's attenuation model. Treisman (1969) did not move the bottleneck further up the chain of processing, in reaction to Moray's and her results. Rather she assigned a new role to Broadbent's filter. Instead of blocking out all incoming messages incompatible with the attended stimulus, her bottleneck weakened the incompatible message, that is, she replaced Broadbent's filter with a Treisman attenuator.

She accounted for the range of available evidence by adopting Morton's (1969) logogen model of semantic memory. Each word has associated with it, in a mental lexicon, a logogen, which is activitated is the presence of the associated word. If the activation achieves a certain threshold, the logogen then "fires" and the word enters consciousness, so to speak. The thresholds of logogens differ. Highly affective words, for instance, have a low threshold, whereas neutral words have a higher threshold. A whispered neutral word may not be heard, but a whispered curse word will.

Consequently, the attenuator assures that only unattended words with low logogen thresholds will be heard. The rest of the unattended message will never achieve a high enough level of activation to fire their respective logogen, Neutral words in the unattended message will never be heard, yet even attenuated curse words will fire the appropriate logogen and be heard.

Late bottlenecks. Several psychologists failed to find Treisman's attempt to accommodate the extant evidence compelling (see Kahneman, 1973, for a review). People seemed to be able to process the unattended message for semantic content, and, to these psychologists, a model of selective attention should face this fact squarely.

The evidence for semantic processing came not only from experiments such as those Moray and Treisman. With an early bottleneck, unattended information should not be processed to a semantic level – even unconsciously. Some lines of evidence support this conjecture. Subjects in a dichotic listening experiment cannot easily detect targets on the unattended channel (Moray and O'Brien, 1967; Treisman and Geffen, 1968), nor can they recall from long-term memory information on the unattended channel (Norman, 1969). Several experiments have shown that the message on the unattended channel is available as long as it rests in short-term memory, but cannot be remembered as soon as subsequent messages displace it from short-term memory (Glucksberg and Cowen, 1970).

Other lines of evidence, however, suggest that unattended information is deeply processed. Corteen and Wood (1972) and Corteen and Dunn (1972), for instance, shock-conditioned subjects to city terms and then, in a dichotic listening experiment, periodically presented these shock-conditioned words on the unattended channel. Although subjects claimed that they did not hear anything on the unattended channel, they nevertheless showed a clear galvanic skin response (GSR) to the unattended city terms. Interestingly, the GSR was found not only for the conditioned words, but also for other city terms. This finding suggests some level of semantic generalization and, according to Corteen and Wood, provides evidence that the unattended message was processed beyond the stimulus level, indeed, probably to a semantic level.

Several other dichotic listening experiments offered the same conclusion. Lewis (1970) showed that the unattended word could interfere with the shadowing of an attended word if the two words were semantically associated. Lackner and Garrett (1972) and MacKay (1973) demonstrated that the paraphrasing of an attended ambiguous sentence could be affected by the semantic content of unattended message. In Lackner and Garrett's demonstration, the attended and unattended messages were, for example:

> Attended: The spy put out the torch as our signal to attack.
> Unattended: The spy extinguished the torch in the window.

Subjects were to shadow the attended message and then paraphase it. Preexperimental analysis established the biases subjects had in paraphrasing the ambiguous attended sentences. Lackner and Garrett found that the unattended message could reverse these biases.

It is difficult to know how to weigh this line of experimentation inasmuch as the experiments do not replicate well. Wardlaw and Kroll (1976) followed Corteen and Woods's procedure and failed to find semantic generalization. They suggest that subjects in Corteen and Wood may have heard the unattended message ever so briefly. Treisman, Nauires, and Green (1974) showed that the Lewis effect is confined to the first few words of the message and that Lewis's results appear more general than they are because Lewis averaged across positions. According to Treisman et al., it takes subjects a few seconds before they can devote their full attention to the message that is to be attended. Consequently, the early words on the unattended channel may be attended to at some level. Finally, Newstead and Dennis (1979) showed convincingly that the results of MacKay also reflects subjects' partial attention to the unattended message. As far as I can tell, Lackner and Garrett is the only experiment of its kind that has not been seriously questioned.

Despite the confusion in this literature, the late bottleneck position has held its attraction to psychologists. Deustch and Deutsch (1963) proposed the first such model.

The Deutsch and Deutsch model. Deutsch and Deutsch (1963) argued that "a message will reach the same perceptual and discrimination mechanism whether attention is paid to it or not; and such information is then grouped or segregated by these mechanisms" (p. 83). Attention is directed to one of these extensively processed messages. Thus, selection occurs only after a message has been processed to the extent that its relation to other messages can be determined. Such processing is clearly beyond the stimulus level and most probably involves some level of semantic analysis. When selection does occur, it is based not on stimulus characteristics of the message, but a more meaning-laden aspect of the message, its "importance." An organism simply selects the most important of the competing messages.

A moving bottleneck. Many psychologists had difficulty accepting the Deutsch and Deustch model (see Kahneman, 1973). An original motivation for Broadbent's (1958) filter model, and most alternatives to Broadbent's model, was to describe a processing system with finite capacity bombarded by an overwhelming quantity of information. Somehow or other, the system must pare down this information to a reasonable size so that the important information can be fully processed. The Deutsch and Deutsch model does not appear to recognize this dilemma. If there is any limitation in the system, it is at the level of response selection or decision making, before that, everything is processed as deeply and fully as it can be. Such extensive processing is a far cry from the limited capacity envisioned by early students of attention.

A model by Norman (1968) attempted to reconcile the desire to limit extensive parallel processing with the finding that unattended messages are deeply processed. According to Norman, at each point in the chain of processing, pertinence values are assigned to each message being processed. Messages that receive a high pertinence value are passed on for further processing. Unless there is something special about it, a message will recieve a low pertinence value. The pertinence value may change, of course, with further processing: A message that may seem pertinent at the beginning of the chain of processing may seem less pertinent as one learns about it. In the end, most messages will have been rejected, and the deepest processing is confined to one or two messages.

This model should not be confused with the model by Deutsch and Deutsch (1963). The Norman model neatly allows for deep processing of more than one message, but does not posit that all incoming messages are fully processed: The system can discard competing messages at an early stage of processing as well as at a late stage of processing, thereby relieving some of the potential processing demands on the system.

The model, however, signals the downfall of the bottleneck metaphor. The neck of any bottle is located at one fixed point. It is a structural feature and certainly would not change from one instant to the next. Consequently, the metaphor of the bottleneck must be stretched beyond recognition if, as Norman claims, the limitation on the system can occur at any point of the chain of processing. Norman is asserting that there is no point at which the limitations of the system imposes itself. Rather, the system is constantly working to limit the demands placed on it. There is no bottleneck, then; just a limited processing capacity that the organism must constantly struggle to stay within. This concept of "limited resources" was to play a central role in the theory of divided attention that Norman helped develop (Norman and Bobrow, 1975).

Extensive or goal-directed processing

One issue that remained rather buried in most of the discussion of attention up to Norman's model has recently surfaced. In the early models of attention – and particularly, in the Norman model – the human information-processing system is goal directed [see Miller, Galanter, and Pribram (1960) for an elegant exposition of this point]. It accepts and further processes information that is important for perceiving and understanding the world, and it rejects the other information. The organism selects among competing messages for at least two reasons: (1) because of any limitations on its processing capacity and (2) because of the need to make sense out of the world.

Attention, then, could guide an organism in building a coherent perception of the world – a position I will call the constructive theory of attention. Recent work of Treisman and Gelade (1980) provides a good example of how this constructive role might be carried out. Any object consists of a complex of features or attributes. A "red F" has at least two features, one concerned with shape, that is, the shape F, and one concerned with color, in this case, the color red. Treisman and Gelade posited that stimulus attributes such as red and "F" are perceived directly and automatically. One does not need to focus on a specific spatial area to perceive them. Focal attention, however, is called for when one perceives the conjunction of these two features, that is, sees not red and F, but red F.

Treisman and Gelade argued for their position by examining conjunction errors. They tachistoscopically flashed an array of three differently colored letters and asked subjects to spread their attention across the array and later to indicate the letters that appeared in the array. Subjects often made errors. If the array consisted of, for example, a blue A, a red F, and a green B, subjects might report seeing, instead of a red F, a blue F, or a red A. However, they rarely saw a black Z, that is, forms and colors that were not originally present. People can perceive the stimulus characteristics, but cannot appropriately conjoin them when their attention is distributed across an array.

If subjects are told to attend to one area of the array, the number of conjunction errors decreases dramatically. People now correctly conjoin features. Attention, from this perspective, serves a important function. If the outside would must be constructed from bits and pieces of automatically processed stimulus features, then attention provides the means of selecting the appropriate features to conjoin. Treisman and Gelade argued that attention is like a flashlight, and inasmuch as objects occupy a discrete space, the appropriate stimulus features will quite naturally be selected when the flashlight is focused narrowly on the space occupied by a single object. Other researchers (Duncan, 1984) have argued that attention is object-oriented rather than spatially oriented. Whichever it is, the role is the same: to guide the construction of a percept from stimulus features by selecting the relevant features and rejecting the irrelevant.

This putative role of attention is mere conjecture, and an alternative, what I will call the exhaustive processing theory of attention, has been forcefully suggested. Instead of actively searching for information to construct the world, the human information-processing system may process everything to some rather deep level and select from this processed information what one wants to place into consciousness. Attention, then, does not play an intimate role in the construction of percepts; incoming information that neither conforms to one's expectations nor fits into the evolving schema of the outside world is not rejected, but is rather fully processed. As Marcel (1983b) recently wrote:

All sensory data impinging however briefly upon receptors sensitive to them is analyzed, transformed, and redescribed, automatically and quite independently of consciousness, from its source form into every other representational form that the organism is capable of representing, whether by nature or by acquisition. This process of redescription will proceed to the highest and most abstract level within the organism (p. 244).

From this perspective, what we process unconsciously is not identical to or closely related to what we are conscious of. When embedded in a sentence, the unintended meaning of an ambiguous word may escape the conscious notice of the reader, yet, the claim would be that both meanings are processed unconsciously. Attention and the associated concept of expectation come only into play in raising one of the various unconsciously processed meanings into consciousness. Similarly, when looking at an array of colored letters, all possible conjunctions of colors and letters are derived unconsciously; attention merely selects from these objects of conjoined features the one that should emerge into consciousness. It does not play any role in the conjunction of the stimulus features.

Moreover, limitations on information processing do not arise from any restriction on the processing system, at least, processing up to the meaning of the incoming information. Any limitation on the system rests in the nature of consciousness. The difference between this position and that of Deutsch and Deutsch (1963) is quite subtle, but important. Deutsch and Deutsch carefully avoided the term "consciousness" and located the bottleneck at the level of "response selection and decision making." Marcel and his associates locate it at the point that information is placed into consciousness. What they mean by consciousness is unclear, but they are insistent that any limitation on the human processing system reflects this limitation on the content of consciousness and not on the amount or nature of processing.

Thus, instead of discussing the location of a bottleneck, students of selective attention can ask what appear to be more substantive questions. Is unconscious processing guided by the expectation and schema that guide selection? Is the apparent limitation on the human information-processing system a reflection of a limit on the contents of consciousness or on the amount of processing that can occur simultaneously within the system? What is the role and function of consciousness?

I have already reviewed some of the evidence for the exhaustive processing theory of attention in the context of the Deutsch and Deutsch model. Other evidence has recently been reported. First, several researchers have shown that multiple senses of words are processed regardless of context. For example, Swinney (1979) asked subjects to listen to an audiotape and simultaneously indicate whether a string of letters presented on a TV screen formed a word. The tape was designed so that on some trials a word semantically associated with the string of letters was broadcasted at the same time or just before the string appeared on the screen. In the crucial condition, the word on the audiotape was ambiguous and only one of its senses was semantically related to the letter string on the screen. Swinney found that when the word on the tape was semantically related to the string on the screen, the lexical decision required of the subjects was faster than when the auditorily presented word and the string were unrelated. More importantly, he also found that this facilitation, or priming, occurred for both senses of an ambiguous word. Even if a sense was counterindicated by the sentence, it nevertheless primed the lexical decision if it was semantically related to the string on the screen. Swinney argued that lexical processing is data driven, that linguistic context and expectation will not affect the processing of lexical information, and that multiple meanings of words are processed regardless of context.

In a recent dissertation, Gildea (1984) questioned the validity of Swinney's claim. Swinney assumed that the priming went from the word on the tape to the string on the screen, but Gildea showed that priming may go in the other direction as well, that is, as subjects process both the word on the tape and the string on the screen, the processing of the string on the screen may guide the processing of the word on the tape. When the string on the screen is a word that corresponds to the inappropriate sense of the word on the tape, it

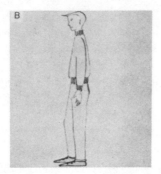

Figure 6.2. Stimuli for Eagle (1959).

may prove to have a powerful enough effect to initiate the processing of this inappropriate sense.

Work on subliminal perception is also offered as support for extensive-processing model (Kolers, 1957; Eagle, 1959; Smith, Spence, and Klein, 1959). The study of subliminal perception has a long and troubled history. In the traditional experimental design, a picture or word is flashed tachistoscopically at subthreshold levels of perception. Subjects simply cannot tell whether something was flashed or the field was blank. Subjects are then given a task, the performance of which can be affected by the subthreshold stimulus. If this is the case, then processing must have occurred that did not surface to consciousness. If it is not, then there may be a limit on the extent to which information can be processed without conscious awareness. For example, Eagle (1959) flashed subliminally either "A_1" or "A_2" of Figure 6.2 and followed it by "B." Subjects were asked to judge the personality of the boy in "B." They judged the boy negatively if A_1 preceded; positively if A_2 preceded. Thus, even though subjects claimed not to see the A picture, it affected their interpretation of B.

Work along these lines, however, did not replicate very well (Guthrie and Wiener, 1966; see Neisser, 1967). Just as we saw in other work on processing

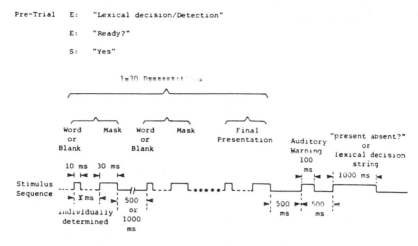

Figure 6.3. Experimental design of Marcel (1983a).

unattended material, questions were raised about the extent to which subjects were indeed "ignoring" the unattended material (Neisser, 1967). It became clear how hard it is to determine the extent to which subjects "saw" the subthreshold stimulus, and interest in the phenomenon lagged.

The situation has recently changed with a series of experiments by Marcel (1983a,b). In the most interesting of his experiments, depicted in Figure 6.3, Marcel flashed, at subthreshold speeds, either a word or a blank field, followed by a mask. The mask was a cross patch of letters designed to prevent the perseverance of the word in iconic memory. One of the two tasks followed the mask. In the detection task, subjects indicated whether a blank field or word had been flashed. In the lexical decision task, subjects were shown a string of letters and asked whether the string formed a word. Response times were recorded. The experiment was designed so that the tachistoscopically presented word was, in some trials, semantically related to the letter string. If the tachistoscopically presented word was processed to at least a lexical level, then one might expect its presence would facilitate the lexical decision task. On the other hand, if the tachistoscopically presented word was not processed that deeply, then its presence should have no effect on the lexical decision task.

The results were quite surprising. Subjects performed at chance levels in the detection task, indicating that they did not see the tachistoscopically presented word. They did, however, demonstrate a clear priming effect: the lexical decision was significantly faster when the lexical string was semantically related to the tachistoscopically presented word than when it was unrelated. These findings, however, have not been consistently replicated. Fowler et al. (1981) replicated them, but attempts in my laboratory and others have failed

[Cheesman and Merickle, 1984; see a discussion of Kahneman and Treisman (1984) on this point]. The dynamics of this effect have not been worked out. The masks from one laboratory to another have differed as well as various timing factors. Moreover, the subception threshold changes with practice, and the ways in which experiments have taken these changes into account have varied.

The Marcel work, however, is undergoing close scrutiny in many psychological laboratories, because it is quite important. When considered with the prior work on subliminal perception as well as work involving dichotic listening, the Marcel work argues strongly for the extensive-processing model. The common assumption that subconscious processing is governed by schemata and expectations may be wrong. Yet before this more traditional model is abandoned, the difficulties that surround all lines of experimentation need to be fully resolved.

Orienting attention

Most of the discussion to this point has involved attention to verbal messages. Few of the experiments involved visual processing, with the exception of Treisman and Gelade's (1980) feature-integration studies and the work on subliminal perception. Visual attention is often confused with foveating. When people want to attend to an object, they shift their head and eyes so that the object fully rests in the fovea. Attending, in these instances, is not different from looking. It is a reflection of an activity rather than an activity itself. To direct "attention" to an object does not involve any special process or mechanism outside those involved in normal perception.

It is possible, however, to attend to an area of space without foveating or "looking directly at it." Posner, Nissen, and Ogden (1978), for instance, asked subjects to push a button as soon as they saw a superthreshold change in the luminance in a dot. Each trial began with an arrow cue, which instructed subjects to shift their attention to a target that would appear 18° from fixation, in the direction indicated by the arrow. When it appeared at all, the target flashed 1000 msec after the cue. The target would appear in the specified location 80% of the time, and, on other trials, in a similar location on the opposite side of fixation. Posner et al. (1978) showed that the arrow cue facilitated detection of the target when it appeared in the indicated location and depressed response time when it was in the opposite location. Importantly, Posner et al. monitored eye movements by measuring electro-ocular movements and found the facilitation/depression effects even when they discarded trials in which eye movements were detected. Consequently, they concluded that it was possible to attend to something that you were not looking at.

From this perspective, attention seems to be something other than merely looking or listening. It clearly involves selection. Subjects are selecting to attend to one area of the visual field over another. What are they training on this selected piece of space? Are they training attention, just as one would train their perceptual mechanism, thereby allowing it to work on one area of the visual field over another? The flashlight metaphor is clearly inappropriate at some level. Attending to one area over another without eye movements may help detection, but it may not enhance acuity (Engle, 1971). Subjects may not be illuminating one part of the visual field over another in Posner et al. (1978), but merely improving detection within a specified area. The results of Posner et al. appear to argue that there are at least two ways one can "see" something. One can merely determine whether something is present or not. This kind of "seeing" is clearly independent of foveating. Alternatively, one can determine the actual shape and nature of the object. This kind of "seeing" appears to require foveating. Such a distinction has been made in several different contexts and, indeed, seems to map onto two distinct neuroanatomically specifiable visual systems.

If one assumes that both systems can process only a small area of the visual field at one time, then the subjects used by Posner et al. are training the detection system on a space that does not correspond to the area at which they are looking. Subjects, then, are not training attention onto a specified corner of the visual field. Rather they are directing a visual system to this area, and their ability to "attend" to this area is constrained by the nature of the visual system and not by any limitation inherent in some reified attentional mechanism.

Posner et al. (1978) provides a means of asking rather precise questions about the nature of selection within the perceptual system. Attention could shift from fixation to target location in one step, or it could sweep across the visual field, much as a flashlight would. Schulman, Remington, and McLean (1979) addressed this issue by varying the location of the target and the time interval between the arrow cue and the presentation of the target. Specifically, in 10% of the trials, the cue appeared midway between the fixation point and the "standard" target location. In this case, they found that the interval between the presentation of cue and target producing the greatest facilitation was about half of the interval found for the standard location of the target. The result indicates that attentional shifts are analog.

Subsequent work indicated that attentional shifts are not completely independent of eye movement. Wurtz and Mohler (1976a,b) speculated that attentional shifts were programs for the movement of the eyes. Klein (1979), however, showed that one could demonstrate an attentional shift in the opposite direction of eye movement. On the other hand, Remington (1980) showed that attentional shifts often preceded eye movements. In reviewing these and other studies, Posner (1980) concluded that the relation between eye movement and attentional shifts was functional: It is usually the case that attention is needed in the same location to which one wants to direct one's eyes. Consequently, in all but the most contrived experimental setting, eye movements and attentional shifts are correlated.

Summary

The reader at this point might feel that after 30 years of effort, the study of selective attention has made little progress: The original filter model failed to account for a wide range of evidence; the metaphor of a bottleneck seemed inappropriate once it is realized that the bottleneck can occur anywhere along the chain of information processing; the relation between unconscious processing and conscious events is unclear; the role of attention in building the world out of stimulus features is also uncertain.

Yet the large question mark that seems to punctuate this discussion need not be viewed pessimistically. Much progress has been made over the last 30 years, and it is worthwhile highlighting this progress.

First, and most importantly, psychologists have developed means of discussing attention that are both precise and raise questions that are empirically tractable. Questions such as the function of attention and the degree to which attention guides unconscious processing could not have been asked, let alone answered, 20 years ago.

Second, psychologists have clearly established that selection can be based on semantic characteristics of the message as well as stimulus characteristics. Exactly when semantic characteristics are relevant and when they are not is poorly understood, but their possible contribution is undisputed at this point.

Finally, the nature of visual attention is beginning to be appreciated. Treisman and her colleagues have shown that whereas primitive features can be processed without attention, the conjunction of these features into objects requires attention. Posner and his colleagues have studied the visual system responsible for detection and find that attentional shifts are analog even though they are not dependent on foveation.

Divided attention

Party-goers can try to divide their attention between two conversations instead of selectively attending to just one of them, but they will find such eavesdropping difficult. Questions about divided attention have occupied as prominent a place in the study of attention as have questions of selective attention. In particular, psychologists have been interested in determining how many things can be done at one time or, to put it another way, predicting the ease with which two tasks can be done simultaneously. Harking back to William James (1890), we find an elegant statement of the sentiment of the field in his time, one that could well serve the sentiments of the present.

If, then, by the original question, how many ideas or things can we attend to at once, he meant how many entirely disconnected systems or processes of conceptions can go on simultaneously, the answer is, not easily more than one, unless the processes are very habitual; but then two, or even three, without very much oscillation of the attention. (p. 409)

A casual reader might think that James's simple statement settles the matter, but a careful reader will notice that it is spotted with terms such as habit, systems, and conceptions. These ambiguous concepts are there because any answer to the "original question" will depend on the experience of the individual, the nature of the tasks, and the relation between the tasks. A young child cannot talk while putting on his coat, but adults can. An adult, on the other hand, might find it impossible to drive and make numerical calculations (Brown and Poulton, 1961). It is easy to tap the right hand in a beat of four while tapping the left to a beat of two, but it is much harder to tap triplets with one hand while tapping doubles with the other. It is easier to attend to an auditory message while also attending to a visual message than to attend to two auditory messages or two visual messages (Treisman and Davies, 1973).

In what follows, I will summarize the attempts of psychologists to model divided attention. If it is any good, a model of divided attention should predict the ease with which two tasks can be done concurrently. Most models of divided attention have had problems meeting this criterion. The history of divided attention is frustrating. One reasonable model is replaced with another equally reasonable model. If anything differentiates them, it is that as the evidence becomes more complex so do the models.

Early models of divided attention

Early models of divided attention took their lead from models of selective attention (Welford, 1968; Broadbent, 1971). The human organism had a limited processing capacity. At some point in the chain of processing a bottleneck occurred, and one message had to be selected for further processing over others. The nature and mechanism of the bottleneck not only determined the basis of selection, but also imposed limits on doing two things at once. From this perspective, one could do more than one thing simultaneously if they did not place undue demands on the limited capacity of the system. The exact nature of these limitations depended, of course, on the location of the bottleneck.

In Broadbent's early filter model, the p system placed the restrictions on the processing (see Figure 6.1). It was like a central processing unit (CPU) in a computer. All information must be processed through the ρ system, or some processing "channel," if it is going to lead to an output, and because of the nature of the mechanism, only so much information can be processed at any one instant.

Making predictions about the ease with which two tasks can be combined is a straightforward if arduous procedure for proponents of this model. One need only determine when any task passes information to the CPU, and then determine whether competing tasks place simultaneous demands on the CPU. If, because of various time factors, or because of various task considerations, concurrent tasks do not place concurrent demands, then the tasks should be easy to combine. If, on the other hand, concurrent tasks place concurrent demands on the ρ system (or CPU, whatever your terminology), the performance in the dual task will be worse than performance of either task done alone.

Resource theory

Early models of selective and divided attention held two assumptions dear: that the human information-processing system has a fixed finite capacity and that this limitation can be viewed as a bottleneck in the information-processing channel. As the bottleneck metaphor became less and less popular, psychologists found themselves left with only the assumption dealing with finite capacity. Resource theory was an attempt to describe processing limitations in a sophisticated manner. In what follows, I will describe the general framework of the theory, articulate some of its mathematical advances, and investigate both its predictive power and study of the nature of resources.

Resource and data limits. Norman and Bobrow (1975) lead the work on resource theory. According to Norman's (1969) model of selective attention, the limits on the processing system could occur at every stage. In their discussion of divided attention, Norman and Bobrow achieved the same flexibility, yet maintain the assumption of fixed limitations of processing by introducing the concept of "resource."

Performance in a task can be limited in two quite different ways. A task may be data-limited, in which case performance depends on the quality of the incoming information and the quality of stored information. Performance in an identification task, for example, will be limited by the quality of the incoming signal and the quality of the mental representation to which the incoming signal must be matched.

A task may also be resource-limited. The intuition motivating Norman and Bobrow is clear. Processing requires resources to function. If we assume that there is only a fixed quantity of resources, then the difficulty in doing two things at once occurs because there are not enough resources to "run" the processing adequately. Norman and Bobrow were vague as to what they meant by resources. Resources, for them, include such things as "processing effort, the various forms of memory capacity, and communication channels" (Norman and Dobrow, 1975, p. 45). As this litany suggests, resources can be treated as fuel that runs processes (fuel such as processing effort); they can also be thought of as cognitive structures (such as short-term memory).

The fuel metaphor seems appropriate because people can often improve dual performance by exerting effort (Kahneman, 1973). Inasmuch as they are doing the same tasks, the processes underlying the tasks and their demands on difficult processing structures probably has not changed with increased effort. What does change, the reasoning goes, is the "amount of fuel" being allocated to the processes.

The "structure" metaphor fits well into component models of human information processing, but it needs more exegesis. The concept of "cognitive structure" is, of course, pervasive in psychology. Fodor (1983), for instance, has proposed that the information-processing system is modular, and consequently can be divided into discrete and independent processing systems. Such modules seem to be good candidates for what attentional theorists mean by "cognitive structures," though much of the attention work predates Fodor. Moreover, as Norman and Bobrow noted, structures such as short-term store and sensory channels are also good candidates. These structures are resources because they have built into them certain fixed limitations. Short-term memory can only hold 7 ± 2 chunks of information at one time, and a sensory channel can only process one message at a time. Thus, the architecture of the structure imposes a limitation on the number of things that a structure can accommodate at one time. This structural limitation differs from a "fuel" limitation. Two factory lines might not function smoothly because there is a limit on the amount of electricity available to run the machines on the line. Alternatively they may run poorly because there is a point where they share a machine for screwing on nuts that can only handle one nut every three seconds.

Performance–resource functions and POC curves. Although the concept of resource is only vaguely defined, and, more importantly, although proponents of resources describe metaphorically the concept of resource in what appears to be distinctly different ways, a rather precise theory of the relation between resources and divided attention can be worked out. The articulation of this theory has been advanced in the main by Norman and Bobrow (1975), Kinchla (1980), and Sperling (1984). This discussion follows Norman and Bobrow (1975). They noted that task performance should improve as more resources are allocated to the task, a relation that can be captured by a performance–

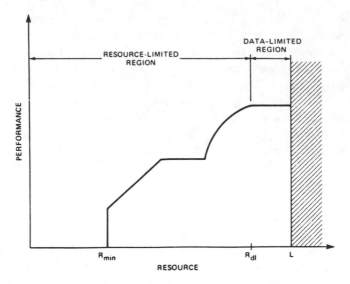

Figure 6.4. An example of a performance–resource function. (From Norman and Bobrow, 1975.)

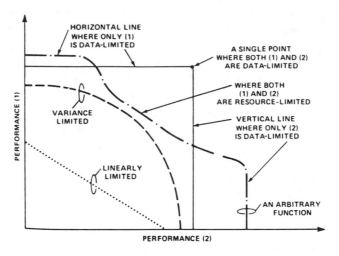

Figure 6.5. An example of a performance-operating characteristic (POC) curve. (From Norman and Bobrow, 1975.)

resource function. The difference between data limits and resource limits can be clearly seen in these performance–resource functions. When a task is resource-limited, one should see a trade-off between performance and resource allocation. As Figure 6.4 shows, the function is curvilinear under these conditions. On the other hand, a task is data-limited when performance is independent of the amount of resources allocated. Here, the performance–resource function is flat.

Two performance–resource functions, corresponding to two different tasks, can be combined to reflect the trade-off in performance when the two tasks are done concurrently, resulting in the dual tasks's performance operating characteristics (POC). As Figure 6.5 illustrates, a POC maps the performance of the primary task into the performance of the secondary task. There are many ways to calculate POC. Norman and Bobrow (see Kantowitz and Knight, 1976; Norman and Bobrow, 1976) suggest changing the priority of the primary and secondary tasks: performance in the two tasks for each priority ratio will yield one point on the associated POC. When the POC is parallel to either the abscissa or ordinate, one of the tasks is data-limited. When the POC has a slope between 0 and ∞, both tasks are resource-limited, with the POC plotting the trade-off in performance.

Performance operating characteristics are almost always discussed in the context of resource theory, but this appears to be an historical accident. Performance operating characteristics are descriptive tools; they are ways of empirically depicting derived data about dual task performance. To be sure, one does not empirically determine every point on a POC. Empirically based points are connected by smooth lines, but these connections could be based more on esthetic considerations than on any knowledge about the way the tasks interact. When viewed in this way, POCs are independent of any concept of resource allocation, and their usefulness is independent of the validity of resource theory. If one is merely interested in plotting the trade-off between two concurrently performed tasks, POCs provide an elegant means of doing this.

Performance operating characteristics seem wedded to resource theory because they are usually depicted as smooth monotonically increasing or decreasing continuous lines, just the form a proponent of resource theory would predict (see Wickens, 1984). The prediction, however, comes not from anything inherent in POCs, but from something inherent in resource theory.

Predicting dual task performance: the principle. If POCs are merely descriptions of data and not tools for predicting dual task performance, how does one make such predictions? As already noted, when resources are thought of as processing structures, prediction involves detailed task analysis. Performance should decline when competing tasks place simultaneous demands on a processing structure. Subjects would have to postpone the processing of one of the tasks, resulting in either a decline in performance or a complete breakdown.

When resources are thought of as fuel, predicting the ease with which two tasks can be combined is more a matter of arithmetic. An individual is endowed with a certain amount of fuel. In order to calculate whether two tasks, A and B, can be performed simultaneously, one must add together the amounts of fuel demanded by tasks A and B and determine whether the sum exceeds the quantity of available resources. If it does, then performance of A and B when performed simultaneously will be worse than performance of either A or B when performed alone. If it does not, then tasks A and B should be performed as well concurrently as they are performed alone.

Central or multiple resources? Although it is clear in principle how to predict dual task performance, it is hard in practice. The structure model requires a task analysis that is fine-grained enough to be impractical. The fuel model rests on the assumption that one knows the resource demands of a task. Various psychologists have tried to determine quasiobjective means of measuring resources, such as by inferring them from related POC curves or obtaining subjective rating of workload (see Wickens, 1984 for a review). As yet, none of these methods have gained wide acceptance, and most have been severely criticized (Moray, 1979).

Moreover, it has become increasingly apparent that a truly predictive science of divided attention is not forthcoming until resources are more clearly specified. Psychologists are not even clear as to whether there is:

1. One central resource – from which all tasks draw
2. Multiple resources – with some tasks drawing from some resource pools and not others, for example, verbal tasks drawing from a verbal resource pool and not a visual resource pool
3. A combination of specific multiple resources as well as a central resource

Wickens (1980, 1984) posited more than one attentional resource. The Broadbent filter model clearly evoked one and just one resource, that is, the ρ system. The extant research argues for a multiple resource approach (see Navon and Gophers, 1979).

McQueen (1917), Sverko (1977), Hawkins and Ketchum (1980), Hawkins, Rodriquez, and Reicher (1982), and Sverko, Jerneic, and Kulenovic (1983) sought evidence for the existence of a "time-sharing capacity," Sverko's term for a central processor or central resource pool. Four simple tasks were paired to create six dual tasks: for instance, a rotary pursuit task and a visual choice reaction time task were paired to create a dual task in which subjects pursued a dot on a rotary while making visual judgments. Sverko assumed that some people must have a larger central resource pool than do others: such individual differences are an inevitable consequence of being human. Thus, if a central resource pool exists, the performance of a particular subject on a one dual task AB should be correlated with his performance on another dual task CD. Sverko was unable to find a significant correlation and attributed his failure to the possibility that a central resource pool did not exist.

Although the work of Sverko, Hawkins, and their associates argue for a multiple resource approach, it does not specify the nature of these resources. Another line of experimentation, which I will call the same versus mixed paradigm, does specify the nature of these resources. In these studies, the simultaneous performance of two tasks with one crucial difference is compared with the simultaneous performance of these two tasks when the crucial feature is equated. For instance, performance on two concurrent tasks presented in different modalities (visual and verbal) can be compared with performance on two visual tasks and performance on two verbal tasks. Except in special circumstances (Wickens and Vidulich, 1982), it was found that two tasks with a crucial difference are easier to perform simultaneously than are the two tasks when the difference is equated. That is, it is easier to do two tasks presented in different modalities than the same two tasks presented in the same modality (Treisman and Davies, 1973; Rollins and Hendricks, 1980).

According to the logic underlying the same versus mixed paradigm, if a task calls upon a central resource, and only this resource, then the dual task in the "mixed" condition should be performed as poorly if not worse than at least one of the dual tasks in the "same" condition. For instance, if a verbal task takes eight units of the available resource units and the visual task takes four, then the visual/verbal dual task, with a demand of 12 units, should be performed worse than the visual/visual task, which only demands eight units. The superior performance in the mixed condition, then, suggests that the verbal and visual tasks must pull resources from different pools.

Wickens (1980, 1984) has gone further than anyone in trying to articulate the exact nature of these resource pools from work on the same versus mixed paradigm. It is not enough to say that there is more than one pool; one must also describe these pools. Wickens has suggested that resource pools can vary on at least four dimensions: (1) stimulus characteristics (visual and auditory), (2) internal codes (visual and verbal), (3) response characteristics (manual and speech), and (4) levels of processing (shallow and deep). Wickens cautioned that this schematization may be incomplete, but suggests that the final version will be close to the present version.

Critique of the multiple resource model. Resource theory is attractive and intuitive. It is not surprising that among some psychologists it is taken as dogma. It has its critics, however. Some critics have specifically attacked the multiple resource version, whereas others have provided a more general critique of resource theory itself. When viewed together, they appear to present a serious threat to this widely accepted theory. In what follows, I will offer three arguments against the multiple resource theory. The first considers the effects of practice on divided attention and problems that any effect raises for resource theories in general. The second argument explores the ontological status of resources. The third argument revisits the same versus mixed paradigm. Although any one of these arguments may not be compelling, together they present difficult problems for any proponent of multiple resource theory.

Practice and divided attention. The first criticism is general and could be applied to both multiple resource and single resource theories. All resource models posit that the limit on attention capacity is fixed. When resources are thought of as processing structures, the architecture of the structure is constituted so that only a fixed amount of processing can go on at one time. Broadbent's ρ system can only process one line of information at a time, and short-term store can only hold about 7 ± 2 chunks of information. When resources are thought of as fuel, the supply of fuel is not only limited, but this limit is also fixed. The storage tank will hold only so much fuel. Moreover, the fixed limit on attentional resources holds firm whether resources are undifferentiated or multiple. When one posits a single resource, the limit on doing two things at once is the limit of this resource, whereas when one posits multiple resources, the limit on divided attention reflects the sum of the limits of each of the many resources.

Psychologists have been impressed by the limited nature of attentional capacity because they rarely study practice in divided attention tasks. In the few studies that exist, an extraordinary ability to divide attention is apparent (Underwood, 1974; Moray et al., 1976; Ostry, Moray, and Marks, 1976). Skilled secretaries, for instance, can type while shadowing (Shaffer, 1975), and proficient cocktail pianists can sight read music while shadowing (Allport, Antonis, and Reynolds, 1972).

Probably the most extensive practice study was undertaken by Hirst, Neisser, Spelke, and their colleagues (Spelke, Hirst, and Neisser, 1976; Hirst et al., 1980). They taught a group of subjects to read prose while writing single words and short sentences from dictation. On the surface, reading and writing appear to be resource-demanding tasks, ones that in combination should exceed available capacity. At the initial stages of practice, the combination of reading and writing did strain the system. Subjects complained bitterly, stating the task was impossible, they would never learn to do it. Nevertheless, by the sixth week, they could not believe that they ever found it hard to read and write simultaneously. It was as if the capacity of their cognitive system had grown with experience.

The subjects were not learning to time share between the two tasks. Extremely rapid time sharing is always a possibility, but it becomes an article of faith unless one can specifically account for how task performance can be maintained in the presence of this alternation. In the reading and writing task, subjects could have taken advantage of the redundancy in the prose, learning to skip over unimportant elements of what they were reading to find time to write. The evidence suggests that most of the subjects did not adopt this strategy, however. Although trained on highly redundant prose such as short stories, they nevertheless proved able to write while reading fairly non-redundant prose such as encyclopedia articles. Such an observation cannot rule out the possibility of time sharing, but it suggests that it is unlikely.

The subjects were also not learning to perform either the reading or writing task automatically. It is widely held that a task no longer requires processing resources when it becomes automatic. Consequently, if either the reading or the writing from dictation became automatic, then reading and writing should be easy to combine. Hirst et al. argued that reading is too complex a process to ever become automatic, no matter how much practice, but admitted that writing from dictation could have, if it did not require semantic processing. However, subjects appeared to be comprehending what they were writing. They made fewer spelling errors when writing meaningful sentences than anomalous sentences. When a sentence contained a homonym, they wrote the spelling consistent with the meaning of the sentence. Most strikingly, they falsely recognized inferences from stories dictated to them.

At some quite sophisticated level, then, the subjects in the experiment by Hirst et al. were learning to do two meaningful tasks at the same time, neither of which are simple enough to permit automaticity. Some researchers have wondered whether this "complexity" definition adequately captures the concept of automaticity (Lucas and Bub, 1981; Shiffrin and Dumais, 1981). However, alternative definitions are equally vague (Neisser, Hirst, Spelke, 1981; Hirst and Volpe, 1984), and until the concept of automaticity is better understood, the abilities of these subjects must be carefully considered.

In particular, the radical change in performance observed in these subjects suggests that the limits on attentional capacity are not fixed. Rather, limits can change with practice and depend on individual abilities. At any particular stage of development, a limit exists as to what a person can or cannot do, but additional practice can change this limit. Trying to find the final limit makes no sense.

Once one allows the capacity of resources to change with practice, the concept itself loses much of its appeal. Early versions of resource theory treated resource as either a reflection of the basic architecture of the mind or the manifestation of an innate reservoir of psychic fuel, but whatever the nature of the resource, their proponents assumed that capacity is fixed. They did so because without a fixed capacity, the capacity of a resource would depend on an individual's experience and abilities. Resources, then, would not reflect basic architecture or innate reservoirs, but the current state of the organism. Indeed, the concept of "resource" that emerges is so different from that described in earlier papers, such as Norman and Bobrow (1975), Hirst et al. (1980) suggested that a more appropriate term might be "skill." The term "skill" captures the flexible nature of dual task performance and easily accounts for variables such as individual differences and practice.

Isolating resources. Navon (1984) has persuasively argued against the onto-logical status of attentional resources. He drew a strong distinction between alterants and commodities. Alterants are states of the organism, whereas commodities are made up of units available to one user at a time. The dis-tinction is best understood through examples. When considering the ease with which two cars can travel down a road, the commodity involved is fuel, in that each car makes exclusive use of the fuel. Factors, such as weather or the road conditions, are alterants, because neither is owned or used by any par-ticular car more than any other. When considering the performance of two concurrent cognitive tasks, locations of working memory are commodities, inasmuch as each location is used exclusively by one of the tasks at any one time. Factors, such as anxiety or the activation of a long-term memory node, are alterants, in that these momentary states of the organism can affect the performance of both tasks.

For Navon, resources, at least as articulated in theories such as Norman and Bobrow (1975), are commodities, not alterants. If so, then problems arise. If one is going to build a theory of divided attention on a concept such as resource, then it is necessary to show that at least in some sense, resources exist and do constrain divided attention. In a series of compelling arguments, Navon shows that there is no unambiguous and straightforward means of determining whether task performance is constrained by something akin to resources. Any set of behaviors traditionally associated with arguments for resource theories could also arise because of the nature of the alterants. For instance, proponents of resource theories explain limits on doing two things at once by appealing to the scarcity of resources. Navon pointed out that tasks may interfere even without competition for scarce resources. Simulta-neous telephone calls may interfere even when their number does not exceed the capacity of the system. Electrical interference, for instance, may produce cross-talk. Irrigation of a plantation may interfere with harvesting in part because of the demand on the scarce resource of labor, but also because the irrigation makes the soil muddy. One task may interfere with another because of the anxiety it produces rather than the demands it places on limited resources.

In exhausting detail, Navon considers a host of possible criteria for un-dercovering resources. One knows that a resource pool has been detected when one can assert that the more difficult a task, the more resources it calls upon; be certain that task priority will affect relative performance of two concurrent tasks by affecting the allocation of resources; determine that task difficulty interacts with the presence of a concurrent task; and ascertain that task difficulty interacts with the difficulty of the concurrent task, especially in instances in which the performance on the secondary task is held constant. Some of these criteria are quite subtle, and their motivation quite complex. Yet in each case, no matter how subtly constructed, the criterion seems to evaporate in Navon's hand when he begins to consider alterants as well as commodities.

Navon found that some criteria may be better indicators of resources than others, but in the end, discovered that none are decisive. Any trade-off in performance in a dual task can be explained equally well by alterants as by resources. For Navon, resources are like a soupstone: folklore dictates that one add it to soup, but the soup is just as good without it. Nevertheless, he admits that he is as captivated by the folklore as anyone. Resources make intuitive sense to him, even though their ontological status is uncertain. Consequently, he argues that discussions of divided attention may still revert to discussion of resources, a conclusion that is too timid for my taste.

Revisiting the same versus mixed paradigm. As already noted, a multiple resource theory is only as good as the description of the resources. Wickens's (1984) model, with its four dimensions, goes a long way toward providing this description. However, as Wickens (1984) himself noted, his description is incomplete. It is possible that as more research is done, the number of resources will multiply. This possibility could open up a Pandora's box for resource theorists. If the number of different resources becomes too large, the theory will lose much of its elegance. It will be cumbersome to predict the ease with which two tasks can be combined. Each calculation will have to consider the way the two tasks call upon, let's say, 200 different resources, an awesome and perhaps overwhelming task.

Recent work in my laboratory (Hirst and Kalmar, 1984) suggests that the number 200 may be much too small. Wickens based his four-dimensional description in large part on results from studies using the "same versus mixed" paradigm. Hirst and Kalmar used this paradigm to examine a wide variety of different tasks. They found that it was easier to spell and add at the same time than to spell and spell or add and add, and that it was easier to spell and categorize at the same time than to spell and spell or categorize and categorize. Most importantly, Hirst and Kalmar examined two detection tasks: one in which subjects had to detect wild animals among a large set of animal terms; another in which subjects had to detect vegetables among plant terms. They found that it was easier to do the animal and plant task simultaneously than either the plant task paired with a similar plant task or the animal task paired with a similar animal task.

This latter finding suggests that semantic features can differentiate resources. Each distinct semantic feature would specify a distinct resource pool. Inasmuch as the number of semantic features is quite large, the findings also suggest that the number of resources is also quite large. Indeed, the number of resources may be so large that predictions about doing two things at once may be difficult, if not impossible. This loss of predictive power sours the intuitive appeal offered by resource theories.

Summary. When considered together, these three criticisms of resource theory appear to be quite compelling. Attentional resources do not have the explanatory power that was initially attributed to them. They do not capture the flexible nature of divided attention, seem to multiply without limit, and in the end, may be unnecessary and cumbersome baggage. As mentioned, Hirst et al. (1980) have suggested that the concept of "skill" may serve as an alternative. I will explore this suggestion in the next section.

Attentional skills

The problem in replacing the concept of "resource" with the concept of "skill" is that one vague concept is replaced by another. Almost anything can be called a skill. Skilled activity involves a coordination of actions with the aim of achieving some specific goal. To be skilled at tennis is to possess the ability to coordinate one's actions in order to play tennis successfully. To be skilled at adding while spelling is to coordinate one's actions in order to add and spell simultaneously. A skill is a catchall term that refers to the cognitive mechanism or structure that guides skilled activity.

Arguing, then, that people can add and spell simultaneously because they possess or have developed the requisite skill does little more than argue in a tautological way that people can add and spell simultaneously because they know how to. Clearly, more meat must be given to the concept of skill if it is going to serve as an interesting theoretical substitute for resource.

Evoking the concept of "skill" is a metatheoretical gesture, in much the same way that early memory theorists' introduction of "schema" into discussions of memorizing and remembering was more metatheoretical than theoretical. The concept of "skill" does not in and of itself explain behavior or even allow one to make predictions. It is too vague and poorly defined. Nevertheless, its adoption as a central theoretical concept of any theory of attention should alter one's perspective on the important questions of attention.

Proponents of resource theory concentrate on the way that people allocate resources to task demands. The processes involved in doing two things at once are of secondary importance. They figure in a discussion of divided attention only in so far as they clarify how competing tasks vie for the same resources. The notion is that there are certain structures or "fuels" that processes share, and that these commodities can be shared by only so many processes at one time. From this perspective, a theory of divided attention seeks two things: a list of the commodities and an understanding of their limitations.

As Navon (1984) argued, however, it is hard to determine whether performance breaks down because of competition for the same commodity or because of changes in the state of the organism. Telephone messages may indeed interfere with each other for reasons other than the capacity of the wire. In part, their transmission will depend on the coding mechanisms at the sending and receiving end of the wire. If the messages are coded so that they are similar, interference is bound to occur. If, on the other hand, they are coded in such a way as to make them as dissimilar as possible, interference will decrease. They ability to transmit more than one message at a time will depend on how one processes them and not necessarily on the capacity of the communication channel.

The same point can be made when considering the flexible nature of limits on divided attention. It is not that the capacity of the available commodities changes with practice, but rather the processing itself. Resource theorists try to accommodate this observation by noting that as tasks become automatic, they no longer require resources. However, this argument seems circular. If commodities cannot change with practice, then what must change is the demands the practiced task places on the commodity. You know, however, that the demand changed because there is no competition for the commodity when originally there was. It seems easier, free from fewer reified entities, to assume that at times people can do two things at once and then to devote one's energies to trying to understand both the conditions that make this possible, and the processes underlying the performance.

Proponents of a skills approach adopt this position. They concentrate on describing what people are doing when performing two tasks simultaneously and what they are learning when practicing dual tasks. They avoid positing reified entities such as resources. For them, resources are a last resort rather than the major theoretical construct.

Such an approach does not condemn one to studying specific skills of specific tasks. It is true that, whatever they are, skills of divided attention are task specific. A person who can play the piano while talking cannot necessarily type while talking, and vice versa. Moreover, people may differ in the way that they divide their attention even when one confines discussion to a single dual task. Their skills will depend on their general abilities, their experience with the task, and variables such as motivation and set.

Nevertheless, there are several general techniques people can use to do two things at once. In the closing section, I will summarize what is known about these techniques. I will focus on four of them: (1) coordinating two tasks into a single higher-order task, (2) practicing one task until it is automatic and then combining it with a second task, (3) learning to keep competing tasks insulated from each other, and (4) finding processing time in one task to share with the other task.

Higher-order structures. People can learn to do two things at once by coordinating the two tasks into a single task. People are constantly doing a multitude of things. When putting on a coat, people are moving their arms, swinging their torso, and wiggling their legs. Somehow, all of these tasks have been combined into a single task, and labeled as such.

Tasks can be organized at either the stimulus or response level. The work of Neisser (1964) and Neisser, Novick, and Lazar (1963) on parallel processing provides a good example of organization at the stimulus level. Subjects were to search a long list of letter strings for the presence of particular letters. Neisser found that with enough practice this search became independent of the size of the list of targets, and suggested that the search must be done in parallel instead of serially. Subjects reported that as they skimmed through the list each target just jumped out. Interestingly, subjects required a great deal less practice when there was a common feature defining the letters, for example, the letters all have curvilinear shapes, as "B" and "P" do (Gibson and Yonas, 1966). When this occurred, subjects could merely search for the common feature instead of the letters themselves. That is, instead of searching for four letters – B, P, C, and Q – they merely searched for curvy lines. By finding a common stimulus feature, they were able to organize the tasks into one.

Seibel (1963) provides another example. Subjects were to push a sequence of buttons in a pattern that mirrored a sequence of light flashes. Initially, response time depended on the number of lights flashed, with a maximum of ten lights, but after extensive practice (75,000 trials), response became independent of the number of lights. Instead of responding to each light, subjects now saw the pattern of lights as a unit and responded with a unitary, integrated action.

Tasks can be organized not only at the stimulus level, but also at the response level. Any beginning pianist has difficulty playing doubles on the right hand while playing triplets on the left hand, yet once this pattern is organized into six beats in a measure, the difficulty disappears. Kelso, Southard, and Goodman (1979) have made a similar observation in their studies of motor skills. Subjects were first asked to move a finger resting on a dot right in front of them to a dot by their side. If the distance from the center to the side was short, it took little time for the finger to traverse the course. If the distance were long, the traverse time was long. However, if subjects combined the two tasks and moved the right finger to a distant dot on the right side and the left finger to a closer dot on the left side, they coordinated their actions so that the fingers arrived at their respective dots at the same time. Even though subjects had the capability of getting to the closer dot sooner, they slowed their traverse to organize the responses into a single unit.

Klapp (1979) also examined response organization. He asked subjects to tap a finger to the flashing rhythms of light, letting the right hand follow the right light display while the left hand followed the left light display. He found that when the rhythms set up by the flashing lights could be easily organized, subjects found it easy to tap both rhythms simultaneously. For example, when one consisted of two beats per second and the other four beats per second, it was easy to tap both simultaneously. However, when the two did not form some straightforward hierarchical pattern, as when the lights appeared randomly, it was quite hard to tap both simultaneously.

If people can learn to do two things at once by organizing stimulus and response, they might also learn to do two things at once by organizing more central processing. For instance, if a process A involves four subprocesses A1, A2, A3, and A4, and process B involves two major subprocesses B1 and B2, do subjects coordinate the subprocesses so that the output of A2 occurs simultaneously with the output of B1? If they do, then one would expect that the reaction time patterns for the simultaneously performed tasks would be different from the pattern established when the tasks are performed alone. Whereas this different pattern may reflect competition for resources or task interference, it also might reflect task reorganization and coordination. As yet, no one has investigated this distinction.

Automaticity. The concept of automaticity has been advanced in discussions of divided attention for many different and sundry purposes (Posner and Klein, 1973; Posner and Snyder, 1975; Schneider and Shiffrin, 1977; Shiffrin and Schneider, 1977; Shiffrin and Dumais, 1981). For instance, as we have already seen, automaticity figured centrally in resource theories. It provided a fairly general explanation for the kind of improvements observed in practice studies such as Hirst et al. (1980) once it was assumed that automatic processing requires little or no resources. This explanation falters, however, when one attempts to specify why it is that automatic activities do not require resources. On the surface, an auditory message should occupy the auditory-processing channel regardless of practice, and a task should require working memory independent of how many times it is done. I am not sure whether this superficial surmise is true, but it is clearly incumbent on resource theorists to explain exactly how the processing changes with practice and why this change results in a lower demand on resources. They have not done so.

Indeed, almost nothing is known about the difference between automatic and effortful processing. It seems premature to enlist these concepts in aid of a theory of divided attention that has, as noted, a host of problems. Rather, it seems more appropriate to study it in its own regard. Clearly, people can

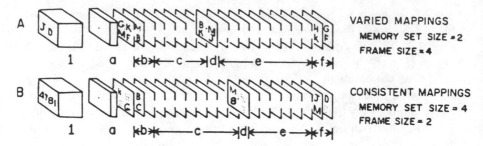

Figure 6.6. Varied and consistent mapping in the Shiffrin–Schneider paradigm (from Shiffrin and Dumais, 1981).

learn to do two things at once by practicing one until it becomes automatic. It is one way to acquire a skill of divided attention. Why not study the nature of automatic processing and the conditions governing the ease with which tasks become automatic? Once these questions are answered, it may be easier to determine whether automatic acts demand resources.

Shiffrin–Schneider paradigm. Ideally, one would like to know how the processes underlying an automatic act differ from those underlying an effortful act. So far, such a detailed description has alluded psychologists. Schneider and Shiffrin (1977; Shiffrin and Schneider, 1977), who have studied automatic processing more extensively than anyone else, have pursued a more conservative track. They have been interested in questions such as: Is automatic processing qualitatively different from effortful processing? What kinds of processes can become automatic? What kinds of learning material and stimulus material facilitate automaticity? What criteria can one develop to determine whether an act is automatic? They pursued these questions within the narrow context of a visual search task, but they clearly hope that their findings will be generalized (Shiffrin and Schneider, 1984).

In the basic Shiffrin–Schneider paradigm, subjects must determine whether an item from a small set of memorized targets is in a complex visual array (see Figure 6.6). There were two conditions. In the varied mapping condition, the target in one trial may act as a distractor in the visual array in the next trial, whereas in the consistent mapping condition, the target and distractors are taken from mutually exclusive sets across trials. Schneider and Shiffrin demonstrated that processing underlying the varied mapping condition does not improve with practice, and they labeled it control processing. They also found that the processing underlying the consistent mapping condition improved markedly with practice until it asymptoted. They labeled this processing as automatic.

Properties of automaticity. Shiffrin and Schneider argued that automatic processing is qualitatively different from control processing. They advanced this argument because their experimental results showed that the processes in their experiments had quite different properties. Three seem worth mentioning.

First, the number of targets had an appreciable effect on performance in the varied mapping condition, but not in the consistent mapping condition. Subjects appear to be searching their memory set serially in the varied mapping condition, whereas in the consistent mapping condition, the search had the earmarks of parallel processing. Schneider and Shiffrin (1977) established this load effect in a large number of experiments, and it is, without a doubt, the strongest and most robust of their findings. It is worth noting, however, that the load effect is tied to the present paradigm. Although it does indicate that subjects in the varied mapping condition are doing something different from subjects in the consistent mapping condition, it may have little bearing on any putative difference between automatic and controlled processing in other experiments or other situations.

Second, concurrent activities will not interfere with an automatic process, but will interfere with a control process. For instance, in Schneider and Fisk (1982), subjects were asked to carry out automatic detection on one diagonal of a four-character display while attempting a controlled search on the other diagonal. The automatic detection involved a manipulation similar to the consistent mapping condition of Schneider and Shiffrin, whereas the controlled search involved stimulus material similar to that found in the varied mapping condition of Schneider and Shiffrin. Performance on this dual search task was compared with performance on a task in which subjects only had to attend to one of the diagonals. It was found that, with sufficient practice, performance in the dual search task was at levels obtained in the one diagonal search task. The automatic detection did not interfere with the control search.

Finally, Schneider and Shiffrin (1977) established that control processes were indeed open to intentional manipulation, whereas automatic processes were not. In several of their experiments, they would reverse the distractors and targets in the visual array. If subjects were in the consistent mapping condition, performance dropped precipitously when the reversal occurred; however, when subjects were in the varied mapping condition, they quickly accommodated themselves to the change. It was as if subjects could direct their processing in the varied mapping, but after extensive practice, the processing in the consistent mapping condition ran itself without the conscious direction of the subject. A similar point has been made by researchers on typing skills, when they note that a novice typist can think about what letter he should type next, whereas similar thoughts on the part of an expert will only interfere with performance (see Shaffer, 1975).

Factors affecting the ease with which a task becomes automatic. Obviously, some tasks require less practice than others to make the transition between controlled and automatic processing. The factors governing this phenomenon are poorly understood, and, outside some work on individual differences, only a few studies exist. Dumais (1979) explored the issue in the context of the Schneider–Shiffrin paradigm. She investigated exactly how consistent the mappings had to be across trials, that is, the degree to which a target in one trial served as a distractor in another. In the original Schneider and Shiffrin work, the mapping was 100%: targets were never distractors. Under these conditions, automaticity was achieved after 1 hr or so of practice. Dumais varied the ratio of target to use to distractor use, using ratios of 10 : 0 (the standard consistent mapping condition), 10 : 5, 10 : 10, 10 : 20, 9 : 61 (the standard varied mapping condition). She found that even after 6000 trials, the 10 : 20 condition resembled the 9 : 61 condition. The first three conditions, 10 : 0, 10 : 5, and 10 : 10, were all superior to the varied mapping conditions. When an item appears as a distractor twice as often as it appears as a target, automaticity may not develop. Moreover, the lower the consistency of the mapping, the slower the development.

Segregation. A task may not be automatic and still be performed simultaneously with another task with little or no interference. For instance, as already noted, there is much less interference between tasks when the tasks have different stimulus characteristics than when they share the same features. Subjects in such tasks seem capable of segregating the two tasks. Allport (1980) noted that

So long as the input–output and translational demands of concurrent tasks can remain adequately insulated from each other so that no stimuli proper to one task activate schemata pertinent to another and no actions in the response domain of one task are required by schemata active in the other, there need be no mutual competition (p. 83).

Navon and Gopher (1979), in their exposition of multiple resource models, also offer a similar conjecture. They posited that divided attention is limited by the demands that competing tasks make on a similar resource and that performance can be improved by distributing demands among resources. As long as processing demands can be segregated into separate resource pools, divided attention is possible. Finally, Navon (1984), in his critique of resource models of attention, noted in a variety of contexts that dual task performance should improve if the tasks can be insulated from each other. It would be easier to transmit multiple messages over a telephone wire if the messages could be kept segregated.

Hirst and Kalmar (1983) discussed segregation in information-processing terms. Any task consists of a sequence of processes, where each process takes an input, transforms it, and produces an output (Miller et al., 1960; Newell,

1972). This stream of processing may in turn consist of subprocesses arranged heirarchically (Neisser, 1963), but these subprocesses also take an input, transform it, and produce an output. For instance, one could, in a simplistic way, view the process of typing a letter as a sequence that (1) takes the letter on the page and transforms it into an internal representation of the letter, (2) takes this internal representation and transforms it into a motor program, and finally (3) takes this motor program and transforms it into observable movements. Each of these steps or processes contains the three necessary parts: input, transformation, and output.

Although human information processing is by no means this straightforward (see Neisser, 1976; Anderson, 1980), this schematic conception nicely highlights one problem people might encounter when doing two things at once. Consider what might happen with typing while shadowing. The two tasks could easily get crossed at the level of internal representation. The internal representation of the letter to be typed could be selected as input for the process that transforms it into a speech program. Alternatively, the internal representation of the letter to be shadowed could be selected as input for the process that transforms it into a motor program. In either instance, performance should decline. The processing may simply fail to run its full course or may produce cross-talk and other errors.

There are a variety of reasons to suspect that something such as segregation may facilitate divided attention. First, several experiments suggest that even complex, nonautomatic tasks, such as semantic processing, can be done without interference indicating that one difficulty in performing these tasks concurrently may be in keeping them insulated from one and other, not merely doing them simultaneously. LaBerge and Samuels (1974) have suggested that some aspects of reading may be free from interference. Work in priming and subliminal perception also suggests that semantic information can be processed without interfering with other tasks (Allport, 1970; Posner & Snyder, 1975; Marcel, 1983a, 1983b). Finally, several researchers have suggested that two lines of processing can occur without interferring with each other as long as the processes involve different cortical structures (Hellige, Cox, and Litvac, 1979; Kinsbourne & Hicks, 1978).

More importantly, work on cross-talk also indicates that people may be learning to direct the appropriate input to the appropriate process when practicing doing two things at once. For instance, Shaffer (1975), in his study of a skilled secretary typing while shadowing, does provide several nice illustrations of cross-talk. He found many examples both at the acoustic and semantic level: the secretary occasionally typed words that were exactly, phonemically, or semantically similar to words that she should have shadowed (and vice versa). In each example, the wrong message appears to serve as input into the wrong process.

Finally, Hirst and Kalmar (1983) showed that people are indeed learning to use phonological features to guide segregating when practicing a divided attention task. They practiced subjects in a selective attention task and then examined the effect of this practice on a related divided attention task. They employed a dichotic listening paradigm, and, in the experimental condition, the two messages could be distinguished from each other on the basis of their distinct phonological features. For example, in one of the messages, all of the words began with fricatives while in the other, the words began with plosives. They found that divided attention improved with an improvement in selective attention. They also practiced subjects in the divided attention task and found that as divided attention improved so did selective attention. They concluded that one skill that facilitates divided attention is a skill that also facilitates selective attention.

The Hirst and Kalmar results only embrace phonological features. At present, it is not clear exactly what features of the message people can use to guide segregation and whether messages must be segregated at all levels of analysis to be free from interference. Shaffer's work on cross-talk indicates that people may be able to use semantic features of competing messages to guide segregation.

Time-sharing. People can also learn to do two things at once by shifting their attention between the two tasks. In the strictest sense, they are not really dividing their attention between the two tasks when this occurs. Rather they are devoting their full attention, first to one task, then to another.

Although the issue is by no means resolved, the current evidence suggests that it takes time to shift attention (Broadbent, 1971). Consequently, if a person wants to do two things at once by time sharing, he must somehow perform one or both of the tasks more efficiently in order to maintain normal levels of performance. Exactly how this can be done will depend on the task and subjects' practice with the task. If the task is already highly practiced, it probably cannot be done more efficiently than it is already being done. However, the processing underlying an unfamiliar task may change substantially with just a little practice. If this is so, it is easy to see how subjects might add a concurrent task, time share between the two tasks, and appear to be doing two tasks simultaneously.

Summary

In their desire to predict the ease with which two tasks can be combined, students of divided attention have devised several intuitively appealing models. In a sense, the models reflect the interest of the researchers and the kind of experiments that they have devised. The resource model was articulated by researchers who viewed the organism as static. These researchers were not interested in how the organism adapts to meet dual tasks demands, rather they were interested in looking at the trade-off between the performance of the two tasks in an organism that did not benefit from practice or experience. Of course, everyone knew that dual task performance could improve with practice, but this improvement was not the starting point for theory development. It was something that would be explained after the theory had been developed.

Proponents of the skills approach examined practice effects on divided attention and hence took as their starting point the changes that occur with practice. Their interest was not in the trade-off between performance – inasmuch as they viewed the processing underlying any dual task as constantly changing. Their interest was in what skills allowed an individual to meet the level of performance that he did and how these skills changed with practice. To this end, they tried to articulate what a person did when doing two things at once, not how he or she allocated resources. In the present chapter, we have discussed what is known about the skills of organizing two tasks into one, of automatizing one of the tasks, of segregating the two tasks, and of time-sharing. The exact nature of these skills is poorly understood, but even the articulation of a taxonomy of attentional skills is a worthwhile endeavor.

7   The neurobiology of attention

David Lee Robinson and Steven E. Petersen

Introduction

Numerous studies of the central nervous system have attempted to identify neural sites, correlates, and mechanisms of attention. The visual, auditory, and somatosensory systems have been explored for attentional effects at peripheral as well as central locations. Figure 7.1 is a drawing of a rhesus monkey brain with labels for many of the areas studied. Since the major emphasis of this chapter is the visual system, these points on the brain are identified extensively.

To date, the major question being addressed is: Where, in the brain, are the structures that exhibit attention-related events? Further questions include: (1) What exactly are the legitimate attention-related phenomena? (2) Is attention a peripheral gating out of irrelevant sensory data, enhancement of relevant data, or both? (3) Is there a single attentional system for all three modalities or are there separate attentional systems for each sensory modality? Some brain areas having attentional effects have been identified, and some study into the neural mechanisms underlying these effects has been performed. We will examine some of these experiments and assess their contribution to the understanding of the neurobiology of attention. Recent reviews (Sprague et al., 1961; Näätänen, 1975; Friedland and Weinstein, 1977; Lynch, 1980; Posner, 1980; Wurtz, Goldberg, and Robinson, 1980; Hyvärinen, 1982; Hillyard and Kutas, 1983) have dealt with other aspects of attention and its neural substrates.

Definitions

Many interrelated factors influence the way in which an organism deals with sensory stimuli, for example, arousal, attention, emotion, and intention to move (Wurtz et al., 1980). In order to study the neurophysiology of attention, it is necessary operationally to define attention so that we can distinguish between attention and other processes currently operating in the organism.

142

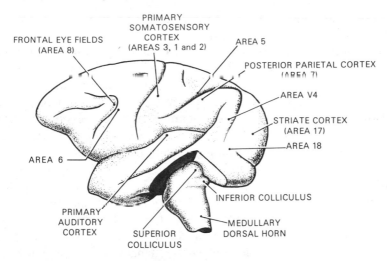

Figure 7.1. Schematic representation of a rhesus monkey brain. The figure is drawn as if part of the inferior temporal cortex has been dissected to expose the brain stem. Dots show the approximate location of brain sites referred to in the text.

We will consider three behavioral processes that interact with sensory signals as they are propagated in the central nervous system: attention, arousal, and intention to move, and we will examine experimental approaches designed to distinguish between them.

Attention is generally agreed to be a selection process: that process by which an organism chooses to deal more effectively with one type of sensory stimulus at the expense of other stimuli. For example, one can focus on auditory events and ignore somatosensory stimuli. We clearly can attend to one sensory modality and not another. Also, we can attend to one attribute in a sensory system, and ignore other characteristics in the same modality. We do this when we search for red-colored clothing or scan only our right visual periphery. In the former we attend to color, in the latter we attend to spatial location. Therefore, any mechanism that is related to attention should have selective characteristics.

In the visual system, the shift of visual spatial attention has a frequent concomitance in the form of rapid saccadic eye movements. We generally make an eye movement to an attended image in order to inspect the image with our foveal vision. However, attention is a process that can be independent of the organism's motor behavior. Although shifts of visual attention and eye movement often occur together, the shift of attention is not dependent on an eye movement. Similarly, shifts of attention in the somatosensory realm can be associated with appendage movement; however, they need not be. Thus, any neural event that is proposed to be related to attention should be dissociable from a specific motor response.

Some sensory events are, in fact, used for the initiation and guidance of movement; one's awareness and use of such stimulation can be functionally different from that which is influenced by attention and arousal. The initiating process for motor acts, or motor set, can modify neural responses to sensory stimulation, but it is not necessarily identical to the process of attention.

This conceptualization of attention suggests a selectively increased (or decreased) ability to deal with incoming sensory information; attention should also be differentiated from nonselective changes in processing that are related to levels of arousal. Levels of alertness drastically alter our ability to deal with sensory events, but this process tends to be general and related to many stimuli and sensory systems. At low levels of arousal, stimuli in all modalities are reduced in their effectiveness; at maximal levels of arousal, many stimuli are effective, and there is little selectivity. Changes in responsiveness to sensory stimuli could be due to several factors, not just attention.

Methods employed

Neuroscientists have used several experimental techniques to determine the loci and neural mechanisms of attention. Together, these approaches have led to advances in our understanding of the neurobiology of attention, but each approach has significant advantages and disadvantages, and a global comprehension of the neurobiology of attention requires a synthesis of the concepts derived from each of these techniques. Our evaluation of these experimental approaches will deal with such features as the control of the subject's behavior, the degree of localization of neural processes within the brain, the time resolving capacity of the technique, the amount of interference with normal functioning, and the amount of neural tissue that can be studied at one time.

Since this chapter is the counterpart of one on the cognitive psychology of attention, it is important to emphasize major differences between cognitive and neurobiological approaches. The psychologist studies the whole brain, from receptors to integrative processes to motor output; the neurobiologist, especially the electrophysiologist, studies only one part of the brain at a time. The psychologist tries to infer the essential functional elements of a process by studying the whole process; the neurobiologist attempts to analyze the process by studying its component parts. What the neurophysiologist observes at one point in the brain can be only a fraction of the mechanism that the psychologist studies as the whole. To date, it is impossible for the neurophysiologist to describe neural substrates that will account for the whole of a perceptual process, just as it is impossible for the psychologist to understand individual mechanisms by testing the whole organism.

1. Single neurons and behavior. One of the approaches to the neurobiology of attention (one with which we are most familiar) has been to record the action potentials from single neurons in awake animals trained on several behavioral tasks. When we use this approach to study the visual system, we teach monkeys to fixate on a spot of light projected onto a screen. Once an animal is trained, we can record the electrical signals from individual cells in the brain and try to present light patterns or behavioral situations that change the activity of neurons in a brain area. During periods of fixation, we flash lights onto different locations on the screen. When we find a location where light excites a cell, we call that area the cell's receptive field. Next we use the lights in the receptive field as cues or targets for different types of behavior to see which behaviors influence the cell's activity. We assume that stimuli or behavior that maximally excite a cell are critical parts of the neural function of the brain area. This approach allows one to test a variety of sensory stimuli on many cells and to determine how various types of behavior influence the responses of such cells to sensory stimuli. It permits the study of rapidly modulating events and allows for comparison of different behavioral effects on these events. By histological study of the brain after an experiment, this approach allows for very precise localization of the neuronal sites of various effects. These techniques permit analysis of a most basic element of neural processing – the single-cell signal.

One of the major difficulties of this approach is the limited communication with an animal. Each animal must be taught each task, and controls must be carefully constructed, to be sure that the behavior measured is relevant to the assumptions of the particular experiments. Whereas one has numerous verbal reports possible from human subjects, there is an uncertainty about exactly what an animal is attempting to do in each task and hence uncertainty about the precise state of its brain. Another limitation is sampling; cells are studied one at a time and there is no way of directly knowing how all of the cells in one area are behaving at one time. Furthermore, only one area of the brain is sampled at a time, and there is no direct way of knowing how other brain areas are functioning while one area is being examined.

2. Regional cerebral blood flow. A recent technique that has promise for understanding the function of the brain is the measurement of regional cerebral blood flow. When an area of the brain is active, it has an increased need for metabolic "fuel." As a consequence of this, the blood flow to this area is increased. For example, there are elevated levels of flow in the auditory cortex when sound patterns are presented to a listening subject. Isotope-

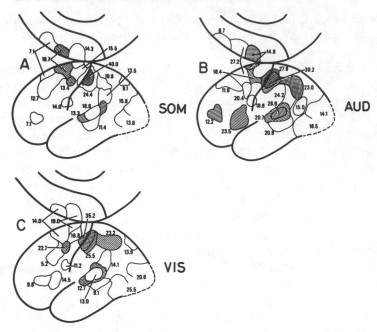

Figure 7.2. Elevations in regional cortical blood flow during the performance of attentional tasks. Subjects were presented with stimuli in the visual, auditory, and somatosensory modalities, but in different tests were required to attend and discriminate in only one modality. When attention was directed toward the visual stimulus (C), there was an increased activity in the posterior parietal area, in the frontal cortex in the area of the frontal eye fields, and in the superior temporal area. Attention to the somesthetic stimuli (A) evoked increased blood flow in the more anterior parietal cortex, the precentral cortex, and a region of the temporal cortex. There were more widespread increments in flow with attention to the auditory events (B), and these were found primarily in the prefrontal cortical areas, the temporal cortex, and a large expanse of the parietal area. (Reproduced with permission from Roland, 1982.)

labeled xenon is injected into the blood supply to the brain, and regional levels of the isotope are measured by a large number of detectors on the scalp. This technique can measure changes induced during a 1-min test period. Figure 7.2 shows the patterns of activity observed in the brains of human subjects who are performing different sensory attention tasks.

This approach has the advantages of easy communication with a human subject and minimal interference with normal functioning. The degree of localization within the brain is much poorer than that in single-cell studies, and the technique cannot be used to analyze the time course of events since a state must be maintained for at least 1 min.

3. Evoked potentials. Electrodes placed on the scalp will record potentials generated from a large number of neural elements beneath it. Evoked po-

Figure 7.3. Demonstration of visual neglect with self-portraits. The German artist Anton Räd-erscheidt made a self-portrait two years before (A), two months after (B), and 5 months after (C) he had suffered a stroke that damaged his right (nondominant) parietal cortex. Prior to the damage the portrait was complete on both sides. Immediately after the damage to the right side of the brain, the self-portraits were severely deficient in representing the artist's left face, the contralateral sensory field. Over time, the left representation improved, but never equaled that of the right and prelesion conditions. (Reproduced with permission from Jung, 1974.)

tentials have been used to understand neural processing in both humans and animals. This approach has the advantage of (1) being usable in humans, which simplifies communication problems, and (2) an excellent time-resolving ability. On the other hand, evoked potential recording requires averaging in order to detect most effects and loses the ability to evaluate moment to moment influences. As in the cerebral blood flow studies, there is relatively poor spatial resolution for cortical effects, and evoked potentials are not regularly useful in humans for measuring effects in subcortical areas. It is also difficult to analyze exactly what a waveform reflects in terms of underlying neural processing.

4. Lesions – experimental and clinical. The oldest approach to the study of the brain is to test for the losses in function following damage to the brain. The observation of deficits in humans after strokes or tumors is the origin of clinical neurology. Figure 7.3 exemplifies the visual perceptual deficit produced in a human as the result of damage to the inferior parietal cortex, a lesion that is believed to remove some attentional function. Although lesion studies could be viewed as the most direct test of the function of the damaged tissue, it does have the limitation of, in fact, demonstrating how well the organism can compensate for the loss of a brain area. An old adage holds that a lesion experiment really shows what the remainder of the brain is capable of doing, rather than what the removed portion did. One can only infer function on the basis of what capacity has been lost. Frequently, deficits are compensated for in a short time, and are not noted with this approach.

A further complication of all studies in which the brain is damaged is that of delimiting and specifying the amount of damage. With humans, lesions seldom limit themselves to one functional area and seldom is histological material available until long after the damage. With both humans and animals, lesions seldom are limited to a single cell group or fiber pathway. Recent studies with chemical lesions, such as those caused by kainic and ibotenic acid, provide hope that experimental animal lesions can be limited to functional groups of cells.

5. Drug injections. Since almost all transmission between neurons is mediated by chemical compounds, changes in these chemical processes can lead to changes in function. With our increasing understanding of the chemicals involved in synaptic transmission, it is possible to modify brain function in selective ways and to observe changes in the behavior of the animal following the change. This approach has the advantage of changing the brain in a relatively nondestructive way. It also has the asset of changing the brain in subtle, more physiological, ways than other techniques, such as lesions or electrical stimulation. This approach is limited by our present understanding of how transmitters function within the central nervous system, and in assessing the functional spread of the injected substance to adjacent areas.

In summary, the various experimental approaches described above are contributing to our understanding of the neurobiology of attention. Studies on humans have permitted a high degree of sophistication in the behavioral aspects of tests, but have been limited in the specificity of their observations on the brain. Those studies that have used animals have been able to localize brain sites and to examine mechanisms quite well, but exact control of the animal's behavior has been limited. Thus, it is with the combined approaches of all of these studies that we have made advances in an understanding of the neurobiology of attention. The balance of this chapter will deal with each of the major sensory systems (vision, audition, and somesthesia) and the progress that each experimental approach has contributed to unraveling the sites and mechanisms of attention.

Sensory systems

Vision – a model system

Within the past decade there has been developed a body of neurobiological data on the role of attention in the visual system. These data have been derived from a variety of approaches and have allowed for specifying brain loci and possible mechanisms. There are several reasons that study of the visual system has facilitated these advances. First, there is a large body of physiological data on the organization of the sensory aspects of this system so that it is possible to move onto the influences of attention with this as a background. Second, in the visual system, as in audition, it is also possible to clearly control and specify the stimuli used in an experiment. Third, some motor responses to visual stimuli (eye movements) are easily measured and controlled by behavioral techniques.

1. Organization of the visual system. The axons that leave the eye send visual information to the lateral geniculate nucleus and to the superior colliculus. All of the cells in the lateral geniculate then send their axons to the striate cortex (area 17). This pathway from retina to the lateral geniculate to striate cortex is extremely important for the fine, detailed aspects of vision. Cells in the striate cortex send information to many places, such as the superior colliculus, part of the pulvinar, and area 18. Although these latter areas are definitely part of the visual system, their contribution to visual behavior is not clear. There are still other parts of the brain, such as parietal cortex and the frontal eye fields, that contain neurons that receive visual information. For these areas, we have little understanding of how they receive their visual information, but we are beginning to appreciate that their function is not in simple visual sensation, but is in the area of active visual behavior.

2. Single cell studies. By comparing the responses of single neurons during the performance of several tasks and in several different areas of the brain, some understanding of how attention and other behavioral influences are organized in the central nervous system has been realized. This section will organize these data on the basis of groups of structures containing similar functional properties as they relate respectively to visual spatial attention, the initiation of movement, and arousal. Finally new experiments dealing with attention to stimulus orientation will be presented.

a. Parietal cortex and pulvinar. (i) Enhancement. Neurons in the parietal cortex (area 7) (Robinson, Goldberg, and Stanton, 1978; Bushnell, Goldberg, and Robinson, 1981) and part of the pulvinar (Pdm) (Petersen, Robinson, and Keys, 1982; Petersen, Morris, and Robinson, 1984; Robinson and Petersen, 1984) have visual and behavioral properties that make them candidates for neural processes of visual spatial attention. Many of the cells in these areas have visual receptive fields (Yin and Mountcastle, 1977; Robinson, Goldberg, and Stanton, 1978). These receptive fields can be demonstrated in awake monkeys by training the animal to fixate on a spot of light. During brief periods of fixation, other lights can be flashed onto the tangent screen, and restricted regions (visual receptive fields) of the screen can be found where light will cause the neurons to discharge. These cells are not as selective

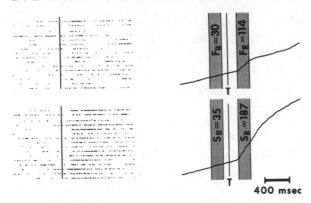

Figure 7.4. Enhanced discharges to actively used visual stimuli. The data on the top show the response of a parietal neuron to a stimulus flashed on the tangent screen while the monkey fixated a spot of light. Each dot represents a spike of the neuron; each row of dots comes from a single fixation trial. The vertical line indicates the time of stimulus onset. Cumulative histograms on the right sum the data in the adjacent raster dot display. For the trials on the bottom there is an increased response of the cell to the same stimulus on this series of trials where the animal makes a saccadic eye movement to the identical stimulus. This type of cell in parietal cortex responds better to identical stimuli when the stimuli are actively used than when they are not. The vertical, shaded columns and numbers on the right are used to quantify the enhanced responses. (Reproduced with permission from Bushnell, Goldberg and Robinson, 1981.)

for the types of stimuli that excite them as are cells in other parts of the visual system. Nonetheless, these neurons do receive information about the presence of light in the external visual environment. More important for attention is the fact that these cells also receive inputs that change their responsiveness to a stimulus, depending on the behavioral context in which stimulation takes place (Yin and Mountcastle, 1977; Robinson et al., 1978; Bushnell, Goldberg, and Robinson, 1981; Mountcastle, Andersen, and Motter, 1981; Petersen and Robinson, 1983; Robinson and Petersen, 1984). Such an effect can be demonstrated in these cortical and pulvinar cells by changing the test conditions. If a stimulus that excites a cell during fixation is used as the target for an eye movement, then many of these cells have more intense activity, an enhanced discharge; these cells respond better to stimuli that are actively used than to stimuli that are not. Figure 7.4 shows the results of such an experiment for a neuron recorded in parietal cortex.

(ii) Selectivity. Additional tests elucidate further important features of this enhanced activity. If the response of an enhanced cell to the same stimulus in the same location is tested during a series of eye movements to different points in space, the enhancement is present only for eye movements directed to the visual receptive field (Robinson et al., 1978; Bushnell et al., 1981; Petersen and Robinson, 1983; Robinson and Petersen, 1984). These tests

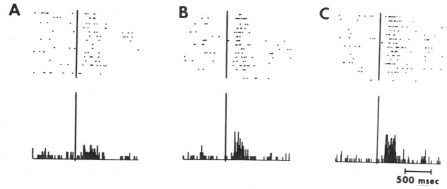

Figure 7.5. Enhanced responses of a parietal neuron with both saccadic and attentional use of a target. Data on the left (A) show the response of the cell to a light flashed on the screen while the monkey maintains fixation. Data in the center (B) show the improved discharge to the same light when it is the target for an eye movement. Data on the right (C) illustrate that the response of the same cell can be enhanced when the animal attentively uses the light without making an eye movement to it. Here the monkey was required to detect the dimming of the peripheral light that occurred at variable times long after the onset of the light. Histograms sum the data in the raster above them. (Reproduced with permission from Bushnell, Goldberg, and Robinson, 1981.)

show that the enhancement reflects a selection process; it is not due to such nonspecific processes as arousal, task difficulty, emotion, the termination of the fixation point, and so forth.

(iii) Movement independence. A final series of experiments links this effect in these areas with visual attention, as opposed to a motor set. If a monkey is trained to use a stimulus in its receptive field in a way other than as the target for an eye movement, then the enhanced response can still be demonstrated (Bushnell et al., 1981; Petersen and Robinson, 1983; Robinson and Petersen, 1984). For example, if the animal must use a peripheral stimulus as a cue for a bar release and is not allowed to make an eye movement to it, the enhancement is still present. Figure 7.5 illustrates the enhanced responses of a parietal cell during both saccadic and "peripheral" use of a light in the cell's receptive field. In this test, the animal is required to respond to the dimming of a peripheral light at some time long after it has been flashed onto the screen. Because monkeys can perform this task well above chance levels, the assumption is that they must be attending to the peripheral stimulus. The enhanced response is present near the onset of the light that occurs significantly before and at variable times before the monkey's response. The results of these experiments show that the enhancement is not functionally related to saccadic eye movements, but is probably related to the shift of visual attention common to both the saccade and peripheral attention tasks.

In an additional set of tests, monkeys were trained to reach with their hand to touch a lighted panel without making an eye movement to the light (Bushnell et al. 1981). Although this task is difficult to teach, monkeys do perform it and enhanced responses are again demonstrable in area 7 cells.

Other experiments by Rolls et al. (1979) show that neurons in area 7 are responsive to stimuli that are aversive as well as those that are of positive significance. It is assumed that the animals attend to the aversive stimuli. These data are consistent with the activity of this area being related to attention rather than to specific behaviors or reward contingencies.

In summary, studies have demonstrated that the enhanced response can be produced in four situations in which the monkey actively uses a stimulus in its visual field: as the target for an eye movement, as the target for a hand movement devoid of an eye movement, as the stimulus for a bar press without any movement toward the light, and as an object of aversion. Since a shift of attention is common to all of these tasks, we conclude that it is attention that is enhancing these responses and not the independent behaviors.

These studies demonstrate several features of neurons in parietal cortex and the pulvinar that tie them to the processes of selective visual attention: (1) The cells receive sensory information (they respond to visual stimuli); (2) they have a selective mechanism (an enhancement that is spatially restricted); and (3) the enhancement is present with shifts of attention regardless of the motor response that indicates to the experimenter the shift of attention.

Work from other laboratories on cells in area 7 of the awake monkey has also suggested attentional properties. Hyvärinen and Poranen (1974) recorded cells that were active with different types of eye movements and reaching movements, and they were the first to propose that this activity had an attentional character. It is hard to be certain that attention was the critical factor of this activity, since the motor set, sensory stimuli, and state of arousal were not controlled in these experiments. Mountcastle and his colleagues (Yin and Mountcastle, 1977; Lynch, 1980; Mountcastle et al., 1981; Andersen and Mountcastle, 1983) systematically studied parietal neurons in awake, trained monkeys and characterized some of the eye and limb movements that activated these cells. Since there were no tests of different motor responses on the same cells, it is difficult to distinguish whether the activity of these neurons was related to the movement or the attention concomitant to the movements. Subsequent studies by this group (Mountcastle, et al., 1981) showed that attentive fixation to a spot of light produced a significant facilitation of the visual discharge when compared to the response without attentive fixation. In a final set of studies they found that the magnitude of the visual response of many of these cells was modulated by the angle of gaze of the monkey (Andersen and Mountcastle, 1983). These data suggest that visually responsive cells have some information about eye position. Such information could be useful in the processes of spatial attention.

b. Superior colliculus and frontal eye fields. Comparable enhancement experiments have been conducted on the cells in other parts of the central nervous system and have produced different results. Many of the neurons in the superficial layers of the superior colliculus, and frontal eye fields (area 8) are visually responsive and also have enhanced responses when lights are used as the targets for saccadic eye movements (Goldberg and Wurtz, 1972; Wurtz and Mohler, 1976a,b; Goldberg and Bushnell, 1982). In both of these areas, the effect is spatially selective (Goldberg and Wurtz, 1972; Goldberg and Bushnell, 1982); the enhanced responses are associated with eye movements into the receptive field but not to other points in the field. So here, as in the other brain areas described, the enhanced response represents a selection process. In contrast to the effects in the visual "attention" areas, the enhancement in these collicular and frontal eye field cells is not demonstrable with attentional behavior (Wurtz and Mohler, 1976b; Goldberg and Bushnell, 1982). Figure 7.6 shows the eye movement enhancement for a frontal eye field cell and lack of enhancement with peripheral attention. These results indicate that the modulations here are a selection process that is specifically related to eye movement and not to spatial attentional behavior. We interpret these observations as indicating that the effect is part of the visual processes for the intention to move.

c. Striate and other cortical areas. Other parts of the brain have been tested for enhanced responses and have produced qualitatively different effects. The neurons in striate (area 17) (Wurtz and Mohler, 1976a) and prestriate (area 18) cortex (Robinson, Baizer, and Dow, 1980) have visual responses that require very selective stimuli to excite them. In this respect, they are different from those in parietal cortex, the pulvinar, the colliculus, and the frontal eye fields. Although a limited number of the visual cortical cells have enhanceable responses, the process is not spatially selective. Furthermore, the effect can be produced with both eye movement and attentional tasks. The results taken together suggest that the processes that are modifying these cells are related to the level of arousal in the animal and are not related to shifts of attention or the intention to move.

Neurons in dorsolateral prefrontal cortex do not have the stimulus specificity of neurons in areas 17 and 18, but respond in both eye movement and peripheral attention tasks (Mikami, Ito, and Kubota, 1982). The neurons were not tested for spatial selectivity, so their relation to attention or arousal is ambiguous.

d. Attention to stimulus properties. All the studies discussed so far deal with the attention to spatial location. Recent neurophysiological experiments have

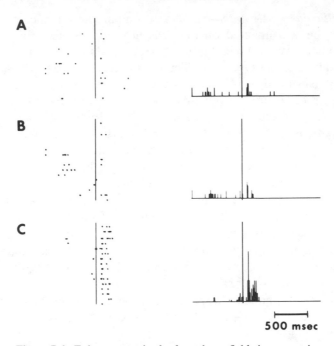

Figure 7.6. Enhancement in the frontal eye fields is present in eye movement trials but not peripheral attention trials. Responses on the top (A) come from trials where the animal fixated while a light was flashed on the tangent screen in the cell's receptive field. Trials on the bottom (C) show the enhanced response evoked by a stimulus when the monkey subsequently made saccadic eye movements to the stimulus. Data in the center (B) demonstrate that cells which have saccadic enhancement in the frontal eye fields do not show enhancement when the monkey attends but makes no movement to the light. Reproduced with permission from Goldberg and Bushnell (1982).

demonstrated modified responses when animals are trained to attend to stimulus configuration (Richmond and Sato, 1982; Haenny and Schiller, 1983). In one of these studies, monkeys were trained to look at a display of stripes that alternated by ninety degrees of orientation. When the regular alternation was changed to a repetition of one orientation, then the animal was required to make a response. Cells in striate and prestriate area V4 were sampled, while monkeys performed this task, and many cells in V4 showed a significant and selective change in their response level when the monkey was attending to a stimulus of a particular orientation. The process involved here could not be related to levels of arousal because the effect is selective for stimuli of a certain orientation. It is also unrelated to any type of motor response since the effect is present long before any motor response and is time-locked to the onset of the stimuli of the correct orientation; it was not synchronized with the actual motor response. Comparable studies of attention to stimulus configuration using cells in the inferior temporal cortex have produced comparable results (Richmond and Sato, 1982). However, when an animal attends to stimulus location, these cells give their poorest responses (Richmond, Wurtz, and Sato, 1983). These studies demonstrate sites for the neural process of visual attention to stimulus configuration and suggest opposing mechanisms for spatial and feature attention. They also indicate that the neural processes of attention can be specific to even a submodality of a sensory system. Thus, attentional effects are related to stimulus form in some regions and to stimulus location in others.

Experiments by Fischer and Boch (1981a,b) have shown spatially selective enhancement in cells in area V4 but have not directly tested for movement dependence, so the relation of V4 neurons to spatial attention or the initiation of movement is still unknown.

3. Regional blood flow. Roland (1982) attempted to localize attentional sites in the human brain by measuring the regional blood flow with isotopes while humans performed various tasks. When subjects were presented with stimuli in the three main modalities but asked to discriminate using only one, Roland found various cortical changes. Attention to the visual modality in this situation produced increased activity in visual association cortex, the frontal eye fields, and the right posterior superior parietal cortex (Figure 7.2). When attention was changed from one modality to another, there was changed activity in the superior lateral part of the prefrontal cortex regardless of the modalities being switched. The primary and immediate associative cortices for each modality were markedly activated in each attentive task. These studies suggest parts of the brain that may be specifically related to attentional behavior and not just the sensory processing associated with any modality regardless of the behavioral state. Furthermore, the data show that at some level the sensory activity which is not being used in attentive behavior still reaches the cortex and is processed to some extent.

4. Evoked potentials. Numerous experiments have attempted to study the relationship between evoked brain potentials in humans and performance on tasks of selective visual attention. Most of these have reported enhanced potentials to attended stimuli (see Näätänen, 1975; Näätänen and Michie, 1979, for review). One of the questions addressed by this approach was whether attention functions as a selective filtering of sensory signals. If this is the case, it is expected that signals of the attended modality are facilitated and those of the unattended modality inhibited. This model would suggest a lowered transmission in the unattended channels (Hernández-Peón, Scherrer, and Jouvet, 1956). However, a recent review of these data has suggested that other, nonattentional, factors may have caused the enhanced potentials (Näätänen, 1975). Possible nonattentional influences include arousal, inadequate control of peripheral sensory stimulation, or nonrandom presentation of stimuli allowing for anticipation by the subjects. Lack of stimulus control could be such that attended and nonattended stimuli were not presented equally at the receptor, and, therefore, stimulation differences could account for the evoked potential differences. Näätänen (1975) and Näätänen and Michie (1979) also emphasized the importance of limiting the analysis to early portions of the evoked potential since one would expect to see early changes specifically related to attention; later changes in potentials could well be related to discriminative decisions or motor responses.

Experiments by Eason, Harter, and White (1969) and Eason (1981) demonstrated that both arousal and attention can augment the potentials evoked by visual stimuli. Aroused subjects showed larger evoked potentials and shorter reaction times, while the effect of attention to one field primarily increased the evoked potential. This work had no control of the subject's eye position so that stimulation of the two visual fields may have been unequal. Furthermore, the subjects could have predicted the stimulus sequence.

Harter, Aine, and Schroeder (1982) found that the early part of the visually evoked potential was well correlated with the location of the attended stimulus, and this potential was most prominent over the hemisphere contralateral to the attended stimulus. A later negativity was more closely associated with the type of stimulus that was attended. In contrast, this negativity was most prominent over the left hemisphere independent of the side of the stimulus. These data suggest hemispheric specialization in attention to pattern versus spatial location. Furthermore, they suggest that attention to location precedes attention to other stimulus attributes.

Van Voorhis and Hillyard (1977) showed that attention to visual stimuli enhanced two components of the evoked potential. The earlier enhancement of the potential was seen primarily over the occipital regions and was suggested to be related to the specific submodality. The later potential, which was enhanced, appeared to be related to the more general aspects of selective attention. They also reported that visually evoked potentials were largest when attention was focused on a stimulus, smallest when attention was focused away from a stimulus, and intermediate when attention was divided between both fields.

In an attempt to determine the spatial frequency tuning of the visual attentional system, Harter and Previc (1978) asked subjects to attend selectively to certain sizes of checkered patterns flashed on a screen. Evoked potentials were measured during the performance of this task. The facilitation of the potential evoked by the flash was greatest when the test pattern exactly matched the standard, and it decreased with difference in size from the attended size.

5. Lesions

a. Human studies. (i) Cortical damage. At present, there are few systematic studies of the effects of brain damage on attention. This is because the systematic study and quantification of attentional behavior is relatively recent. One way of circumventing this problem is to hypothesize how an "attention-less" individual would behave, and then if patients with brain lesions are found to exhibit such behavior, it would be possible to attribute an attentional function to the damaged area.

If the attentional system is involved in some scanning mechanism that searches the peripheral visual system, it might be expected that its loss would lead to a type of neglect. The presence of this neglect is also called hemi-inattention. Without the ability to scan the periphery, an organism would be expected not to ulitize visual data coming at its visual system from the con-tralateral fields. Therefore, for the present, we will discuss situations in which neglect is present and deal with this in terms of damage to an attentional system. It is unlikely that all cases of neglect are due to change in an attentional system. For example, if the superior colliculus is related to the visual initiation of movement (and not true attention) then damage to this structure could lead to a lack of visuomotor responses. This would be a type of neglect, but not an attentional one from our frame of reference. Also, damage to an arousal system (Heilman, Pandya, and Geschwind, 1970) would lead to a lack of responsiveness to sensory stimuli. This would be classified as neglect, but would not be a true attentional problem.

It should be noted that there are probably many processes of attention and many related brain areas. Damage to any part of the system or malfunction of any process could lead to neglect. We will exclude from our analysis any lesion that produces a simple sensory loss because, at best, this confounds any problem and makes the analysis of attentional problems confused. Fried-land and Weinstein (1977) have observed that:

Patients with only dense homonymous or bitemporal hemianopias may exhibit no evidence of visual hemi-neglect, and are generally able to draw and read without any asymmetry in their performance. Patients with even severe hemisensory deficit or hemiparesis may also show no evidence of hemi-inattention and may remain well aware of the presence and disability of the affected extremities.

When there is damage to a primary sensory system, there is often an awareness of the loss. Individuals with neglect most often are unaware of their difficulties.

One of the most dramatic demonstrations of neglect produced by damage to the human brain results from destruction of the inferior parietal cortex (see Friedland and Weinstein, 1977; Lynch, 1980; Hyvärinen, 1982, for re-views). Humans with such damage have a profound neglect of their contra-lateral sensory space: visual, somatosensory, and auditory. In the visual modality, these people ignore and do not spontaneously use data from their contralateral visual field. Such individuals are capable of detecting and re-sponding to stimuli presented in this area, but they do not do so spontaneously, and they have great difficulty dealing with images presented simultaneously to their ipsilateral and contralateral visual fields. When stimuli are presented to ipsilateral and contralateral fields, only the ipsilateral stimulus is identified. This phenomenon is termed extinction. One interesting feature of this problem is that the patients are unaware of, and frequently even deny, the problem. Such effects are most common when the lesion is in the nondominant (right) hemisphere. The effects are less common with left-sided lesions. Unfortunately for those interested in studying these problems, neglect is generally of brief duration, appearing suddenly after a lesion and disappearing within a few weeks. Neglect is most often found with rapidly developing damage to the brain such as a stroke or infiltrating tumor.

Posner et al. (1984) have proposed three processes that are probably involved in attention; any one or all of these may be lost in patients with neglect. The first process is disengagement from the present image of interest. The second is the actual shift of attention to the new object of interest. The third is the engaging of the new image. We would propose an additional step that should precede this triad and that would be the selection of a peripheral image to which attention might be shifted.

In a series of experiments on normal humans, Posner (1980) demonstrated that people respond more quickly to the onset of a spot of light in the visual periphery when the location of the target is preceded by a cue at a location near to the spot than when the spot is preceded by a cue at a distant location. A similar result is obtained when the site of the target is indicated by a centrally located arrow pointing to a location. Posner's (1980) hypothesis is that the cue shifts the person's attention to one side or the other, thereby facilitating the individual's response to targets appearing at that side.

Because humans with damaged parietal cortex are presumed to have damage to their attentional system, Posner and colleagues (Posner et al., 1984) tested such patients on this task. These patients could perform rather normally in situations where they were presented with valid cues to either side, whether to the intact side or the neglected (contralateral) side, although they were slightly slower on the contralateral side. They were also able to respond reasonably well when the cue was in their affected field but the target was in their good field. However, they had severe problems in the situation where the cue was in their good field and the target was in the affected field. They frequently took several seconds to respond or did not respond at all. In this situation, it appears that when the cue shifts their attention into their good field, they have tremendous difficulty in shifting attention back into their affected field. Posner et al. (1984) proposed that parietal patients have lost the ability to disengage their attention from images of interest, although they are able to shift attention and engage it once shifted. Studies by Holtzman et al. (1981) of humans with section of the corpus callosum show that the attentional mechanism of a single hemisphere has control of attention in both visual half-fields.

In an additional set of studies, Posner, Cohen, and Rafal (1982) tested patients who had brainstem lesions that impaired their ability to make saccadic eye movements. When tested with the cue task, these people were clearly able to shift their attention in directions in which they were unable to make eye movements, although they were slightly impaired in the speed with which they responded. The interpretation of these experiments is that there is some interaction between the systems that generate eye movements and those that shift attention, but that a functioning oculomotor system is not necessary for attentional shifts.

Heilman and Valenstein (1972b) have reported that people with damage to their frontal lobe have neglect. The patients studied had lesions of the frontal lobe on the nondominant side, and the lesions were either on the medial surface or the dorsolateral surface. The observed neglect was unilateral, and there was no primary sensory problem. These people also showed extinction.

(ii) Subcortical lesions. Earlier we suggested that subcortical as well as cortical areas may function in an attentional system. Several reports have noted neglect after damage to the basal ganglia (Hier et al., 1977), thalamus (Watson and Heilman, 1979; Zihl and von Cramon, 1979), and mesencephalic reticular formation (Watson et al., 1974). Infarcts that damage the putamen in humans have been shown to produce a unilateral neglect. The cases reported all involved damage to the nondominant side. Since some of these individuals had visual field defects and others did not, the effect is not dependent on simple sensory problems. Thalamic damage of the right side (nondominant) has been reported to cause contralateral neglect (Watson and Heilman, 1979). In addition, these individuals have visuospatial disorders. The authors of these reports proposed that the problem produced with these lesions could be related to damage of an arousal system. Patients who have suffered brainstem lesions and show signs of damage to their centrencephalic system frequently show signs of inattention to one side of their sensory fields (Weinstein, Kahn, and Slote, 1955). Such individuals do not scan their affected field. These people also tend to be withdrawn from much interaction, suggesting that they have more severe emotional problems. There is one report of a patient with a pulvinar lesion who had visual neglect and extinction (Zihl and von Cramon, 1979). Although the functional deficits are very well studied and documented, the patient also had parietal damage that could account for the deficits.

b. Animal studies. (i) Cortical lesions. Studies on experimental animals looking for attentional deficits are limited in number. Monkeys with damage to cortical area 7 have some signs similar to those of human neglect. These include misreaching to visually presented targets as well as increases in the latency to initiate responses (Faugier-Grimaud, Frenois, and Stein, 1978). Heilman, Pandya, and Geschwind (1970) produced unilateral neglect in monkeys by parietal lesions. The animals responded to ipsilateral stimuli when both fields were tested simultaneously. They only responded to food presented in their ipsilateral fields. These problems lasted 2–4 weeks.

The frontal cortex may have an attentional function as suggested by the work of Kennard (1939) and Welch and Stuteville (1958). They showed that lesions of the frontal lobes produce a lack of responsiveness to visual stimuli presented in the contralateral visual field, especially in the periphery. The deficit is not a simple sensory loss and is accompanied by deviations of the head and eyes. Reeves and Hagamen (1971) demonstrated neglect to stimuli of many modalities following frontal cortical lesions in cats. Area 8 was thought to be most important in producing this deficit, but the problem may actually be an oculomotor, rather than attentional deficit.

(ii) Subcortical lesions. Sprague, Chambers, and Stellar (1961) found that lesions of the lateral brainstem in the cat led to deficits in most sensory modalities. In some respects, these animals were inattentive, but we are not including these as attentional deficits because the animals have primary sensory difficulties. Reeves and Hagamen (1971) found multimodal neglect from lesions of the midbrain reticular formation, but here, too, there were sensory deficits which complicate the interpretation. Multimodal attentional problems have been described in rats with lateral hypothalamic damage by Marshall, Turner, and Teitelbaum (1971). These animals do not orient to any sensory stimuli. These rats could produce autonomic responses to contralateral stimuli, suggesting that the effect was not totally sensory. Ungerstedt (1974) chemically destroyed the nigrostriatal system in rats and observed multimodal unresponsiveness, although it is possible that these animals had such a severe sensory loss that the problem is not attentional.

Subsequent studies by Sprague and Meikle (1965) have demonstrated more attentionlike problems from damage to the superior colliculus in the cat. Animals with unilateral damage have some visual capacity in their contralateral fields, but have problems in making localizing responses to images presented here. These deficits were especially clear when stimuli were simultaneously presented in the ipsilateral and contralateral visual fields; response was only to the ipsilateral stimulus. Auditory and somesthetic stimuli also evoked irregular responses. One of the issues that needs to be clarified with these experiments is whether the animals have trouble in shifting their attention to contralateral images (covert orienting) or whether the problem is in initiating motor responses in that direction.

The rostral subthalamic area has been suggested as an attentional area by

Adey (1962) because cats with unilateral damage to this region neglect the contralateral visual space when trying to locate food rewards. However, they are capable of performing other visual functions in this part of space. They are aware of objects in either half of the visual field. These animals appear to be easily distracted in visual task situations, suggesting that they may be "overattentive." The deficit lasted only 1 week. Bilateral lesions produced longer lasting effects that were present in both visual fields.

Orem, Schlag-Rey, and Schlag (1973) have proposed an attentional contribution for the intralaminar nuclei of the thalamus. Destruction of this area in cats leads to difficulties in orienting to visual stimuli in the contralateral field and other visual field problems. These problems are limited to the contralateral visual field and have a short duration. In addition, there is a deficit in the optokinetic system.

Several experiments have shown that damage to the pulvinar leads to behavioral deficits that could be interpreted as attentional. Ungerleider and Christensen (1977, 1979) reported that damage to the inferior as well as adjacent lateral pulvinar leads to animals that have prolonged periods of fixation even while performing visual discriminations. Chalupa, Coyle, and Lindsley (1976) demonstrated that monkeys with damage to the inferior pulvinar are severely impaired in making visual discriminations when the images are presented tachistoscopically. Such a deficit may be related to the loss of the attentional scanning system for viewing and scanning the peripheral visual fields. Although these data are consistent with an attentional function for the pulvinar, it is possible (even likely) that these effects are related to damage to the corticotectal pathway and are not pulvinar effects at all (Leiby, Bender, and Butter, 1982).

6. Drug injections. Electrophysiological studies in our laboratory have suggested that a subnucleus of the pulvinar may be involved in selective visual attention (Petersen and Robinson, 1983; Robinson and Petersen, 1984). In an attempt to support and clarify this function, we tried to modify the normal functioning of this area and thereby change an animal's attentional behavior. Monkeys were trained on a task similar to Posner's in which the reaction times to the appearance of validly or invalidly cued targets were measured. Monkeys show the costs and benefits from invalid and valid cues that are similar to human values. When drugs, which changed the levels of gamma-aminobutyric acid (GABA) (a known neurotransmitter in the pulvinar) (A. Hendrickson, personal communication), were injected into the pulvinar, there were clear changes in the animals' reaction times that we interpret to reflect changes in attentional behavior (Petersen, Morris, and Robinson, 1984; Robinson, Morris, and Petersen, 1984). Elevated levels of GABA made it difficult for the monkey to shift its attention to the contralateral visual field, while lowered GABA levels facilitated the attentional shift. These studies support our hypothesis that the pulvinar is involved in selective visual attention.

Audition

Although there has been an increase in work on the central processes of auditory attention, it is difficult to synthesize these data. There have been very few systematic studies of areas of the auditory pathways using several behavioral tasks, so that all of the relevant controls are conducted for each physiological effect.

1. Single cell studies. Studies from several different laboratories have attempted to determine if attention to an auditory event can change the response of single neurons at various levels of the auditory system. Hubel et al. (1959) were the first the describe "attention" cells in auditory cortex. They recorded from neurons in primary and secondary auditory cortex of awake cats and found that 10% of the cells did not respond to any auditory stimuli that they could present unless the cat looked at the source. They concluded that the animal's attention was a necessary factor for activating these neurons. While it is possible that it is attention that is affecting these cells, it is also possible that any of several other factors is causing the response. The response may be related to head and eye movements toward the sound source, to the act of fixating the object, a change in arousal induced by the auditory stimuli, or combined visual–auditory stimulation. Nonetheless, these studies were the first to propose that neurons in a primary sensory system may be under nonsensory influences.

Neurons in the auditory cortex of awake monkeys can be influenced by behavioral conditions (Miller et al., 1972). Cells recorded while an animal is using an auditory stimulus in a behavioral task are much more responsive to the stimulus than to the same stimulus presented when the same animal is not performing a task. Similar effects were also noted by Hocherman et al. (1976). Furthermore cells recorded from untrained monkeys appear to be even less responsive to a stimulus, and even when they respond to a stimulus, their discharge reliability is quite poor. These data suggest that behavioral state can influence neurons in auditory cortex.

Beaton and Miller (1975) trained monkeys to perform a reaction time task using an auditory stimulus as the cue for responding. The auditory cues were presented in two different situations, one where all auditory stimuli were reinforced if they were responded to and another condition where only one of the various frequencies was reinforced. The monkey's reaction times were slower and more variable when the auditory cues had to be discriminated than when all tones were appropriate for a response. Neurons were tested using their best response frequency as well as adjacent frequencies. A quarter of the neurons in auditory cortex were differentially affected by this task. The weakest response to an auditory stimulus was found in the condition where all tones could be used for the response. More vigorous activity was seen in cells when the monkey had to discriminate the tone, but paradoxically this facilitation was not present when the "selected" tone for the discrimination was at the neuron's best frequency. The task produced increased evoked activity for stimuli only when they were discriminated and unreinforced. Attention away from the best frequency of a neuron apparently facilitates the response of auditory cortical cells.

Ryan and Miller (1977) trained monkeys to perform a reaction time task that used an auditory stimulus as the cue for a motor response. While monkeys performed this task, cells were recorded in the inferior colliculus. They found that the majority of neurons in this subcortical structure were more responsive to a stimulus in the context of the behavioral task than when the same stimulus was presented at times when the monkey was not engaged in a task. The effects could be either an increase in the magnitude of the evoked discharge or a reduction in the amount of suppression caused by the stimulus. Another frequent effect was a lengthening of the latency of the stimulus-evoked response. Performance of the task did not have a consistent effect on the background activity of the population of cells. Off responses were frequently attenuated by the task. The work could be interpreted to show that these effects are related to the attention that the monkey must have directed toward the auditory stimuli. Unfortunately, other processes, such as arousal or the intention to move, could be active in this experiment, and it is not possible to make a clear functional interpretation.

Hocherman, Itzhaki, and Gilat (1981) trained monkeys to push one lever if a tone was presented and another lever if white noise was presented. After training, the sounds were preceded by valid or invalid cues. They found that the animals' reaction times were faster for validly cued trials than for invalidly cued trials. While monkeys performed this task recordings from neurons in the auditory cortex were made. It was found that about half of the cells tested responded differentially depending on the validity of the cue. The response changes could be either facilitations or suppressions of the baseline response. The focus of these studies was to determine whether such activity could be related to the neural mechanisms of prediction, but it is possible that the influences acting on these cells might be related to the processes of selective attention to frequency in the auditory system.

Benson and Hienz (1978) trained monkeys to respond to tone pulses presented to only one selected ear. The selected ear was indicated to the animal by illuminating the response button to be pushed. In this situation, it was assumed that the monkey attended to only one ear because its level of performance was extremely high. They found that for a small fraction of neurons recorded in primary and adjacent auditory cortex, there was a significant facilitation of the response of the cell to stimuli presented in the attended ear at the cell's optimal frequency. This facilitation was compared to the response of the same cell to the same stimulus in the same ear when the animal was not attending to that ear. These data suggest that selective auditory attention can modulate the activity of cells in primary auditory cortex. These experiments also showed that there were significant changes in the response of neurons for the identical stimuli when presented in the behavioral test situations and at other times outside of the test situation. These latter changes are of two types: a modulation of the magnitude of the discharge of the cell and a decrease in the latency of the response. These studies show that attention to the spatial location (ear) from which an auditory stimulus is presented produces a change in the response of auditory cortical cells.

Hocherman et al. (1976) trained rhesus monkeys to push a lever to one side or the other depending on (1) the side of an illuminated light or (2) the nature of an auditory stimulus. Trials were run in blocks where either the auditory stimuli or light stimuli were the relevant cue for indicating the direction of the lever movement. While the monkeys were performing this task, neurons were recorded from primary and adjacent auditory cortex. The neurons studied in these conditions were modulated by the behavioral conditions of the task. Some of the affected cells responded better to identical auditory stimuli when they were the cue for directing the monkey's behavior than when the relevancy cue was presented in the visual modality. The other affected cells had weaker responses when the auditory relevancy cues were in the auditory modality. Most of the change in the discharge pattern could be seen in the earliest components of the response. In those situations where both visual and auditory cues indicated the correct direction of lever push, the discharges were generally better than when the combined stimuli were in the opposite direction. These experiments indicate that attention to auditory stimuli can modulate responses of auditory cortical neurons and that these influences can be independent of the direction of the motor response that indicates the attentive use.

M. H. Goldstein, E. Baadia, D. A. Benson, and R. D. Hienz (personal communication) trained monkeys to respond to visual or auditory stimuli that were presented at several different locations around the animal. The response was either a "localizing" one in which the monkey contacted a switch at the locus of the stimulus, or a "detect" task where the animal contacted a central switch regardless of where the stimulus originated. They recorded from neurons around the arcuate sulcus in the frontal cortex. Some of the neurons recorded in this region responded weakly to passive auditory stimuli, also poorly to the same noise burst when it was merely detected, but gave much better responses when the same sound was the target for a localizing movement of the arm. For most of these cells there was no activity when the stimulus as a visual one. Other cells had clear localizing responses that were independent of the modality of the stimulus initiating the response. Such cells would discharge to the same arm movement whether it was made to an auditory target or a visual target. However, the same cell did not respond with the same movement if it was made as part of a "nonstimulus" response. In this task, the monkeys were trained to respond with movements in one direction even though the location of the sound source was varied. These experiments indicate that the activity of these cells in frontal cortex is related to a sensory localization process; it is neither sensory nor motor. As such, this could be a component of the auditory spatial attention system.

2. Cerebral blood flow. In the study by Roland (1982) described previously, he tested humans who are being presented with visual, auditory, and somatosensory stimuli simultaneously. When subjects in this experimental situation are required to discriminate only the auditory stimuli, there is increased blood flow in the auditory association cortex, the midtemporal cortex, Broca's area, and the frontal eye fields (Figure 7.2). Whenever attention is shifted from one modality to another, regardless of the modalities involved, there are also changes in the superior lateral part of the prefrontal cortex.

3. Evoked potential studies. Tests of auditory attention conducted by Hillyard et al. (1973) while recording evoked potentials from the surface of human scalps have demonstrated potentials that specifically reflect selective attention and not nonspecific preparatory behavior or reactive changes subsequent to attentional shifts. Subjects were presented with a rapid sequence of tone pips in both ears and required to discriminate a pattern presented in one ear only. In this situation, the subjects had to attend to only one ear and the potentials recorded from the subjects showed that the N_1 potential, which peaks at 80 to 110 msec, was selectively suppressed on the unattended side and selectively enhanced on the attended side. The authors emphasized the importance of the rapid stimulus presentation rate and the ease with which the relevant and irrelevant stimuli could be identified. Without these features, the clear potential differences would not be demonstrable. These data show that there are cerebral events that are related to attention and do not reflect sensory processing or motor response. Because the effects occur at such short latencies, Hillyard et al. propose that the attentional process is a tonic influence on the incoming auditory information. Because these studies deal with undetermined central effects, there is no reason why the attentional mechanisms cannot move very rapidly, and that this effect does not represent a rapid transient process. A late positive potential that occurs at 250 to 400 msec was selectively present only on the trials where the recognized tone was presented to the attended ear, suggesting that some cerebral events are specifically related to the recognition of attended stimuli. Although these experiments cannot localize the source of these potential changes, they do demonstrate that there are specific potentials measurable during attentional tasks.

Schwent, Hillyard, and Galambos (1976a) suggested that "information load" (i.e., the rate of information presented to the system) was important in being able to demonstrate changes in the N_1 potential. Subsequent studies (Schwent, Hillyard, and Galambos, 1976b) demonstrated the N_1 potential was only enhanced when stimuli were presented at rapid rates. This high rate was also associated with the most accurate target detectability scores. The authors suggest that a large information load makes it very difficult for the subject to process information in the unattended channel and must commit all his resources to the attended channel. Also such a high density of stimulation may make it easier to maintain attention to one channel. Alternatively, the shift of attention may be so fast that tasks such as this are required to control attention consistently enough to be measured.

Additional experiments (Schwent et al., 1976b) showed that the attentional enhancement of the N_1 potential was minimal for loud tones and increasingly greater for lower tones. Furthermore, tones that were obscured by masking noises were also associated with enhancement of N_1. These studies suggest that the brain produces greater attentional potentials in those situations where there is the greatest demand for a selection process to the sensory channel.

Hillyard and colleagues (Van Voorhis and Hillyard, 1977; Hansen and Hillyard, 1983; Hillyard et al., 1984) conducted experiments to determine if attention to spatial location in one modality would influence the sensory-evoked potentials generated by stimuli of other modalities at that location. Subjects were instructed to respond only to the appropriate modality at the attended location. When stimuli from the wrong modality were presented at the attended location, there was some modulation of the evoked potential, but not the full modulation seen when that was the modality of choice. The results are intermediate between the two hypotheses for these experiments.

Knight et al. (1981) attempted to determine the contribution of the frontal cortex to the selective potential changes recorded during an auditory attention task. Control subjects showed a change in the N potential when they attended to tones. Lesions of the frontal lobes reduced this attention-related potential and produced poorer behavioral performance. These potential reductions were the same whether they were recorded over intact or damaged cortex, suggesting that the site of recording was not the site of the neurophysiological effect. Lesions of the left frontal cortex tended to modify the attention-related potentials for either ear, whereas right lesions only affected attention potentials for the contralateral ear. Possibly the left frontal cortex is related to global attentional processes, whereas the right deals with more localized processes. These experiments show that the frontal cortex may mediate the attention-related potentials but suggest that these changes are probably not originating in the frontal cortex.

4. Animal evoked potential studies. The data just reviewed deal with the neural effects of attention directed toward auditory stimuli. It has been shown that attention to one modality reduces the effect of stimuli in other modalities (Hernández-Peón, Scherrer, and Jouvet, 1956; Oatman, 1971, 1976). Hernández-Peón et al. (1956) were the first to suggest the neural mechanisms of this process. They showed that slow wave potentials evoked in the dorsal cochlear nucleus to auditory stimuli were greatly attenuated when the animal was attending to stimuli in other modalities (visual, somatic, or olfactory). This study had the problem that there was little control of the auditory stimulation because the animal was free to change head position and thus the orientation of its ears. Oatman (1971, 1976) used auditory stimulation that remained constant with changes in the animal's head position and showed that the transmission of auditory information throughout the auditory system is attenuated. Cats were trained to make a visual discrimination in the presence of behaviorally irrelevant auditory stimuli. During the time period of the visual attentive behavior, potentials evoked in the round window, cochlear nucleus, and auditory cortex were attenuated compared to other conditions. This effect was not due to the action of the middle ear muscles and was most prominent for auditory stimuli of low intensities. It was proposed that this effect was mediated by the olivocochlear bundle. These data confirm the original hypothesis of Hernández-Peón et al. (1956) that there is a screening out of unattended sensory information at a very peripheral level of the sensory system.

5. Lesions – human and experimental. Damage to the parietal lobe, particularly the dorsal portions, has been shown to lead to visual and somatosensory neglect, but little study has been directed to audition. Denny-Brown, Meyer, and Horenstein (1952), followed by Bender and Diamond (1965; Diamond and Bender, 1965), described humans with unique auditory problems. These individuals had no auditory acuity deficits but could not deal with bilateral simultaneous stimuli. In such a situation, they would not respond to stimuli contralateral to the lesion even though they could respond to the same tone when presented alone. Even though they were capable of detecting such unilateral stimuli, they mislocalized them within their contralateral auditory space. These people had an auditory neglect. Heilman and Valenstein (1972a) have observed that the parietal lesions in humans that extend laterally to part of the temporal lobe produce a neglect of auditory stimuli originating from the contralateral side. The damaged cortex appeared to be the caudal portion of the dorsolateral surface of the temporal lobe. To date this has been observed only with unilateral lesions on the nondominant (right) hemisphere. Heilman et al. (1971) have also demonstrated that monkeys with damage to the posterior parietal and preoccipitotemporal cortex have auditory extinction. When presented with stimuli simultaneously on the ipsilateral and contralateral sides, they respond only to the ipsilateral ones. This is in spite of the fact that they can respond to the identical stimulus in the ipsilateral field when it is presented by itself.

Other studies (Welch and Stuteville, 1958; Heilman and Valenstein, 1972a,b) have shown that damage to the frontal lobes yields a type of auditory neglect in man and monkey.

Somesthesia

1. Single-unit studies. In the somatosensory system, as in the others, a common theme in the neurobiology of attention is in the enhancement or modulation of a sensory response in different behavioral conditions. In studies on somatosensory cortex, both Hyvärinen, Poranen, and Jokinen (1980) and Nelson (1984) found that the responses of neurons in primary somatosensory cortex to tactile stimulation could be modulated by behavioral set. The modulated responses were common in the posterior part of primary cortex (area 1) in both studies and were absent, for the most part, in the anterior portion (area 3b).

In Nelson's study, the animal was trained to move his hand in response to a vibrotactile stimulus on a block of trials. The stimulus was presented to the responding hand. In subsequent or preceding blocks of trials, the animal was presented with the same vibrotactile stimulus but was unrewarded for the movement. The responses of neurons during trials with movement in the rewarded condition were compared with responses without the movement in the no task condition, and differences were noted. There is an enhancement of the evoked activity in the active condition, but the basis of the enhancement (simple arousal, intention to move, or selective attention) was not addressed.

Evidence that part of the enhancement is not solely dependent on the intention to move the stimulated hand comes from the work of Hyvärinen et al. (1980). In these experiments, an animal was trained to respond with the unstimulated hand in some trials and no movement in others. Some enhancement was found, indicating that the facilitating process was not dependent on a particular movement, but something common to both movements, possibly arousal or selective attention. Experiments by Rizzolatti et al. (1981) have shown that for neurons responsive to tactile stimulation of the mouth, enhanced responses can be seen when the tactile stimulation is followed by a related movement. Further experiments to determine more precisely the conditions necessary for this enhancement (e.g., attention alone) were not conducted. In the second somatic sensory area, Poranen and Hyvärinen (1982) found that multi-unit responses could be modulated by changing the behavioral state of the animal, but the possibility that a nonspecific arousal could account for the enhancement was not ruled out. There is no real evidence of selectivity in these experiments, so the modulations could be due to a non-specific arousal or to attentional factors.

There is, however, evidence for attentional enhancement at the level of the medullary dorsal horn (Bushnell et al., 1984), a much earlier processing station in the somatosensory system. Neurons in this structure, when presented with nociceptive stimulation in several different behavioral conditions exhibited response enhancement. The enhancement took place only when the animal was actively discriminating the somatosensory stimulus, not when the same stimulus was presented in a no response condition and not when the animal made the same motor response as in the discrimination task. If the motor response was made to a visual stimulus, and not the simultaneously present somatosensory stimulus, there was no enhancement. Some of the attentionally enhanced neurons were shown by antidromic stimulation techniques to project to the ventrobasal thalamic nuclei that in turn project to the cortical regions of primary somatosensory cortex. These studies represent a clear demonstration of attentional processing in the somatosensory brain.

2. Regional blood flow. The work of Roland (1982) on selective regional cerebral blood flow failed to implicate any region in an attention task when the person was to pay attention to the somatosensory modality. The region defined by Roland as somatosensory association cortex showed no change from control during the performance of the task. However, the technique is unable to measure blood flow in the secondary somatosensory area, so attentional changes here could have gone undetected.

3. Evoked potential studies. Somatosensory evoked potentials exhibit the same modulation that are seen in the other modalities. Desmedt and Robertson (1977) found an early negative enhancement of a somatosensory-evoked potential when the person was attending to an attribute of the stimulus. This early enhancement was most obvious when the task was difficult in that there were weak stimuli and rapid rates of presentation; in other words, where the information rate was high and the stimuli difficult to discern.

4. Lesions. In humans, lesions of the parietal lobe have produced defects in attention in the somatosensory realm as well as those described for other modalities. These defects are again subsumed under the umbrella of neglect and comprise a large number of related clinical observations. Some of the most dramatic examples of neglect are in the somatic realm (see Friedland and Weinstein, 1977; Lynch, 1980; Hyvärinen, 1982). People with parietal lesions often fail to acknowledge all or part of the contralateral side of their bodies. They will fail to dress half of themselves, will leave their contralateral foot out of a car, will often claim that the affected limb is someone else's. The attentional nature of the effect is shown in somatosensory extinction, similar to visual extinction. The person is often capable of responding to pressure or touch on an affected limb; when the unaffected limb is stimulated at the same time, there is no report of the stimulation on the affected side.

Monkeys also exhibit somatosensory extinction after restricted lesions of parietal cortex (Eidelberg and Schwartz, 1971). Reeves and Hagamen (1971) reported extinction following lesions to the midbrain reticular formation in cats and to the cingulate gyrus in monkeys. These are rather compelling examples of an attentional deficit. Rizzolatti, Matelli, and Pavesi (1983) showed that damage to the periarcuate cortex of the frontal lobe produces a contralateral somatosensory neglect. Cells in this area have receptive fields on the mouth, and animals with damage here fail to use the mouth to grasp food. Although this deficit is termed neglect, the problem is most likely related to the sensory initiation of movement rather than being related to attention. Neglect was reported for lesions to several areas, including hypothalamus (Marshall, Turner, and Teitelbaum, 1971), mesencephalic reticular formation (Watson, Miller, and Heilman, 1978), and frontal cortex (Watson, Miller, and Heilman, 1978), but these reports often do not distinguish between arousal, detection, and attention.

Watson et al. (1973) have found unilateral neglect from damage to the cingulate cortex of the monkey. Sensation was intact in all three modalities, but the animals demonstrated extinction to somatosensory stimuli. When both legs were stimulated simultaneously, the animals responded as if the ipsilateral leg alone was touched. This deficit was interpreted as a problem in intention to move since the monkeys were capable of responding to stimuli presented on the ipsilateral and contralateral legs; they had difficulty moving the leg contralateral to the lesion when it was the response indicator in this task.

Conclusions and future work

One of the primary efforts of neurobiological research in the area of attention has been to localize areas of the brain that are sites for attentional activity. Once found, it is then appropriate to determine the mechanisms that the structure employs in aspects of attention. Neurophysiological work on awake, trained animals, regional blood flow studies in man, and tests of brain-damaged humans and animals have been productive in identifying new sites for such activity. By focusing on the visual system and limiting ourselves to spatial attention, we have been able to identify two attentional loci: posterior parietal cortex and a region of the pulvinar. These areas have visual properties, a selection mechanism, and an independence of the attentional process from any form of motor response. Subsequent drug-infusion studies have supported the attentional function of the pulvinar by showing that changes in transmitter levels in this area can change an animal's attentional behavior. Damage or alteration of these areas leads to neglect and other attentional disorders. In each of these areas, the mechanism of action of the brain is to intensify the neuronal discharge to the attended stimulus. Although there appear to be no published single-unit observations of reduced responses when attention is diverted, we have seen examples in the pulvinar. A shift of attention from a visual stimulus sometimes produces a weaker response than when a shift does not take place. New studies demonstrate that attention to stimulus attributes other than spatial location produces modulations of sensory responses, and this opens other parts of the sensory pathways to testing for this type of effect and a search for its neural mechanism.

Earlier in this chapter, we listed four steps in the shift of spatial attention: selection of a new target, disengagement from the present target, shift of attention, and engagement with the new image. It is entirely possible that future work in cognitive psychology will provide additional steps in the attentional shift that will lead to new neurobiological approaches. Posner's studies of humans with parietal lesions suggest that their deficit is in disengagement from the present image of interest. Future work will be required to determine the neural mechanism of this process, if parietal cortex functions in other stages of attention and if other identified attention areas operate in this and other stages of attention.

Once an attentional site has been clearly identified and its neural mechanism studied by normal physiological techniques, it is appropriate to evaluate some of the neuropharmacological processes involved. Such studies would be able to identify the chemicals used in one brain area and also identify some of the functional subsets of the mechanism. For example, the sensory data sent to an attentional area may be mediated by one transmitter while the attentional input (or inputs) may utilize other transmitters. By using selective transmitter blocking agents during the course of recording from neurons in an attentional task, it may be possible to dissect the functional mechanisms that combine to produce the attentional activity of cells.

The past decades have brought new approaches for studying the human and animal brain. These developments will allow for more direct study of the neural bases of higher cognitive functions. As a result, many of the hypotheses proposed by cognitive psychology should soon be directly testable.

8   A neurobiological view of the psychology of attention

David Lee Robinson and Steven E. Petersen

The areas of overlap between neuroscience and cognitive psychology can be seen in the similarity of two major questions of these disciplines: For integrative neuroscience, how can we understand the organization of behavior from the study of neural processes? In cognitive psychology, how can we understand the organization of behavior from the study of behavior? Because the levels of observation are quite different, and because the approaches utilize different types of analysis, some common ground must be established. To relate these two disciplines and to explicate the differences and similarities of the two approaches, we will examine a simplified model (Figure 8.1).

The model consists of three basic components with information flowing from top to bottom. The first steps begin with the input box of sensory stimuli, instructions, and training. These are next processed in the large black box of the brain and its associated peripheral sense organs. Finally there is the output in terms of measurable behavior. Inside the processing box, we have represented in a very simplified fashion several stages. It is known that sensory modalities are carried over separate pathways for several synapses into the central nervous system, so we have shown them to be separate in the diagram. Within modalities, there are complicated sets of structures that are interconnected in various ways. In this diagram, we have shown areas connected in both parallel and serial paths. As we progress further into the brain, the processing connects various modalities and submodalities, and we have subsumed this higher-order processing under a box labeled sensorimotor integration. Finally, the integrated bits of information are used in the organization of behavioral outputs, the generation of which is performed by various motor plants, such as skeletal, ocular, and verbal behavior. These outputs are rendered as behavior, and aspects of behavior are measured by practitioners of both subfields. Using this model, we can examine some of the ways that cognitive psychology and neuroscience are different, and areas in which they are similar and could interact on the subject of attention.

We would like to thank J. David Morris for his contributions to the simplified model presented in Figure 8.1 and for discussions of multiple attentional processes.

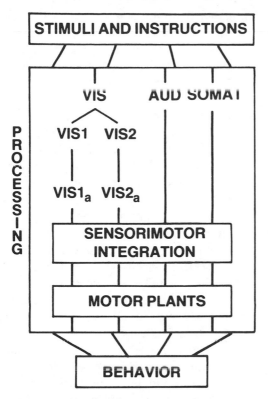

Figure 8.1. Outline drawing of some of the stages between stimuli and behavior. External events that influence and direct behavior are represented at the top as stimuli and instructions. The multitudes of processing that eventually lead to behavior are included in the central box. Within this section, the three major modalities are initially separated, and these individual modalities are even further divided. Whatever the within and without modality processing, there is eventually some conversion of these data toward responding, and this is referred to as sensorimotor integration. This integration leads to the motor plants for eye, limb, and speech movements. Finally, these motor plants lead to observable behaviors that the scientists measure.

Basically, the cognitive psychologist employs an outside–in approach, treating all of processing as a black box and trying to infer its workings without penetrating it. The neurobiologist employs a building block approach by formulating wiring diagrams within the black box using neuroanatomical techniques, studying subunits using neurophysiological techniques, and attempting to ascribe functions to subunits by assessing behavioral deficits following damage to parts of the brain. These basic differences in approach lead to many specific differences in what the respective subdisciplines can study and what they can tell us.

Cognitive psychologists have control of the stimuli presented to the subject and can measure the concomitant behavior. Through manipulations of stimuli presented to animals and humans, the approach attempts to measure and limit behavior so that it reflects the underlying organization of the brain. Since the psychologist is dealing with the entire set of processes, the stimuli he can present can be complex and can cross sensory modalities or submodalities. For the psychologist, the stimulus and instructions are the measurable inputs. On the output side, the psychologist has access only to the behavior of the organism. Because the psychologist has access to the entire organism, often with humans, the behavior measured can be complex. As a result of all of the above attributes, the psychologist can study higher-order processes and uses the behavior to infer the organization of the underlying neural substrate.

The neurobiologist has similar control of the stimuli and instruction set and can make observations on the behavior. He also has access to parts of the black box itself. Because the neurobiologist generally studies animals, he has less control over instruction, can present similar stimuli, and is limited in the forms of cognitive behavior that can be measured. For the neurophysiologist, since he is studying a small subset of the black box, he must use stimuli that are appropriate to the function of the area under investigation. If he is studying a visual structure of the brain, the most likely stimuli to use would be visual. The neurobiologist has a measure of neural processing that the psychologist can only infer. The neurobiologist has access to the behavior and to the physiological monitor that he is employing. The neurobiologist can infer some nonsensory inputs to the brain area recorded by monitoring the behavior of the animal. For instance, in the case of parietal cortex, if the animal is performing a simple fixation task, the response of a cell is less than if the animal responds to a peripheral target. Although the stimuli are the same, the response of the cell is different because of the two different behavioral conditions. Finally, the neurobiologist must have simplified responses to assess; the use of animals precludes verbal or conceptual responses. The neurobiologist studies finer details and simpler behavior, and can infer functional segregation to a particular area of the brain. These and other differences between these approaches are listed in Table 8.1.

An outcome of the neurophysiological approach to attention is that activity related to attention can be seen to be a rapid, transient process. The neural processes that we study and those measured by evoked potential techniques last only fractions of seconds. The experimental psychologist generally assumes that attention is a process that focuses on an area and can be made to stay there. It would appear to be worthwhile to envision attention as a continually moving process (like the direction of gaze), and one that is most naturally studied in a dynamic condition rather than being forced to focus in one and only one locus.

There are several ways in which the disciplines can and have interacted to good use. Cognitive psychology has defined attentional behavior using par-

Table 8.1. *A comparison of the neuroscience and cognitive psychology approaches in studying attention*

	Neuroscience	Cognitive psychology
Stimuli and instructions	Less complex, must be tailored to system being studied	More complex, tailored to experimental question
	Most often single modality	Often crosses modalities
	One of several potential inputs	Only input
Processing	Measurable components of processing	Must infer processing by measuring behavior
Behavior	Less complex behavior is measured (e.g., no semantic behavior)	Access to complex behaviors
	One of several outputs	Only measurable output
General	More often studies in animals than in humans	More often studies in humans than animals
	Access to neural events in 10 msec resolution	Less access to fine-grain resolution of time course
	Access to finer details of processing segregation	Includes higher-order processes

adigms that can be exploited by neuroscientists to investigate neural functioning. Cognitive studies have provided results indicating that attention uses separate behavioral channels and then gives clues to the way that function is segregated in the brain. Neurobiologists by collecting and organizing data on structures in the brain can provide clues as to what behaviors might prove fruitful in an analysis of the brain. Hirst, in Chapter 6, talks of performance operating characteristics, alterants, resource pools, and other concepts that in the realm of psychology are theoretical constructs to organize what we know about behavior. Perhaps by utilizing the data from neurobiology, we could find correlates between the constructs and neurophysiological events or neuroanatomical structures.

Questions concerning attentional bottlenecks and their central or peripheral placement are similar to questions of behavioral modulation at more or less central locations in the brain. The disciplines come to similar answers: the notion of a bottleneck and its placement is too shallow a concept. The bottleneck seems in some cases to be central, in some cases peripheral; in other words, the bottleneck is in some way task specific, and therefore not really a bottleneck at all. Likewise, behavioral modulation can be seen in the cells of the superficial layers of the superior colliculus, a region that gets a direct input from the retina, and thus could be considered to be more peripheral than cortical area 7. Area 7 exhibits behavioral modulation under different

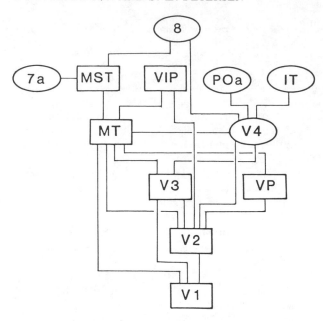

Figure 8.2. Hierarchical organization of visual areas in the macaque. Each block represents a functional area within the cerebral cortex and lines connecting them represent pathways. All visual information for this plan begins in V1; from here it goes to V2, V3 or MT. Subsequent processing of information takes place by various interconnections among these areas. (Reproduced with permission from Van Essen and Maunsell, 1983.)

conditions than that in the superior colliculus, and is several synapses further from the sensory periphery.

An issue raised by Hirst that neuroscience can address is the problem of single versus multiple attentional processes. Cognitive psychology has shown that tasks performed simultaneously in two different modalities interfere with each other less than two tasks performed in the same sensory modality. This is not surprising since anatomically and physiologically the sensory modalities are kept relatively separate from one another through several synapses. It is also interesting that some same-modality tasks are easier to perform simultaneously than others. By employing physiological recording and anatomical tracing techniques, the neurobiologist can begin to put together a wiring diagram of the functional areas of the brain. By studying the properties of single cells in these areas, one can begin to assess the functional specializations of particular areas. Within the visual system, there is a hierarchy of visual areas such that certain pathways and sets of areas process certain aspects of vision (Figure 8.2). The pathway from V1 to MT appears, from several lines of evidence, to process visual motion information, while ignoring the color or form of the stimulus. Another pathway from V1 to V2 to V4 to IT appears to process form information and possibly color information, and these areas are less interested in the direction of motion of stimuli. It would be interesting to know if a set of simultaneous tasks that utilized the separate pathways, such as a motion versus form test, would exhibit less interference than if the tasks utilized the same pathway, such as a form versus color test. As the physiological properties of all the levels and pathways shown in Figure 8.2 are more fully understood, then the flow of useful information from neurobiology to cognitive neuroscience can be expected to increase.

Another line of neurological study suggests that there are several areas contributing to separate attentional processes. The clinical neurological literature has shown that the inferior parietal lobule is in some way related to attention. Humans with parietal damage frequently neglect their contralateral visual and/or somatosensory fields; only with different lesions are problems in the realm of audition present. Even though such lesions are large and crude, they clearly suggest that the neural substrates for visual attention are not those acting in audition. Furthermore, the intensity of the visual and somatosensory deficits are seldom the same, suggesting that different regions of the brain have different contributions to visual and somatosensory attention.

A topic to which neuroscience can contribute only sketchily is in the area of skill and learning. Although experimentation to determine the neural sites and mechanisms of learning is making advances, neurobiological examples of plasticity in complex behavior are not numerous. However, it is a very frequent observation that damage to the brain leads to clear changes in behavior, but it is rare that such changes are permanent. Almost always there is some recovery of function that suggests that another part of the brain compensates in order to mediate the changed behavior. Although the brain does not have unlimited capacity to accommodate to damage, it is clear that it is not a static structure that is etched in stone during development and incapable of modification. It is unlikely that the parts of the brain that are easily compensated for are close to the peripheral sensory or motor systems. Generally, damage at these levels produces devastating and permanent deficits.

It appears to us that one of the major focuses of visual neuroscience is in determining the ways in which visual images are broken down in the brain, and the specific types of behaviors (eye movements, attention, arousal, etc.) that influence visual responses. Although not specifically directed toward the questions of attention, they do provide data on how vision is fractionated and synthesized, and these concepts should be important in determining the different aspects of visual attention, etc. The psychologist sometimes addresses the basic elements of a sensory system. Thus, when attempts are made to determine the constraints of an attentional system, some of these may be related to the organization of the sensory system and not neccessarily attention. This dilemma is most likely when complex sensory stimuli are used in testing situations, especially when such stimuli may have learned significance (words, letters) or affective value. Here the analysis is complicated by issues that are beyond the sensory and attentional questions of the experiment.

In summary, there are issues where cognitive psychology has contributed to neuroscience and vice versa. As developments in each discipline are brought to the attention of the other, this cross-fertilization will continue and a true cognitive neuroscience of attention could emerge.

A psychologist's reply

William Hirst

I am pleased to see that many of correspondences that I noted in my discussion were also noted by Robinson and Petersen in Chapter 7. We both observed that cognitive psychologists use much more complex stimuli than do neuroscientists. We both wondered whether a correspondence could be found between the multiple resources posited by some cognitive scientists and various brain structures, and we both agreed that the bottleneck metaphor has only limited utility.

Nevertheless, in searching for similarities and differences, I think that one must be cautious that one is not comparing apples and oranges. In Table 8.1, Robinson and Petersen compare the study of processing in the two fields and note that neuroscience studies measurable components of processing whereas cognitive psychologists must infer processing by measuring behavior. As Table 8.1 presents it (and as the text itself confirms), Robinson and Petersen fail to acknowledge that the processing studied by neuroscientists and cognitive psychologists are of different kinds. Neuroscientists study brain functioning. When they refer to "process," they are discussing the chain of excitation and inhibition in neural systems. Cognitive psychologists, on the other hand, study mental functioning. When they refer to "process," they are discussing computation at a mental level somewhere between behavior and physiology.

Although the computer metaphor can be overused, it nicely illustrates the differences between the two uses of process. When neuroscientists investigate process, it is like computer scientists exploring the functioning of the hardware of a computer. To assess the hardware of the computer, computer scientists may use electrical probes or various "direct measurements" of the electrical flow through the computer just as neuroscientists use both electrical and chemical probes to study neural action. When cognitive psychologists discuss process, it is like computer scientists talking about the functioning of the software of the computer. Only indirect measurements will reveal the nature of the software from the study of behavior.

Thus, it is misleading to write, for instance, that neuroscientists directly measure processing, whereas cognitive scientists only indirectly measure it. It should be clear that neuroscientists are not directly measuring mental processes, nor are cognitive psychologists indirectly measuring brain processes.

The lack of correspondence between the vocabulary of the two fields does not mean that conversation between their respective researchers is doomed to failure (although it does suggest that Table 8.1 should be more carefully worded). After all, the editors assembled this book because they thought that there was a basis for discourse. Indeed, the research suggests that measurements of neural processing may be used to infer mental processing, and vice versa. One must still determine how and under what circumstances one can make the necessary inferences. However, books such as the present one may help set up the ground rules.

9 A psychological view of the neurobiology of attention

William Hirst

After reading Robinson and Petersen's chapter, I take heart that there is genuine ground for discourse between neuroscientists and cognitive psychologists. What follows is speculative, and the careful reader may find it short on details, indeed, often neglectful of details. However, if a discourse between neuroscience and cognitive psychology is to begin, a little speculation, vagueness, and generalization may be in order.

On definitions

Let me begin by observing that practitioners in the two fields appear to be studying the same thing. Whenever one deals with mental phenomena, it is always possible that shared terms refer to quite different things. Yet there appears to be much agreement about what should be studied, both in terms of definitions and experimental realization of these definitions. Robinson and Petersen began their chapter with a good description of the phenomenon of attention. For them, attention is "generally agreed to be a selection process; that process by which an organism chooses to deal more effectively with one type of sensory stimuli at the expense of other stimuli." Whereas this definition does not have the literary flare of James, and it certainly is not burdened with the phenomenological concepts with which the definition of James is, it captures James's essence and would satisfy any cognitive psychologist.

Moreover, in expanding on this definition, Robinson and Petersen are careful to distinguish attention from arousal and from an "intention to move." If they did not, it would be hard to relate the research that they reviewed with the interests of cognitive psychology. Cognitive psychologists have clearly established that arousal and "intention to move" are not the same as attention. As I noted in Chapter 6, there is ample evidence that people can attend to, for example, a spatial area, without moving their eyes to that area or even intending to move their eyes to the area (Posner, 1980). Furthermore, whereas an increase in arousal often accompanies more concentrated attention, this is not always the case. At certain levels, an increase in arousal may actually

180

make it harder for a person to concentrate his attention (see Kahneman, 1973, for a discussion of this point).

Thus, both neuroscientists and cognitive psychologists treat attention as a process of selection, a process distinct from arousal or foveation. Moreover, when it comes to translating this shared understanding into experimental paradigms, both of them appear to adopt the same strategy. In both fields, the subject is cued to attend to a spatial location or to a particular message and is then asked to respond to a target or message in a variety of different ways.

These similarities in experimental methodology should not mask substantive differences, however. First, researchers in the two fields are interested in different dependent measures. The cognitive psychologist confines his interest to the subject's performance, that is, how well the subject responds to the target. In a dichotic listening experiment, the cognitive psychologist may monitor shadowing rate or target detection. Neuroscientists, on the other hand, are chiefly interested in changes in brain activity. They are looking for a change in brain activity that corresponds to a change in performance in the attention task.

In addition, the experiments in the two fields differ in the complexity of the stimuli that they present to the subject. Nothing inherent in either discipline limits or furthers stimulus complexity; the differences in experimental methodology merely reflect the present state of the two fields. Cognitive psychologists have generally been interested in determining what characteristics of competing messages can serve to guide selection. Consequently, they have varied the stimuli, examining competing messages that differ in modality, timbre of voice, spatial location, phonological character, or meaning. Neuroscientists appear to have contented themselves, at least initially, with a more modest goal. They merely want to find some area of the brain whose activity appears to change with a shift in attention. Thus, in their review of selective attention to visual material, Robinson and Petersen rarely detail the nature of the stimuli, and when they do, the stimulus often turns out to be a simple dot of light. When the stimuli are treated more seriously, the results are interesting. Brain structures that show increased responsiveness when attention is directed to one class of stimuli show no change in responsiveness when attention is directed to another class of stimuli. A careful exploration of this phenomenon and its implications does not seem to have been worked out in detail, however.

Neuroscientists' unsophisticated treatment of the stimuli is important to cognitive psychologists, it seems to me, because it limits the range of questions that neuroscientists can ask. Consider, for instance, questions about the basis of selective attention. Early work in cognitive psychology suggested that selection is based on the stimulus characteristics of the message; later work suggested that, in some instances, selection can be based on the meaning of the message. Can neuroscience provide data that might help differentiate between these two positions? It is doubtful unless both the experimental material and subjects' responses are complicated. Studies of Hillyard, for instance, found an evoked potential N_1 quite early in the chain of processing (see Chapter 7). Although this result would appear to support the claim that selection is based on the "unprocessed" stimulus characteristics of the message, its generality is questionable. Is there an N_1, or some other evoked potential, when selection is based on the meaning of the message, for instance? Robinson and Petersen do not review any work addressing this question. Moreover, based on the chapter by Robinson and Petersen, it would appear that the vast body of work using single-unit recording, cerebral blood flow, and lesioning does not address this issue either.

Attention and detection

Despite these limitations, when neuroscientists do narrow in on an attentional phenomenon, they raise substantive cognitive questions. Consider the work on the effect of attention on detection and concentrate on the single unit recording technique as paradigmatic. Two salient facts emerge:

1. The responsiveness of certain cells to the presence of a visual stimulus will change as a function of an attentional cue.
2. Statement (1) is true of some cells and not other perceptually relevant cells.

These facts complement the work done by Posner and his colleagues (see my discussion in Chapter 6), and they do so in interesting ways. At one level, they confirm Posner's finding that there can be attention to a spatial area without fixation or even the intention to fixate. More importantly, they provide a nice means of mapping the cognitive theorizing of Posner and his colleagues into theorizing at the level of brain function. In particular, they suggest that when people shift their attention to a spatial area, they increase the responsiveness of certain cells to visual stimuli.

This explanation does not, however, account for the analog sweep of attention that Posner finds. One might conjecture that the "analog sweep" arises because those cells that increase in responsiveness seem to be confined to those cells that will play a role in the detection and perception, but the line of reasoning here is tentative and entails a great deal of hand waving.

Nor do the explanations or experimental data at present provide a clear picture of what causes an increase in responsiveness, as far as I can tell. The change in cellular responsiveness could be viewed as a consequence of the mechanism of attention rather than the mechanism itself. A single brain structure may be responsible for increasing or decreasing the responsiveness of various brain cells. Alternatively, the mechanism for changing the responsiveness of a cell may be tied specifically to the functioning of that cell or cellular system. In this case, the mechanism for detection could not be easily separated from the mechanism for attention. There would be no single brain structure corresponding to "attention," but a host of structures corresponding to different kinds of attention. For instance, there may be a different structure for attending to visual stimuli and a different structure for attending to auditory stimuli, a situation suggested by the work on cerebral blood flow.

Thus, a clear and more complete articulation of the physiological mechanism of attention is of particular interest to cognitive psychologists. Cognitive theorists have vacillated when faced with the issue of determining whether "attention" is an actual mechanism. Broadbent's filter model and Treisman's attenuation model treated attention as a mechanism, for instance. Resource theories, however, have no particular mechanism of attention – there is only the allocation of resources. Single resource theories would predict that resources come from a single specifiable source. Multiple resources theorists, on the other hand, would expect a multitude of separate sources of resources.

Although it is by no means necessary, it might be expected that cognitive mechanisms of attention should be reflected in some way in brain structures. If neuroscientists did find one or even a few structures responsible for the increased responsiveness they have observed, then they would be supporting cognitive theories of attention that posit filters, attenuators, or some other mechanism. On the other hand, if they found that the brain mechanism for the observed increase in responsiveness could not be separated in any reasonable way from, let us say, the brain mechanisms responsible for detection or perception, they would be supporting those cognitive models that failed to treat attention as an actual, specifiable mechanism.

The important point here is not that neuroscience has failed to provide an answer, but that it has the potential to constrain cognitive theorizing. We shall see some of this potential actualized in the next section.

Neglect

In Chapter 6, I raised questions about the extent to which information is processed without rising to consciousness: Is unattended information processed at a semantic level, or is processing confined to stimulus attributes? The evidence from the cognitive psychology laboratories seemed confusing, at best, because of replication problems. The study of patients with neglect may be relevant because these patients are not "aware" of objects that are clearly within their visual field. Are these patients processing these objects at all? And if so, to what extent?

As Robinson and Petersen noted, the work to date suggests that patients with neglect process incoming information of which they are unaware. Volpe, LeDoux, and Gazzaniga (1979) dramatically illustrated this point in a study on patients with lesions of the parieto-occipital cortex who exhibited a related phenomenon known as extinction. Here, the presentation of a single stimulus in any area of the visual field results in its accurate description, but lateralized simultaneous presentation of two stimuli, one in each field, results in the verbal description of only the stimulus in the right visual field. Volpe et al. asked their patients to say whether the two objects presented simultaneously were the same or not. Although the question did not make sense to the subjects, inasmuch as they only "saw" one object, subjects could respond quite accurately. The basis of this response is not clear from Volpe et al. Their subjects could have been responding on the basis of the stimuli's physical features or on the basis of the stimuli's meaning. Nevertheless, the results fit nicely into the literature that I reviewed that argues for rather deep processing without awareness.

Resources

Whereas I was pleasantly surprised at the number of points of common interest, I was disappointed that little work was done on divided attention. In particular, I would have liked to see what a neuroscientist makes of the idea of resources. Clearly, cognitive psychology is confused at present over the ontological status of resources. If neuroscientists could make sense of it in physiological or neurological terms, it would clearly cast some light in the present discussion among cognitive psychologists.

Several researchers have recognized the importance of developing a neuroscience of resource and have suggested that "resources" may be equated with metabolism level or electrical activity. It is tempting, for instance, to think of an increase of an evoked potential as an allocation of more resources (Wickens et al., 1983), but proponents of this position have failed to take into account many of the arguments against resources that I outlined in Chapter 6.

Holtzman and Gazzaniga (1982) have provided the best neurological data in support of resource theory, in my mind. They examined a split brain patient, V.P., whose commisurrotomy prevents her from transferring information from one hemisphere to another. Holtzman and Gazzaniga speculated that interference between two concurrent tasks can arise because of perceptual interference or because of competition for shared resources. If the source of task interference is solely perceptual, then a split brain patient should find it easy to do two things at once when each task is presented to one and only one hemisphere. Moreover, a change in the demands of one task should have little effect on the performance of the other. On the other hand, if both tasks call upon a common resource pool, then a split brain patient should find it difficult to do the two tasks simultaneously, even if they are segregated across hemispheres. More decisively, the performance of one task should be sensitive to any changes in the demands presented by the other. Holtzman and Gazzaniga verified this latter prediction and, consequently, argued that resources must either reside in subcortical structures or be transferred between the hemispheres via subcortical pathways.

This conclusion raises many questions, especially as one attempts to specify more concretely where the "resources" reside subcortically. Moreover, there are alternative explanations for the finding. For example, it is well known that anxiety can interfere with the performance of a task. The increase in the demands in the primary task in Holtzman and Gazzaniga could have increased the anxiety of V.P. If so, then V.P. may not have been transferring resources subcortically, only her anxiety.

Whatever the eventual disposition of Holtzman and Gazzaniga, the study clearly demonstrates how neuroscience and cognitive psychology can and do interact. With more discussion, and a more concerted attempt of both fields, many of the issues and problems confronted in Chapters 6, 7, and the present discussion may find a happy resolution.

A reply from neurobiologists

David Lee Robinson and Steven E. Petersen

It was heartening when we read Hirst's chapter on attention to realize that there are issues and experiments that are shared between neurobiology and cognitive psychology. It is even more encouraging that the discussions of both chapters tend to focus on the same issues, for example, stimulus complexity, resource allocation, and the different types of attention. In this section, we will clarify some of the issues raised by Hirst and suggest other approaches to addressing important questions about attention.

One of the criticisms raised by the Hirst discussion is the limited range of stimuli used in the neurophysiological studies. The majority of stimuli used in the attentional studies we described are rather simple, while there are numerous other stimuli available. The reason for using such stimuli in experiments to date has been that these are stimuli that reliably excite the cells in the area being studied. We feel that it is important to use this type of stimulus because an enhanced response evoked by a less optimal stimulus could be ambiguous. The nervous system would have to determine if a strong response is evoked by an optimal stimulus or is an enhanced suboptimal stimulus. More recently, there are many new areas of the visual brain that have been discovered, and these frequently are organized to respond to more complex stimulus configurations. Hence, it is now possible to study cells in these areas for the effect of attention on sensory responses to more complex stimulus parameters.

Hirst asks whether neuroscience can address the question of the basis of the selection for attention: Is it determined on stimulus characteristics or stimulus meaning? Although there are few data that directly address this question, some physiological studies provide suggestive cues. As was mentioned in our discussion of Hirst's chapter (see Chapter 8), neurons at several levels of physiological processing show behavioral modulation of their responsiveness. The basis for selective attention could be either a stimulus characteristic or stimulus meaning at different levels of neural processing.

An approach to attention that has not been used in single-unit physiology is that of divided attention. Frequently in cognitive studies, divided attention has been tested with stimuli from different modalities. This would pose serious problems for the neurophysiologist because it would require determining the sensory responses of each cell to stimuli of different modalities and then determining the effects of divided attention on these responses. The technical demands of this would be quite high. In addition, it would require studying cells in an area that responds to stimuli of different modalities and training an animal to respond to variation in these different stimuli. To date there are few areas of the brain that have been carefully studied for their multimodal properties. A more feasible approach to the neurobiology of divided attention might be to test the effects of transmitter-related drugs on performance of such tasks, or evaluating the effects of local lesions on such behavior. These techniques would eliminate the need for detailed and complicated electrophysiological testing.

A final issue raised by Hirst is whether attention is an actual mechanism or simply the reflection of the limits of the brain's resources. If it is a mechanism then he proposes that it should be possible to locate a limited number of brain areas responsible for attentional effects. If attention is the reflection of brain organization and capacity, then attentional processes should not be dissociable from those responsible for "detection or perception." Although it is straightforward to test many of the visual characteristics of cells, it would be premature to ascribe a specific cognitive function to such activity at this point in time. Because we cannot state the sites required for a single sensory perception, we cannot yet know if attentional effects can be dissociated from them.

Part III

Memory

10   The psychology of memory

Daniel L. Schacter

Experimental psychologists have been concerned with the analysis of memory ever since the publication of Ebbinghaus' (1885) groundbreaking monograph 100 years ago. The numerous experimental studies published since that time have examined a multitude of questions concerning memory, but it is possible to discern two relatively distinct phases in the development of the field. The first phase was initiated by the publication of Ebbinghaus' experiments, and continued until the early 1950s. The dominant characteristic of this first phase was the strong influence of the then prevalent doctrine of behaviorism, which attempted to account for phenomena of memory solely in terms of stimulus–response contingencies, without regard to intervening processes. The behaviorist approach to memory was outlined and endorsed by McGeoch and Irion (1952), whose monograph represents the culmination of this first phase of memory research. McGeoch and Irion contended that "we are interested primarily in the responses made by the learner. In view of this, it is most practical to consider learning in terms of the relation between stimuli and responses" (p. 570).

The second phase of memory research began shortly after the publication of McGeoch and Irion's textbook. It was stimulated by the new information-processing approach that was sweeping many areas of experimental psychology (for historical review, see Palermo, 1971; Lachman, Lachman, and Butterfield, 1979). A number of key publications ushered in this new era of memory research, including Miller's (1956) classic article on chunking, Broadbent's (1957, 1958) publications on attention and short term memory, and Sperling's (1960) experiments on iconic memory. However, no single publication captured the spirit of the emerging cognitive approach to memory as well as Neisser's (1967) text. Neisser contended that memory, like perception, is an active or *constructive* process, not a passive one. It is not sufficient,

This work was supported by a Special Research Program Grant from the Connaught Fund, University of Toronto, and by Research Grant No. U0361 from the Natural Sciences and Engineering Research Council of Canada. I thank Carol A. Macdonald for help with preparation of the manuscript.

argued Neisser, to view the organism as a receptacle of stimuli that impress themselves directly on the mind and lead to predictable responses. Rather, he contended, it is necessary to examine how environmental stimuli are comprehended, elaborated, and transformed in light of the knowledge, skills, and intentions of the organism. Curiously, such a view had been articulated many years earlier by the British psychologist Bartlett (1932), but perhaps because of the dominance of behaviorism, had been ignored by memory researchers. Like Bartlett and Neisser, contemporary cognitive psychologists have not been satisfied with a description of memory purely in terms of stimulus–response bonds. Instead, they have formulated theories and models that attempt to account for the internal processes and structures that transform external stimuli into mnemonic representations, and that permit the organism to reconstruct these representations when access to them is required. It is this concern with the understanding of mental processes and structures that separates the approach to memory described by Neisser from the one described by McGeoch and Irion.

The present chapter represents a partial progress report on the cognitive psychology of memory that has two principal purposes. First, I hope to illustrate some of the methods and findings of cognitive psychology for those who have little or no acquaintance with the field. Second, in view of the cross-disciplinary nature of this book, I will focus on three conceptual themes that have emerged from cognitive research and that, in my own estimation, constitute worthwhile lessons concerning the nature of memory that may have interesting implications for neuroscientists. These themes are general ones in the sense that they are not tied to any one specific theory or model; they concern, respectively, the qualitative nature of encoding operations, the role of retrieval operations, and dissociations among different types or forms of memory. Following a brief discussion of the methods that are used in the cognitive psychology of memory, I will present and discuss data that illustrate the significance of these three themes.

I emphasize that the chapter represents a *partial* progress report on the cognitive psychology of memory, because it would be impractical – if not impossible – to review exhaustively the vast empirical and theoretical literature that has accumulated over the past quarter century. Moreover, there are a number of excellent reviews that include considerably wider coverage than will be attempted here (Baddeley, 1976; Crowder, 1976; Wickelgren, 1977; Craik, 1979; Klatzky, 1980).

In view of the necessarily limited scope of the chapter, I should mention explicitly some of the important topics that will not be covered. First, this chapter will focus on long-term memory (i.e., memory over minutes, hours, and days), and hence will not cover the extensive literature on immediate and short-term memory (see the reviews cited above for discussion). Second, most of the experiments discussed here focus upon memory for recently experienced events and facts – what has been called episodic memory (Tulving, 1972) – and are not concerned with the utilization of well-established knowledge and skills – what has been called semantic memory (Tulving, 1972). The interested reader is referred to the paper by Kintsch (1980) for an informative overview of semantic memory research. Third, because a main objective of the present chapter is to portray general themes that are not specific to any one particular theory or model, many of the well-developed theories that have emerged from the cognitive psychology of memory will not be covered. In lieu of explicit discussion, I will refer the interested reader to a sampling of models that illustrate some of the main approaches in contemporary theory (Anderson, 1976; Anderson & Bower, 1973; Ratcliff, 1978; Mandler, 1980; Raaijmakers & Shiffrin, 1981; McClelland & Rumelhart, 1981; Murdock, 1982; Eich, 1982; Hinton & Anderson, 1981; Tulving, 1983; Anderson, 1983).

Some comments on research methodology

Although the cognitive psychology of memory shares little conceptually with its behaviorist predecessors, it has inherited and used many of the basic methods and techniques that were pioneered by early researchers, such as Ebbinghaus and G. E. Müller, and that were developed further by others during the first part of the twentieth century. Perhaps the most ubiquitous characteristic of early memory research, as well as much of the recent cognitive research, is a reliance upon what will be called the *list-learning paradigm*. Experiments that conform to the list-learning paradigm are relatively simple to describe and to perform: Subjects are shown a series of discrete units, such as nonsense syllables, single words, paired associates, or pictures and, after some variable delay, are asked to recall or recognize the studied items on any one of several types of memory tests. Memory performance is measured either by evaluating the *accuracy* with which subjects produce or identify the list items on the critical test, or by measuring their *latency* to respond to various list and nonlist items (for discussion of issues related specifically to latency measures, see Taylor, 1976; Ratcliff, 1979). The list learning paradigm has several attractive features: The experimenter has control over the to-be-remembered (TBR) units, a large number of parameters in the experimental situation can be varied quite easily, and it is a relatively straightforward matter to determine how many of the TBR units are remembered in different conditions. These may be some of the reasons why the list-learning paradigm has been used so extensively throughout the history of experimental memory research.

There are, however, a number of shortcomings inherent in this approach that have been acknowledged by an increasing number of cognitive psychologists (see Puff, 1982, for review and discussion). First, because the TBR units employed in most list-learning studies (words and pictures) are relatively simple, the view of memory that is obtained from such experiments may be narrow and even distorted. Several psychologists have suggested that to understand memory, it is necessary to sample a broad range of situations and to examine how people remember more complex materials, such as sentences, stories, and sequences of events. Indeed, this point had been made forcefully by Bartlett (1932), who argued that reliance on nonsense syllables prevented psychologists from observing a host of crucial features of memory that emerged only when meaningful materials were used. The past decade has witnessed an increasing trend toward the use of richer and more elaborate materials, such as sentences and lengthy prose passages (for review, see Kintsch and van Dijk, 1978; Franks, Bransford, and Auble, 1982; Voss, Tyler, and Bisanz, 1982).

A second, related objection to the list-learning paradigm is that it bears little or no relation to the situations that involve memory in everyday life. One of the most forceful proponents of this view has been Neisser (1976, 1982), who contends that it is desirable either to abandon the list-learning approach altogether, or at least to supplement it with a more naturalistic orientation that focuses on how memory operates in everyday life. This movement toward *ecological validity* (Neisser, 1976; Bahrick and Karis, 1982) has become increasingly influential and has spawned a variety of studies characterized by a radically different methodology than the traditional list-learning experiment. For example, several studies have used cueing procedures to examine the content, frequency, and temporal distribution of autobiographical memories reported by people from different periods of their lives (e.g., Crovitz and Shiffman, 1974; Robinson, 1976; Teasdale and Fogarty, 1979; Schacter et al., 1982). Other studies have provided information concerning phenomena, such as long-term retrieval of the names of high-school acquaintances (Bahrick, Bahrick, and Wittlinger, 1975), retention of distinctive events known as flashbulb memories (Brown and Kulick, 1977; Winograd and Killinger, 1983), and recollections of politically significant conversations (Neisser, 1981); still others have included actual on-line observation of mnemonic processes in various everyday contexts (e.g., Linton, 1975; Kintsch and Bates, 1977; Zola-Morgan and Oberg, 1980; Schacter, 1983b).

My own approach to the issue of methodology in memory research tends toward a kind of situational relativism: The appropriateness of experimental methods and materials can only be gauged relative to the questions that are posed by the experimenter. Clearly, it is not possible to do justice to the richness and diversity of memory by endless studies of paired-associate learning or categorized word lists. When we want to understand how memory works in everyday life, observational studies are necessary, if only to provide a basis for formulating testable hypotheses; when we want to understand how people process and retain complex sequences of ideas and events, studies that use stories and texts are necessary. At the same time, however, experiments that conform to the list-learning paradigm may be useful for investigating specific problems or for formulating tests of alternative theoretical positions. Unfortunately, because of the ease with which such experiments are executed, the cognitive psychology of memory has too often been plagued by the "functional autonomy of methods" (Tulving and Madigan, 1970) – experiments are performed simply because they can be performed. Thus, cognitive psychologists have often been confronted with an outpouring of data that reveal little that is new about memory, and simply provide information about the experimental paradigms that are used to study memory. However, the recent debate concerning this problem (Allport, 1979; Baddeley, 1976; Tulving, 1979; Neisser, 1982) leaves some room for optimism that the cognitive psychology of memory is entering a phase in which it will be the master, and not the prisoner, of its research methodology. At the very least, researchers who tolerate and perhaps endorse an eclectic methodology will be in a position to determine whether the same principles of mnemonic function hold across materials and situations.

Three emergent themes in memory research: some experimental evidence

Encoding operations and memory performance

One of the central expressions of the emphasis in cognitive psychology on the constructive nature of memory is the large amount of work that has been devoted to the understanding of *encoding operations*. The phrase "encoding operations" refers to the processes by which stimulus inputs are transformed into mnemonic representations. Concern with the construct of encoding operations was initiated by the seminal work of G. A. Miller (1956) and Miller, Galanter, and Pribram (1960), and assumed a central role in the subsequent development of organization theory (e.g., Tulving, 1962; Mandler, 1968). It would perhaps be only a slight overstatement to suggest that the construct of encoding operations constitutes the single most distinctive characteristic of a cognitive approach to memory and distinguishes the memory research of the past 30 years from the earlier work reviewed by McGeoch and Irion. To illustrate the significance of the notion of encoding operations and demonstrate how cognitive psychologists have investigated it, I will discuss representative experiments from several different yet related research domains.

One area of research was sparked by the levels of processing approach to memory advocated by Craik and Lockhart (1972). These investigators suggested that memory could be viewed as a by-product of the processing activities that are applied to incoming information. They argued further that high levels of memory performance are attributable to "deep" or semantic processing activities, whereas low levels of performance are attributable to "shallow" or nonsemantic processing activities. Although the particulars of this approach have received much criticism (e.g., Nelson, 1977; Baddeley, 1978), leading to substantial changes in the "levels" position (Cermak and Craik, 1979), Craik and Lockhart's (1972) emphasis on the importance of the qualitative nature of encoding operations has had a lasting influence. To illustrate what is meant by the expression "qualitative nature of encoding operations," consider one of the experiments reported by Craik and Tulving (1975). Subjects were presented with a series of common words, and answered different questions about them that were intended to induce different processing operations. Some questions focused on the *semantic* features of the target word (e.g., Is the word a type of fish? SHARK), others concerned the *phonemic* features of the word (e.g., Does the word rhyme with weight? RATE), and still others pertained to the *structural* features of the word (e.g., Is the word in capital letters? TABLE). After answering a long series of such questions that had both Yes and No answers, the subjects were given recall and recognition tests. Craik and Tulving found that the type of encoding task had large effects on memory performance: Recall and recognition were consistently highest in the semantic encoding condition, were intermediate in the phonemic condition, and were lowest in the structural condition. The most striking feature of these results is that the traditional sources of variance in memory experiments, such as the frequency, meaningfulness, and concreteness of the target words, were held constant across conditions, but there were still large differences in memory performance as a function of the different encoding tasks. Craik and Tulving attributed the variations in performance to the different encoding operations in which subjects engaged when they processed the critical materials.

Craik and Tulving examined the effects of varying encoding operations among semantic, phonemic, and structural domains. Matthews (1977) has demonstrated that varying encoding operations *within* the semantic domain affects memory performance. In one of his experiments, subjects studied word triplets that consisted of a category label (e.g., *a mammal*) and two other words that had one of three relations to the category: (1) *both* words were category members (e.g., LION, WHALE), (2) one of the words was a category member (e.g., ELEPHANT, TRAPEZE), or (3) neither of the words were category members (e.g., STEEL, PLATINUM). The experiment was designed so that all target words appeared equally often in each of the three conditions. The subjects' task was to judge which of the words belonged to the category. Matthews found that this manipulation yielded large differences on a subsequent recall test: Level of performance decreased systematically from the first to the third conditions noted above. Again, it is important to emphasize that even though the design of the experiment ensured that the nominal properties of the target items were held constant across the three study conditions, memory performance was influenced markedly by the different types of encoding activity.

A second area of research that provides insight into the importance of encoding operations has focused upon the effect of prior knowledge of a particular topic on subjects' ability to remember new information related to that topic. The general theme of this line of experimentation is that existing knowledge plays a large role in determining the nature of the encoding activities in which people can engage. A good example of the influence of prior knowledge is provided by De Groot's (1965) work concerning memory for chess positions in masters and novices. When confronted with a meaningful configuration of 32 pieces on a chess board, masters could reproduce virtually the entire configuration after a brief exposure of several seconds, whereas novices could remember the location of only a few pieces. By contrast, when they were presented with random or nonmeaningful configurations of pieces on the board, the masters recalled no more than did the novices. Thus, the chess masters did not have "better memories," in any general sense, than the novices had. Rather, the data suggest that when confronted with a meaningful configuration, the masters could use their rich and well-differentiated knowledge of chess to create a qualitatively different kind of encoding of the TBR materials than could be achieved by the novices.

Similar conclusions concerning the importance of prior knowledge are suggested by research concerning memory for complex verbal materials. Spillich et al. (1979) exposed subjects to a fictitious account of segments from a baseball game. Some of the subjects possessed extensive prior knowledge of baseball; others possessed little or no baseball knowledge. On a subsequent test for recall of the passage, high-knowledge subjects remembered more of those parts of the passage that were relevant to the structure and goals of the baseball game than did low-knowledge subjects, whereas low-knowledge subjects remembered more of those parts of the passage that were irrelevant to the structure of the game itself. As in the case of the chess masters, prior knowledge seemed to influence subjects' ability to encode meaningful information about the game. Similar evidence on prior-knowledge effects has been obtained in many other studies (e.g., Chiesi, Spillich, and Voss, 1979; Schustack and Anderson, 1979; Chase and Ericsson, 1981; Morris et al., 1981).

An interesting extension of research concerning prior knowledge and encoding operations is found in recent work concerning memory for information about the self. Rogers, Kuiper, and Kirker (1977) suggested that "the self" could be viewed as a well-organized and differentiated body of knowledge that people possess about their own attributes. They reasoned that this extensive prior knowledge would aid memory performance when subjects encode new information with reference to the self. To investigate this issue, Rogers et al. used "self-reference" orienting tasks in which subjects judged whether different adjectives (e.g., shy, outgoing) described themselves. The subjects were also given semantic and nonsemantic orienting tasks similar to those used by Craik and Tulving (1975). Rogers et al. found that recall of the adjectives was higher following the self-reference task than in other conditions (see also Bower and Gilligan, 1979; Keenan and Baillet, 1980). Further information concerning the relations among self, prior knowledge, and encoding operations has been provided by research that has adapted the Rogers et al. procedure to the study of clinically depressed patients. Several investigators have suggested that depressed individuals have extensive, well-organized knowledge of their own negative attributes (cf. Beck, 1976). Derry and Kuiper (1981) investigated the effect of this "negative prior knowledge" on memory performance by using an orienting task in which subjects judged whether positive words (e.g., wonderful) and negative words (e.g., failure) described themselves, and then attempted to recall the words. When they made self-reference judgments, depressed patients remembered more negative words and fewer positive words than did nondepressed patients; no differences in recall were observed when subjects made semantic or structural judgments. The results suggest that the depressed patients' well-differentiated knowledge of their negative attributes provided the basis for a rich and distinctive encoding of the negative words. The results with the depressed patients are similar to those obtained with chess and baseball experts insofar as they illustrate the influence of prior knowledge on encoding processes; they differ, however, because the kind of knowledge that underlies the performance of the former group of "experts" is somewhat less adaptive than the knowledge possessed by the latter two groups.

A third source of evidence concerning the nature of encoding operations derives from research on the role of *inferences* in memory performance. A number of studies suggest that information that is not presented explicitly, and hence is not part of the nominal TBR item, can nonetheless influence memory if it is inferred by subjects at the time of study and hence encoded as part of a TBR unit. Consider an experiment reported by Till (1977). Subjects studied sentences such as (a) *The secretary circled the dates* and (b) *The proofreader circled the dates*. Till contended that people often infer that an event involving a *calendar* is described in the first sentence, and that an event involving a *manuscript* is described in the second sentence. If these inferences are encoded as part of subjects' memory of the sentence, Till reasoned, then *calendar* should be an effective retrieval cue after studying sentence A, but not sentence B, whereas *manuscript* should be an effective cue after studying sentence B, but not sentence A. Precisely this pattern of results was observed in Till's experiment: Cues were effective following encoding conditions in which it was plausible for subjects to infer them at the time of study (for similar data, see R. C. Anderson et al., 1976). It would be difficult to account for such an outcome without a concept such as "encoding," because there is nothing in the nominal properties of the input sentences that provides a basis for predicting the observed pattern of results.

An experiment that demonstrates nicely the joint influence of inference making and prior knowledge on encoding operations has been reported by Arkes and Freedman (1984). Like Spillich et al. (1979), their study included people who had much or little prior knowledge of baseball. The subjects heard several passages about imaginary baseball games and were later tested for recognition of some of the key words in the passages. The critical manipulation involved the distractor items on the recognition test. Some of them represented actions that had to do with subtle features of the game that were likely to be inferred during study by the high baseball knowledge subjects, but not by the low-knowledge subjects. Arkes and Freedman suggested that high-knowledge subjects would be likely to commit a false alarm when an inferred action was included as a distractor on the recognition test, whereas low-knowledge subjects would not be likely to commit a false alarm to the same action because they had not inferred it at the time of study. The data supported this suggestion: The experts' false alarm rate exceeded that of the nonexperts on the critical items. Here, the concept of "encoding" is absolutely essential to understand how prior knowledge of a particular subject can *impair* memory for new information about that subject.

The three aforementioned kinds of evidence, then, highlight one of the central themes to emerge from recent cognitive psychology: To understand memory, it is necessary to do more than describe the characteristics of the input to the system; one must also examine how incoming information is related to, integrated with, and transformed by existing knowledge structures. This is what cognitive psychologists attempt to do when they study encoding operations.

The product or outcome of the encoding process is an hypothesized entity that is commonly referred to as a *memory trace* or an *engram* – the stored representation of an experience that is preserved over time. Most cognitive psychologists do not view the memory trace as a literal "copy" or "snapshot" of an event. A more commonly accepted idea is that a memory trace is a fragmentary and often distorted representation that does not bear a direct relation to an external stimulus. However, there has been a great deal of debate among cognitive psychologists concerning the most useful way of conceptualizing the nature and composition of memory traces. Some theorists have suggested that memory traces can be viewed as organized bundles of features or attributes (e.g., Bower, 1967; Underwood, 1969), whereas others portray memory traces in terms of altered connections between nodes in an associative network (e.g., Anderson, 1976, 1983). Still other researchers have questioned whether anything is to be gained by postulating the existence of a representational entity such as a memory trace (Bransford et al., 1977; Kolers, 1979). Instead, it has been suggested that the effects of experience are preserved by alterations in the actual processes or procedures that are used to encode incoming stimuli (Kolers, 1975, 1979), that is, there is no "record" of experience as such, only a system whose processing tendencies are altered by encounters with the environment.

Questions concerning the nature of stored representations are clearly fundamental ones, but relevant discussions have tended to generate more heat than light; indeed, it is unclear whether the conceptual and methodological tools of cognitive psychology are sufficient to resolve the issue (cf. Anderson, 1978). Although it is easy for researchers to agree that the memory system can be changed by perception and encoding of an event, it has proved difficult for cognitive psychologists to describe the exact nature of this change.

Retrieval processes and study–test interactions

The emphasis on the qualitative nature of encoding operations is a relatively recent development in the cognitive psychology of memory. A second, equally significant, recent development is the increasing concern with the qualitative nature of retrieval operations – the processes by which, and the conditions under which, stored information is elicited on a memory test. It is interesting to note that there is a long history in psychological research of attempting to describe memory purely in terms of events that occur during the input stage (what we would now call the encoding stage) of the memory process. For example, early twentieth-century memory researchers identified variations in memory performance exclusively with variations in the strength of associations that are produced by study-phase manipulations, such as number of repetitions, amount of study time, and the like (Schacter, 1982). The general view of these researchers was that the probability of remembering a particular event or fact is high when a strong association has been established, and low when a weak association has been established, regardless of the conditions under which retrieval is attempted. It is not difficult to discern the residue of such an assumption in many sectors of modern memory research (Tulving, 1976, 1983).

One scientist who did argue that a proper understanding of memory required careful analysis of retrieval conditions and processes was the nineteenth-century German zoologist, Richard Semon. He developed an elaborate theoretical model that foreshadowed many contemporary ideas on retrieval, but was ignored entirely by his contemporaries (see Schacter, Eich, and Tulving, 1978; Schacter, 1982). A key insight that was articulated by Semon – one that has been confirmed and extended in contemporary cognitive psychology – is that the *relation* between encoding conditions and retrieval conditions is of crucial importance for memory performance, and that memory is not determined solely by the strength of memory traces. One of the first experiments to provide evidence for this view was reported by Thomson and Tulving (1970). Subjects in their experiments studied common words, and then attempted to recall them in the presence of either "weak" or "strong" associative cues. Weak cues were those that elicited the targets on a free-association test about 1% of the time (e.g., *hand* – MAN); strong cues, in contrast, elicited their targets about 50% of the time on a free-association test (e.g., *woman* – MAN). A critical manipulation in the experiment was that subjects studied the target items in two encoding conditions: alone (e.g., MAN), or in the context of the weak associate (e.g., *hand* – MAN). Not surprisingly, after studying the targets alone, subjects recalled many more of them to the strong cue (68%) than to the weak cue (43%). The unexpected finding was observed following study of the target in the context of the weak cue: The targets were recalled much less frequently to the strong cues (23%) than to the weak cues (82%). These data, then, demonstrate a strong interaction between encoding and retrieval conditions; they indicate that it is incorrect to speak of one encoding condition as "better" than another irrespective of retrieval conditions, and that it is also incorrect to speak of one type of retrieval cue as "better" than another irrespective of encoding conditions.

One might protest that the Thomson and Tulving finding is attributable to some specific feature of their experimental procedure and hence is of questionable generality. However, a study of sentence memory reported by Barclay et al. (1974) indicates that the Thomson and Tulving finding is not an isolated one. Their subjects studied sentences that emphasized one of two properties of a target noun. For example, subjects in one group studied a sentence such as "The man tuned the PIANO," whereas subjects in a second group studied a sentence such as "The man lifted the PIANO." On a subsequent recall test, subjects in both groups were given two different cues for the target noun (e.g., *something heavy* or *something with a nice sound*). Depending upon which sentence had been studied, one of these cues matched the meaning of the target noun emphasized by the sentential context, and one did not. Barclay et al. found that the performance of subjects in the two conditions depended critically upon type of retrieval cue presented on the recall test. For example, the cue *something heavy* elicited PIANO much more frequently for subjects who had studied "The man lifted the PIANO" than for subjects who had studied "The man tuned the PIANO"; the opposite was true of the cue *something with a nice sound*. In this study, then, it was not possible to describe the effects of the encoding conditions independently of the retrieval cues that were available to the subjects.

The intimate relation between encoding operations and retrieval operations has been further confirmed in studies that have compared memory for semantic information and phonemic information. It was noted earlier that research in the levels-of-processing tradition indicated that semantic encoding yields higher memory performance than does phonemic encoding. Several experiments, however, have demonstrated that the usefulness of semantic and phonemic encoding depends critically upon the conditions of retrieval. Consider, for example, one of the experiments reported by Fisher and Craik (1977). Their subjects studied paired associates in which the cues and targets were associated semantically (e.g., *dog* – CAT) or rhymed with each other (e.g., *hat* – CAT). Memory for the target items was probed on a cued recall test in which subjects were given either rhyme cues or associative cues. Fisher and Craik found that following associative encoding, associative cues were more effective than rhyme cues, whereas following rhyme encoding, rhyme cues were more effective than associative cues. Thus, one could not argue on the basis of these data that associative encoding is "better" than rhyme encoding; the usefulness of the encoding depended upon the type of retrieval cue given to the subject. Similar conclusions are suggested by the work of Morris, Bransford, and Franks (1977). Subjects in their experiments were given two types of encoding tasks: (a) a *semantic* task in which they judged whether a target word fit in a particular sentence frame (e.g., The ____ had a silver engine. TRAIN), or (b) a *rhyming* task in which they judged whether a target word rhymed with another word (e.g., ____ rhymes with legal. EAGLE). When retention of the targets was tested on a standard Yes/No recognition task, performance was higher in the semantic condition than in the rhyme condition. However, a different pattern of results was observed when recognition was tested by showing subjects new words that either did or did not rhyme with a target item, and asking them to indicate which words rhymed with a study item. Under these conditions, recognition performance was higher in rhyme study conditions than in semantic study conditions. Once again, the usefulness of a particular encoding activity depended upon the retrieval cues given to the subject at the time of test.

An intriguing extension of this general line of research is found in recent work concerning the effects of subjective states in memory, such as those related to drug effects or mood changes. It has been demonstrated, for example, that when a subject studies a list of words in a particular drug-induced state, recall of the words is higher when retrieval is attempted in the same state than in a different state (see Eich, 1980, for review). Similar effects have been observed when subjects' mood state is altered through hypnosis or other psychological manipulations (Bower, 1981), or when patients with cyclical mood variations are tested in the same or different mood states on various occasions (Weingartner, Miller, and Murphy, 1977; Clark and Teasdale, 1982). These results indicate that memory for recent events depends not only on the specific information provided to subjects as a retrieval cue, it also depends upon "retrieval conditions" in a broader sense – the extent to which the general cognitive state of the subject at the time of retrieval matches the state that prevailed at the time of encoding.

The results of the type of research described in this section (for more extensive review, see Craik, 1979, 1983; Tulving, 1983) have had a pronounced effect on recent theorizing about the nature of memory. One of the most influential positions has been advocated by Tulving and his associates (e.g., Tulving and Thomson, 1973; Tulving, 1976, 1983) under the general rubric of *encoding specificity*: "The encoding specificity principle is a general assertion that remembering of events always depends on the interaction between encoding and retrieval conditions, or compatibility between the engram and the cue *as encoded* . . . " (Tulving, 1983, p. 242). This view stresses the close relation between encoding and retrieval factors and discourages attempts to explain memory solely in terms of one of the two constructs. A similar notion has been advanced by Bransford and Franks and their colleagues under the rubric of *transfer appropriate processing*: " . . . the value of particular types of acquisition [encoding] activities must be defined relative to the type of activities to be performed at the time of test" (Morris, Bransford, and Franks, 1977, p. 531). The assumption of strong interdependency between encoding and retrieval processes can also be found in a number of quantitative models of memory that differ in other respects (e.g., Jones, 1976; Ratcliff, 1978; Eich, 1982).

Although cognitive research has been relatively successful in demonstrating the critical importance of retrieval conditions, other aspects of the retrieval process have been relatively neglected. For example, as Baddeley (1982) has pointed out, next to nothing is known about how appropriate retrieval cues are generated and used in unconstrained recall conditions. It is evident from everyday experience that a critical aspect of remembering centers on the attempt to find the "key" – the retrieval cue – that will fit the "lock" – the sought-after memory. However, cognitive psychologists have just begun to study this phenomenon. Some of the initial attempts have been made by Williams and his colleagues (Williams and Santos-Williams, 1980; Williams and Hollan, 1981), who have taken an observational, rather than an experimental, approach to the issue. Subjects in these studies were instructed to try to remember the names of high school classmates and to think out loud while doing so – in some cases, for more than 8 hours spread across a number of different sessions. Although most subjects claimed that they had exhausted their fund of relevant names after only several minutes of retrieval, Williams and his colleagues observed that with encouragement from the experimenters, subjects added many more new names across the subsequent hours and days of the task. More importantly, they observed that subjects used a wealth of different strategies in their attempt to generate useful cues and facilitate recall of the names. One of the most common and effective strategies was to generate a specific environmental context that was salient during the relevant temporal interval, and to search for candidate names within that context. For example, some subjects reported imagining a class and "scanning" each row of desks, or imagining their neighborhood and moving mentally from house to house. These reconstructive activities proved essential for retrieving new names as the task progressed. Although this work is largely descriptive and not analytical, it does provide a useful beginning for examining the hitherto neglected aspects of retrieval that are involved in what could be called "the search for effective cues."

The foregoing research also highlights another aspect of the retrieval process that is receiving increasing attention from cognitive psychologists: the subjective feeling that one either will or will not be able to gain access to an unrecalled bit of information. In a relatively unconstrained recall task, such as the one used by Williams, it is important to decide whether or not to continue searching for an inaccessible name. However, how does one know that a name, or any other type of information, is available in memory if one cannot retrieve it? People often report that they have a *feeling of knowing* that they will be able to retrieve unrecalled information under specific conditions. A number of studies have established that feelings of knowing are accurate, at above chance levels, in several different situations. For example, in an experiment by Hart (1967), subjects studied paired associates that consisted of stimulus words and response trigrams. When, at the time of recall, the stimulus word failed to elicit the correct response, the subjects judged whether or not they would be able to *recognize* the response on a subsequent forced-choice test. Hart found that subjects recognized more of those unrecalled items that had been assigned a feeling of knowing "Yes" than those that had been assigned a feeling of knowing "No." Similar evidence for feeling of knowing accuracy has been provided by others (e.g., Blake, 1973; Koriat and Lieblich, 1974; Nelson and Narens, 1980; Nelson et al., 1982; Schacter, 1983a). Given that feelings of knowing are fairly accurate, a natural question concerns the basis of the phenomenon: What gives rise to the feeling of knowing that one will be able to recognize an unrecalled item? Although a firm answer to this question is not yet known, recent research has provided some suggestive clues. Several studies have indicated that subjects are more likely to make a "Yes" feeling of knowing judgment about an unrecalled item when overall level of recall is high than when it is low. For example, Nelson et al. (1982) found more frequent positive feeling-of-knowing judgments as a function of higher degrees of overlearning of study pairs, whereas Schacter (1983a) found more positive feeling-of-knowing judgments following a slow presentation rate than a fast presentation rate, and after a short retention interval than after a long retention interval. These findings suggest that feelings of knowing are determined to some degree by the same processes that mediate level of recall. Consistent with this idea, several experiments have shown that the feeling-of-knowing judgments depend upon recall of attributes of the inaccessible item (Blake, 1973; Koriat and Lieblich, 1974; Schacter and Worling, 1985). However, if researchers are to unravel the complex retrieval processes that are involved in reconstructive tasks such as the one used by Williams, it will be necessary to develop a much deeper understanding of the basis of feelings of knowing and how they influence reconstructive activities.

Priming effects and implicit forms of memory

There is a common theme that ties together most of the research described in previous sections. In virtually all of the cited studies, subjects' retention of TBR information was assessed by asking them to recollect, in a conscious and deliberate manner, events and information that were part of a learning episode. For example, when subjects were given a free- or cued-recall test for a list of studied words, they are being asked to recollect which target words were presented at a particular time and place. Similarly, when subjects are shown a word or a picture, and are asked to decide whether or not it appeared on a previously studied list, their task is to report whether or not they explicitly remember seeing the word or picture during a specified episode in the recent past.

As indicated by the foregoing examples, a common assumption that characterizes a great deal of cognitive research is that "memory" entails an ability to recollect deliberately and consciously what transpired under a specific set of contextual conditions. Indeed, such a notion is reflected in everyday language use of terms such as "memory" and "remembering": We speak of remembering the joke that we heard at a party last week, remembering the most exciting play from last year's World Series, or trying (and sometimes failing) to remember the name of the person that we met at a recent conference. In these everyday examples, the terms "memory" and "remember" imply conscious access to explicit knowledge concerning the content and context of an event.

In contrast to this commonly accepted conception of memory, a growing number of recent studies have provided evidence that the effects of recent experiences can be expressed *implicitly* on retention tests that do not demand explicit recollection of a prior learning episode. The results of these studies suggest that the type of memory that most cognitive psychologists have been concerned with – and the type that most of us think of when we use the term "memory" – may in fact represent just one of several distinct *kinds* or *forms* of memory. Consider, for example, an experiment reported by Tulving, Schacter, and Stark (1982). In this experiment, subjects studied long lists of low-frequency words such as ASSASSIN and BEESWAX. Retention of the words was tested after a 1-hr or 1-week delay in two different ways. On a traditional Yes/No recognition test, subjects were shown some of the target items along with an equal number of lures, and were asked to state whether they *remembered* the occurrence of the word on the study list. By contrast, on a *word-fragment completion* task, subjects were given incomplete fragments of words and were told to try to complete each fragment with an English word. Some of the words represented by the fragments had appeared previously on the study list, the recognition test, or both (e.g., the fragment for ASSASSIN was A--A--IN), whereas others appeared for the first time on the fragment-completion test (e.g., the fragment for LACROSSE was -AC-OS--). The important point about this task is that subjects are not told to try to *remember* the previous episode; they are simply told to do their best to fill in the missing letters of the word fragments.

The results of the experiment revealed that exposure to a word on the study list exerted a marked effect on fragment-completion performance: Baseline probability of correct fragment completion for new words was 31%, whereas probability of completion was 46% for words that had appeared just once in the study list. The difference between completion of new and old words is referred to as *direct* or *repetition priming* (e.g., Cramer, 1966; Cofer, 1967). Two outcomes of the experiment suggested that repetition priming might be mediated by a quite different kind of memory than the one that underlies recognition performance. First, the effect of retention interval was quite different on the two tasks: Accuracy of recognition declined substantially from 1 hr to 1 week, whereas performance on the fragment-completion task remained unchanged over the same interval. Second, performance on the word-fragment completion task was statistically independent of performance on the recognition task: Subjects were equally likely to complete fragments of words that they had recognized as they were to complete fragments of words that they had failed to recognize. If the recognition and fragment-completion tasks were tapping the same kind of memory, subjects should have completed more fragments of previously recognized words than of previously nonrecognized words, that is, performance on the recognition and completion tasks should have been positively correlated, but it was not.

A number of other studies have provided evidence that performance on word-completion tasks can be dissociated from performance on the more traditional tests of recall and recognition. Graf, Mandler, and Haden (1982) manipulated encoding processes by having subjects perform either semantic or nonsemantic orienting tasks on a series of common words. In the semantic condition, subjects rated the words according to how much they liked them, whereas in the nonsemantic condition subjects judged the number of vowels in each word. On the ensuing free-recall task, Graf et al. found that words from the semantic condition were remembered much more frequently than words from the nonsemantic condition. This effect of encoding condition on explicit recall is consistent with the results of studies described earlier (see section on encoding operations and memory performance). However, a quite different outcome was observed on the word-completion task, in which subjects were given the initial three letters of old and new items and were asked to produce the first word that came to mind. There was a reliable priming effect on this task: After one exposure to an item, significantly more of the three-letter fragments were completed with the target than in a baseline condition. The critical finding, however, is that completion performance was virtually identical following semantic orienting and nonsemantic orienting. In this study, then, type of encoding task had a powerful influence on the subjects' ability to *remember* target words on the recall task, but had no effect on the magnitude of priming on word completion. Further data that confirm and extend these results have been reported in a more recent study by Graf and Mandler (1984).

The dissociations between remembering and priming reported in the foregoing experiments are not confined to word-completion tasks. Several researchers have reported quite similar patterns of data using a variety of other tasks that are sensitive to repetition priming effects. For example, Jacoby and Dallas (1981) compared the effects of exposure to a list of words on recognition memory performance and on *word identification* performance. On a word identification task, words are exposed for extremely brief durations (on the order of 35 msec) and subjects attempt to identify them. Priming occurs on the word identification task when subjects "see" more studied words than new words. As on the word-completion task, subjects need not make explicit reference to the study episode in order to demonstrate enhanced performance as a function of prior exposure. Jacoby and Dallas found that a single study exposure to an item yielded enhanced word identification performance relative to baseline conditions. Moreover, they also observed sharp dissociations between recognition memory and word identification: Variables such as level of processing (semantic versus nonsemantic), retention interval (immediate test versus 24-hr test), and sensory modality at study and test (same versus different) all had substantial effects on recognition memory, but had little or no effect on word identification. Although subsequent research indicates that some experimental variables affect the two tasks similarly (see also Jacoby, 1983a), the most striking result of this study is that variables that influence subjects' ability to remember explicitly the occurrence of a word on the study list need not affect the magnitude of priming on word identification – exactly as was observed in the previously discussed studies of word-fragment completion.

Another task that has been used to study repetition priming effects is known as a *lexical decision* task: Subjects are presented with letter strings and are required to state whether the string constitutes an English word (e.g., fruit) or a nonword (e.g., froit). Scarborough and his colleagues (Scarborough, Cortese, and Scarborough, 1977; Scarborough, Gerard, and Cortese, 1979) have found that exposure to a word on a list affects the latency of a subsequent lexical decision about that word: Subjects are faster to make a lexical decision about previously studied words than about nonstudied words. Significantly, Scarborough et al. (1977, 1979) observed that the magnitude of the priming effect on the lexical decision task is independent of the length of the study–test interval. Carroll and Kirsner (1982) have also compared priming effects in lexical decision with recognition memory. In accordance with many previous studies, they found that the semantic relatedness of word pairs at the time of study affected subsequent level of recognition memory: Subjects recognized more related pairs (e.g., gold–silver) than unrelated pairs (e.g., gold–sorrow). By contrast, the relatedness variable had no effect on subsequent lexical decision latency; the magnitude of the priming effect was about the same for related and unrelated pairs.

Priming effects have also been detected using a homophone spelling task devised by Jacoby and Witherspoon (1982). In this study, subjects answered questions that contained a low-frequency homophone (e.g., What is an example of a reed instrument?). Priming effects were examined on a "spelling test" in which words are read aloud and subjects are asked to spell them. Jacoby and Witherspoon found that subjects tended to provide the low-frequency spelling of the homophone (e.g., *reed* vs. *read*) more frequently as a function of the prior exposure. Once again, however, the occurrence of the priming effect was independent of recognition memory: Subjects provided the low-frequency spelling of the homophone about as often when they recognized that it had occurred earlier and when they did not. Using a similar task in a divided attention paradigm, Eich (1984) has furnished even more dramatic evidence on the independence of priming and recognition. Eich's subjects took part in a dichotic listening task in which the critical items were presented on the unattended channel, while subjects monitored a spoken passage on the attended channel. The critical items were word pairs in which the second member was a homophone and the first member was a word that biased the low-frequency interpretation of the homophone (e.g., *taxi*–FARE).

When tested later for recognition, subjects remembered little or nothing about the previous occurrence of the unattended words. Yet, in spite of the virtual absence of conscious recollection, they demonstrated a substantial priming effect on the spelling test. Subjects tended to give the low-frequency spelling of the homophone (e.g., *fare* vs. *fair*) more frequently as a function of the previous unattended exposure.

The dissociations between conscious memory and priming that have been described were all obtained with normal college students. An important line of converging evidence for the separability of the two kinds of performance is provided by studies that have examined brain-damaged patients who are afflicted by the *organic amnesic syndrome*. The amnesic syndrome occurs after various types of neurological dysfunction, including bilateral lesion of the temporal lobes and hippocampus, closed-head injuries, Korsakoff's syndrome, early stages of Alzheimer's disease, encephalitis, anoxia, and damage to the dorsomedial nucleus of the thalamus (for review, see Schacter and Crovitz, 1977; Whitty and Zangwill, 1977; Squire, 1982; Cermak, 1982; Hirst, 1982). Most amnesic patients have relatively preserved intellectual function, use language normally, demonstrate appropriate social skills, retain much of their premorbid knowledge, and have little difficulty on immediate memory tasks such as digit span. However, these same patients remember little or nothing when queried about events that occurred just minutes or hours ago, perform disastrously on standard laboratory tests of conscious memory, such as free recall, cued recall, and recognition, and have little or no ability to remember ongoing events in everyday life (Schacter, 1983b). Consideration of amnesic patients in the present chapter is particularly relevant, because the study of organic amnesia currently provides one of the most productive links between cognitive psychology and neuroscience (e.g., Shallice, 1979; Cermak, 1982: Schacter and Tulving, 1982a; Schacter, 1984). In the present context, the critical phenomenon that has been revealed in studies of organic amnesia is that in spite of their inability to remember recent events and new information, amnesic patients demonstrate relatively normal repetition priming effects on tasks such as word completion, lexical decision, and homophone spelling. In fact, it may be fair to say that a good deal of the recent interest concerning priming in word-completion tasks was sparked by what has turned out to be a landmark finding reported by Warrington and Weiskrantz (1970, 1974). In their studies, amnesics and controls were shown lists of common words, and were then tested on a Yes/No recognition task and a word-completion task. Not surprisingly, the amnesics were severely impaired relative to controls in the recognition test; they had great difficulty remembering whether or not a word had appeared during the study episode. On the word-completion task, however, the amnesics demonstrated just as large a priming effect as the controls did. Graf, Squire, and Mandler (1984) replicated and extended this dissociation, and have also demonstrated that the instructions given to the subjects at the time of test exert a powerful influence on the pattern of results. In their experiments, amnesics and controls performed semantic and nonsemantic orienting tasks on lists of common words. Word-completion performance was evaluated by giving subjects the first three letters of the critical items and asking them to complete the fragment with "the first word that comes to mind." Under these conditions, word-completion performance of amnesics and controls was equivalent. A different outcome was observed when the test was changed to one of cued recall by asking subjects to use the three-letter fragments as cues for remembering the words that had been presented earlier. Although the fragments were nominally identical to those in the word-completion condition, the amnesics were significantly impaired under cued-recall instructions. Controls apparently could use the cues to help reconstruct the study context and hence facilitate retrieval of additional words, but amnesics could not.

Dissociations between recognition and priming in amnesic patients have also been observed with the lexical decision task (Moscovitch, 1982), and the homophone spelling task (Jacoby and Witherspoon, 1982): The magnitude of priming effects is virtually identical in amnesics and controls, even though the amnesics have little or no recollection of the learning episodes that underlie their facilitated performance. In addition to exhibiting normal priming effects, amnesics can also acquire some motor skills (e.g., Milner, 1962; Milner, Corkin, and Teuber, 1968) and cognitive skills such as mirror reading (Cohen and Squire, 1980; Moscovitch, 1982) and puzzle solving (Brooks and Baddeley, 1976). The relation of preserved skill learning to preserved priming effects is as yet uncertain and will not be pursued further here; relevant discussion can be found in Tulving (1983), Cohen (1984), Moscovitch (1984), and Schacter (1985b).

The data concerning repetition priming effects in normal subjects and amnesic patients are reasonably clear-cut insofar as they indicate that the effects of recent experiences can be expressed in an *implicit* mode that is dissociable experimentally from the *explicit* mode that entails conscious access to information concerning the occurrence of recent events (cf. Graf and Schacter, 1985; Schacter, 1985b). If, however, one attempts to go beyond such a general descriptive statement and seek more specific hypotheses concerning the mechanisms involved, a good deal of uncertainty and disagreement is found. Although it is beyond the scope of the present chapter to consider the strengths and weaknesses of the various theoretical interpretations of repetition priming effects (for discussion, see Feustel, Shiffrin, and Salasoo, 1983; Jacoby, 1983b, 1984; Tulving, 1983; Moscovitch, 1984; Schacter, 1985a, 1985b), it is worth distinguishing among several ways of approaching the issue. One general approach emphasizes the differences between priming and conscious recollection. For example, several researchers have suggested that fundamentally different forms of memory are involved in priming, on the one hand, and in conscious recall and recognition, on the other; only one of these forms of memory is thought to be damaged in amnesia (e.g., Tulving, Schacter, and Stark, 1982; Squire and Cohen, 1984; Tulving, 1983; N. J. Cohen, 1984; Graf, Squire, and Mandler, 1984; Schacter, 1985a, 1985b; Schacter and Moscovitch, 1984). Other investigators have argued that priming effects are based upon the automatic activation of old, well-integrated knowledge, whereas conscious recollection depends upon new knowledge that is acquired through elaborative or strategic processing during a study trial (e.g., Wickelgren, 1979; Mandler, 1980; Warrington and Weiskrantz, 1982). A somewhat different approach emphasizes the continuity between conscious recollection and priming (Craik, 1983; Jacoby, 1983a, 1983b, 1984). Here, the general idea is that priming effects, like conscious recall and recognition, are based upon newly established memories, not merely upon the activation of old knowledge. Differences between priming and recollection are accounted for in terms of the differential effectiveness of different cues in guiding or constraining retrieval processes. By this view, an amnesic patient might establish a representation of a recent event, but only would be able to gain access to it when retrieval processes are sufficiently constrained by available cues at the time of test.

It is not yet possible to distinguish between the foregoing interpretations, because the phenomenon of repetition priming is a relatively new one and decisive data do not yet exist. Several recent studies, however, have provided evidence concerning one of the central questions that will have to be addressed by any attempt to account for priming phenomena: Are priming effects based entirely upon the activation of old and well-integrated knowledge, or do priming effects also include new, contextually sensitive knowledge acquired during a particular learning episode? All of the priming effects described so far could be accounted for by postulating an activation of old knowledge. The results of several studies, however, suggest that priming reflects more than an activation of old knowledge and is influenced by new information that is encoded during a study episode. Some evidence along these lines has been provided by Jacoby (1983b) in a study that examined contextual influences on word identification and recognition memory. Jacoby's experiments included three different encoding conditions in which subjects either studied a word alone (e.g., COLD), studied it in the context of an antonym (e.g., *hot*–COLD), or were asked to generate the target antonym in the presence of the context word (e.g., *hot*–?). Jacoby found that the magnitude of priming on the word identification task was affected by the contextual manipulation. Identification of the target word (i.e., COLD) was most accurate when the target had been studied alone, was significantly less accurate when it had been studied in the context of an antonym, and was still less accurate when it had been generated in response to the context cue; precisely the opposite pattern of results was observed on a recognition test. The critical point about the word identification data is that they indicate that priming is sensitive to contextual conditions at the time of study.

It should be noted, however, that the context effects observed by Jacoby were obtained with materials that have a preexperimentally established relation with one another. These data thus do not indicate whether *new* relations among previously unrelated materials can be revealed through priming. However, converging evidence from several different priming tasks suggests that new associations established during a study episode can influence the magnitude of priming effects. McKoon and Ratcliff (1979) had subjects study word pairs that had either weak preexperimental relations (e.g., *city*–GRASS) or strong ones (e.g., *green*–GRASS). Subjects then performed a lexical decision task in which the presentation of target words was preceded by its list cue or by a word that had not been part of the study pair. McKoon and Ratcliff observed that latency of lexical decision was facilitated when a target was preceded by either a related or an unrelated list cue. Moscovitch (1984) has employed a different paradigm to examine contextual influences on priming. Normal and amnesic subjects studied weakly related word pairs, and were later asked simply to read over lists that contained either the same pairs, new pairs, or repairings of the old words. Moscovitch found that both normals and amnesics read the old intact pairs faster than the new pairs. Moreover, he also found that both groups of subjects read the old intact pairs faster than the old broken pairs. Schacter, Harbluk, and McLachlan (1984) found that amnesic patients sometimes retained new facts about well-known and unknown people, even though they had no explicit recollection that an experimenter had told them the facts just seconds or minutes earlier.

Graf and Schacter (1985) have examined whether priming effects on a word-completion task are enhanced when a context word that was paired with a related or unrelated target word on the study list is also paired with the target fragment on the completion task. In one of their experiments, amnesic patients, matched control subjects, and college students were shown lists of related pairs (e.g., *window*–GLASS) or unrelated pairs (e.g., *window*–REASON), and were asked to encode the two words with respect to one another by associating them in a brief sentence. A word-completion task was then administered under the guise of a second filler task whose purpose was to obtain some normative data. The subjects were given a sheet containing word pairs in which only the first three letters of the second item were present. In the *same context* condition, the pairings were intact with respect to the study list (e.g., *window*–REA___), and in the *different context* condition, the pairings were broken with respect to the study list (e.g., *lake*–REA___). Subjects were instructed to read over each pair and to complete the fragment with the first word that came to mind. After this task was administered, subjects were told that their memory for the word pairs presented earlier would be tested by giving them the first member of the pair and asking for recall of the second.

The word-completion data yielded a striking pattern of results. Probability of completing a fragment with the target word was significantly higher in same context conditions than in different context conditions for all three groups of subjects, and the magnitude of the context effect in amnesic patients did not differ from that of the control groups. Moreover, the context effect was present for unrelated word pairs as well as for related word pairs. These data suggest that newly established associations between previously unrelated pairs of words do influence priming effects in both amnesics and normals. In contrast, however, when words were presented as cues to help *remember* the target item, the amnesics' performance dropped dramatically, whereas controls recalled more items than they had produced on the completion task. However, subsequent research suggests that the associative effect on completion performance may be observed only in patients with mild-to-moderate amnesia and not in patients with severe amnesia (Schacter and Graf, in press).

The results of the foregoing experiments indicate clearly that priming effects are sensitive to contextual influences and that newly acquired knowledge can be revealed through priming, hence discouraging the notion that priming is confined to activation of old knowledge. Consistent with this notion is the finding that magnitude of priming effects is sensitive to study test changes of sensory modality (Morton, 1979; Jacoby and Dallas, 1981) and physical parameters of the stimulus (Winnick and Daniel, 1970; Scarborough, Gerard, and Cortese, 1979). At the same time, however, several other studies highlight the crucial role that is played by preexisting knowledge in direct priming effects. Diamond and Rozin (1984) found that amnesics revealed no priming effects when the study materials were composed of pronounceable nonwords (e.g., NUMDY) and the initial three letters were given as fragments at the time of test. The same patients demonstrated substantial priming when familiar words were used. Cermak et al. (1985) observed similar results on a perceptual identification task. Jacoby and Witherspoon (1982), who found independence between recognition and perceptual identification when familiar words were used, observed dependence between the two tasks when nonwords were used. Feustel, Shiffrin, and Salasso (1983) compared lexical decision priming of words and nonwords in college students. They found some priming for both types of materials, but also uncovered a number of qualitative and quantitative differences between priming effects obtained in the two cases. Feustel et al. argued that many of the observed differences could be attributed to the presence of unitized, preexisting representations for familiar words that are absent for nonwords.

Further evidence concerning the importance of unitized representations in priming effects is found in a study by Schacter (1985a). In this experiment, amnesics, matched controls, and college students were exposed to two-word adjective–noun phrases. In the *unitized* conditions, the phrases consisted of idioms such as *sour grapes, small potatoes, black magic,* or *big deal.* Idioms were used because they constitute examples of well-integrated, unitized representations (Horowitz and Manelis, 1972). In the *nonunitized* condition, the components of the idioms were rearranged to form phrases such as *sour potatoes* and *small grapes.* These phrases do not have a preexisting unitized representation. Priming was evaluated by including the first word of each unitized and nonunitized phrase in a long list of words, and asking subjects to write down next to each item the first word that came to mind. Following exposure to the unitized pairs, all three groups of subjects revealed an enhanced tendency to produce the target on the completion test; the magnitude of priming did not differ significantly among amnesics, matched controls, and college students. However, exposure to the nonunitized pairs yielded no evidence of priming in any of the groups. By contrast, when subjects were subsequently given the same cues and were asked to try to remember the second word of the phrase, both groups of controls recalled a large number of items in the unitized and nonunitized conditions. The amnesics, however, failed to recall a single nonunitized item, and did not produce any more unitized items than they had already produced on the completion test. In this study, then, the existence of a well-established unit was crucial for priming, but was not necessary for cued recall.

The patterns of data discussed in this section indicate that although priming effects are sensitive to contextual influences and can provide a basis for acquiring new associations, they are also critically dependent upon whether or not part of a preexisting representation is included on a priming test. A challenge for theorists, who are attempting to understand the relatively new and intriguing phenomenon of priming, will be to account for both of these experimental facts in a single theoretical model. An additional challenge will be to describe exactly how the implicit form of memory that is tapped on a priming test is related to the more conscious and explicit forms of remembering that have long constituted the "bread and butter" of memory research. It is relatively safe to conjecture that such issues will be investigated thoroughly by cognitive psychologists during the coming years.

Concluding comments

Has cognitive psychology advanced our understanding of memory during the past 30 years? Researchers who have addressed this question divide readily into two groups, the pessimists and the optimists. Pessimists portray a field that is enslaved by a restrictive methodology that has yielded an assortment of laboratory curiosities of dubious significance (Newell, 1973; Allport, 1979; Neisser, 1978), and that is dominated by theories that disguise old ideas in fashionable terminology or premature formalism (Tulving, 1979). Optimists point to the existence of diverse and replicable empirical phenomena (Baddeley, 1976), to new theoretical models that are characterized by breadth as well as mathematical precision (Wickelgren, 1977; Anderson, 1980), and to the emergence of a relatively coherent general picture of mnemonic function (Craik, 1979). Depending upon whom one reads – an optimist or a pessimist – one could come to quite different conclusions regarding the current status and future prospects of the cognitive psychology of memory.

In my own estimation, the arguments advanced by the optimists and the pessimists are not mutually exclusive; both have some validity. On the one hand, when we consider that the cognitive psychology of memory represents a relatively recent development, there are grounds for concluding that progress toward the understanding of memory has been made. Although the rate of progress may not be exactly overwhelming, it is useful to keep in mind that concepts such as "qualitative nature of encoding operations," "encoding–retrieval interactions," and "explicit and implicit forms of memory" were nowhere to be seen on the psychological landscape when McGeoch and Irion published their text in 1952. The fact that these and other concepts are now supported by a reasonably consistent body of experimental observations provides grounds for some measure of optimism. On the other hand, it must be acknowledged that many current concepts in cognitive psychology, including the foregoing ones, are disturbingly vague, circular, and difficult to test unequivocally. Indeed, cognitive theorizing about memory is sufficiently "flexible" that even highly sophisticated and detailed models can be patched up quickly when apparently contradictory evidence is produced, hence avoiding the necessity to change any basic assumptions of the models (cf. Tulving, 1979). The vagueness of many basic cognitive concepts, and their frequently tenuous relation to experimental data (Roediger, 1980), represent obstacles to progress that raise questions concerning just how successful a cognitive approach will prove to be in unraveling the ultimate secrets of memory. Indeed, it has been suggested that cognitive psychologists may be, in principle, unable to resolve theoretical debates concerning fundamental issues such as whether a memory representation can be described independently of the processes that act on it (Anderson, 1978). The limits of a cognitive approach, however, may also encourage the building of bridges with sectors of memory research that can contribute to the solution of seemingly intractable problems. The growing influence of artificial intelligence models on memory research (cf. Norman, 1981) represents one example of this trend; the increasing interaction between cognitive psychology and the neuropsychology of memory disorders (cf. Cermak, 1982) represents another. My own suspicion is that the neurosciences may prove to be a valuable ally when the resolving power of cognitive concepts and theories is simply too coarse to provide closure concerning a theoretical issue. If basic cognitive concepts could be instantiated in a model of brain function, it might be possible to obtain answers to previously elusive questions through advances in understanding of neuronal function. Of course, to achieve any kind of equivalence or meaningful relation between cognitive constructs and neurophysiological ones will require extended multidisciplinary interaction between cognitive psychology and neuroscience. Hopefully, this book represents a start in that direction.

11   The neurobiology of memory

Michael Gabriel, Steven P. Sparenborg, and Neal Stolar

Prologue

Evolution and learning

The mechanisms of natural selection and genetic transmission are insufficient to specify the full set of behaviors needed for survival. These mechanisms are thus supplemented by learning processes, which finetune the behavior of individuals so that it may meet specific environmental demands too capricious and subtle to be anticipated by the genetically based adaptive mechanism. The importance of learning processes is thus comparable to that of other very fundamental attributes essential to survival. Because learning processes are so important, and also fascinating, there has for many decades been an active science concerned with their biological bases. The principal objective of this research is to discover how learning works. What are the biological changes that represent stored information, and what functions are performed by different systems of the brain in the acquisition and use of information?

Reductionism and functionalism

Research on the biological bases of learning processes has been conducted in two fundamentally different ways. One of these, labeled here the reductionist approach, tries to identify the biophysical, biochemical, and ultramorphological changes that occur in the nervous system when something is learned. The second, the functionalist approach, attempts to discover the ways in which the functioning and interaction of brain circuits and systems bring about learning and information storage. It also tries to link the systems and their interactions with various categories (e.g., skill learning) and components (e.g., long-term memory) of learning. Central neural cytoarchitecture, connections, and impulse conduction define the neural levels of discourse addressed by the functionalist approach.

It may be helpful to regard the reductionists as conducting research that moves from an acquired change in behavior (i.e., learning), or a change in neuronal activity (neural plasticity), "vertically downward" to more molecular causal substrates of the change. In contrast, the functionalist approach is concerned with identification of the "horizontal" causal relations among interacting neural systems involved in mediating neural and behavioral changes, as well as the upward vertical causal flow from the systems to the behavior.

The contrast between the reductionist and the functionalist approaches can perhaps be facilitated by making an analogy between the study of neural activity and the study of electricity in physics. Reductionism seeks to explain certain forms of activity [long-term potentiation (LTP)], or behavior, in terms of molecular constituents. It can thus be likened to research to identify protons and electrons, the particles that comprise electrical current. The functionalist approach is based on the premise that one must understand how neural plasticity functions in the context of a larger system. Thus, it can be likened to studies in physics that led to Ohm's law, the equation that describes the relationships of current with resistance and voltage. In this instance current is a "given," and what is sought is an understanding of its relationship with other important variables. Here then, the stress is placed on the horizontal flow of causality from variable to variable, rather than on the vertical reduction of a particular variable to its more molecular constituents. The analogy with the study of electricity also informs us of the obvious, that both enterprises are of enormous value.

In this chapter some of the goals, assumptions, methodologies and contributions of the two approaches are described. To accomplish this description, we have chosen to present in some detail a relatively few instances of current research, rather than to give an exhaustive survey. The instances presented were chosen because we felt that they approximate prototypes that in sum represent a good part of the variety of reductionist and functionalist research.

Reductionist studies

Rationale

Wax, soap, clay, and plastic are metaphors for neurobiological substrates that change when memory is formed. This is so because hard objects pressed into such materials leave behind rather permanent replicas of their surface geometries that persist until erosion occurs, or until some other impression is imposed. The basic question addressed by reductionist studies concerns the identity and composition of the waxy substance of the nervous system, upon which replicas of our worlds are impressed. Which property manifests the essential plasticity? Is it the shapes or branching of neurons, the number of synapses, the relative volume of certain neurotransmitters in synaptic terminals, or should one look at even more molecular levels, such as the synthesis of proteins, that subtly alter synaptic activities of neurons without noticeably changing their shapes, numbers, or the amount of available neurotransmitter? These guesses as to the critical substance are intended as a few tenable examples, but not as an exhaustive list. The alternatives are also not intended as a set of completely independent possibilities. Presumably any learning-related change in the nervous system is complex, represented by alterations at many biological levels. Diverse forms of assay may tap different manifestations of the same fundamental change, and changes at one level (e.g., synapse formation) should be translatable to changes at adjacent levels (e.g., altered neuronal activity).

Historical background

During the past two decades there has been a substantial narrowing of the range of procedures used in studies in the reductionist tradition. There is probably a good reason for this. The application of the reductionist strategy to learning mechanisms is a relatively recent enterprise, originating in the 1950s with studies of the effects of drugs and other chemicals (such as antibiotics that inhibit the synthesis of proteins), on the acquisition of learned behavior in mammals. The goal was and is to find the "magic bullet" (i.e., the single biochemical substrate that is the essential substance of the learning process). This objective was approached with some abandon, and neurotransmitter agonists and antagonists, protein assays, and protein synthesis inhibitors were administered to mammalian animals in a wide array of behavioral tasks. After two decades, a great deal had been learned about the effects that such drugs and biochemicals have on the performance of learned behavior. However, no magic bullet had turned up. Furthermore, the work had gained something of a negative image because it seemed unable to escape certain problems of interpretation. Thus, in studies that attempted to impair or facilitate learning through administration of neurochemicals, one had always to contend with the question of whether observed behavioral alterations were due to specific interaction of the agent with the learning mechanism or with some other process, such as arousal, attention, motivation, interoceptive experience, motor impairment, somnolence, and so forth. Similarly, studies employing assay of the nervous system for morphological or biochemical changes produced during acquisition of learned behavior were subject to the possibility that any observed change reflected some process correlated with learning, such as use-related hypertrophy of tissue or arousal, rather than learning per se. An excellent discussion of these issues, and citation of several reviews of the earlier studies are provided by S. P. R. Rose (1981).

Given these problems, the dominant impact of the early work was to refocus the research effort upon simpler and more uniform laboratory preparations – preparations permitting precise and unfettered measurement and control of learned behavior, as well as the neural substrates presumed to mediate it. One manifestation of this refocusing has been a concentration of much recent work on studies of something called long-term potentiation (LTP), a relatively permanent form of stimulation-induced change in neuronal excitability that can be measured at certain synapses. A second manifestation is the concentration of effort on certain elementary forms of behavioral plasticity (habituation, sensitization) in invertebrate (very often molluscan) animals. Knowledge of these two approaches will take one a long way toward appreciation of the current thrust of the reductionist studies.

Long-term potentiation

Electrical stimulation of certain axonal pathways in mammalian brains at high frequency enhances the efficiency of neural transmission across synapses accessed by the stimulated axons (Figure 11.1). The proper choice of stimulation parameters can cause efficiency increases that last for several days, leading to speculation that this sort of synaptic alteration (i.e., LTP), contributes to learning and memory processes [see Swanson, Teyler, and Thompson (1982) for a general review of the work on LTP]. This interpretation, though unproven, is bolstered by the widely held belief that synapses of the brain's hippocampal formation yield more dramatic and enduring varieties of LTP than that observed in other brain systems, and it is the hippocampus and related limbic system structures that have been implicated by studies of brain-damaged humans in the mediation of important aspects of memory (see section on historical background of functional approach). It is argued that if the hippocampus is indeed a critical neural substrate of memory, then surely one will understand something important about the biology of memory if one can completely understand how hippocampal synapses are potentiated. These ideas provide the rationale for the rather massive research effort currently being directed at the identification of the molecular substrates of LTP, work that addresses many levels of synaptic functioning, including changes in synaptic morphology, receptors, neurotransmitter levels, and molecules, brought on by potentiating stimulation.

The most prolific and highly integrated program of research on this issue has been carried out by Dr. Gary S. Lynch and his colleagues at the University of California at Irvine. This work has culminated in a detailed hypothesis about LTP, which incorporates many findings at the structural and biochemical levels (Lynch and Baudry, 1984). The hypothesis states essentially that LTP in the dentate gyrus of the hippocampal formation results from a chain of events initiated by the intraneuronal influx of calcium triggered by poten-

MEASUREMENT of LTP

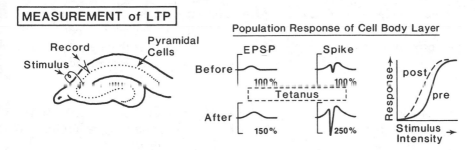

Population Response of Cell Body Layer

ELICITING CONDITIONS

- 15–100 Hz for 1–10 sec (100 Pulse Total)

- Stimulation of: ... Elicits LTP in:
    - Perforant Path .... Dentate Granule Cells
    - Mossy Fiber CA3
    - Schaffer Collaterals CA1
      (stratum radiatum)
    - Stratum Oriens Fibers CA1

MAGNITUDE of LTP

	Average Increase (%)	Range (%)
EPSP	50	0–100
Spike	250	0–120

TIME COURSE of LTP

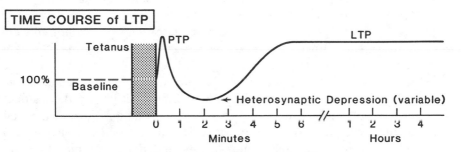

Figure 11.1. Long-term potentiation (LTP) properties, including method of measurement, conditions for elicitation, magnitude, and time course. (Adapted from Swanson et al., 1982.)

tiating stimulation. Calcium activates an enzyme (a neutral thiol proteinase, calpain) that breaks down proteins that normally block the access of the neurotransmitter glutamate to glutamate receptors. It is the increased availability of glutamate receptors that accounts for the enhanced synaptic response in LTP.

Pivotal to the development of the hypothesis was the discovery that agents that inhibit the activity of calpain block an observed calcium-induced increase in glutamate receptors. The most convincing blocking effects were obtained with leupeptin, a tripeptide that potently inhibits calpain.

An important feature of this account is the explanation that it provides for the relatively long duration of the LTP effect. Thus, in vitro studies indicated that the calcium-mediated increase in glutamate receptors is essentially irreversible in the sense that it persisted after calcium was removed by bathing the substrate in calcium-free fluid, or with ethylene glutamine tetraacetic acid (EGTA), a calcium chelator. The irreversibility is an unusual property, but it is quite reminiscent of the action of proteinases, which cause the irreversible cleavage of peptide bonds.

The hypothesis of Lynch and Baudry is compelling because it bridges completely the input–output gap from stimulation to potentiation. Yet it remains very much an hypothesis. Other explanations of the LTP effect are quite possible and even plausible. This is so because many consequences, in addition to those singled out by Lynch and Baudry, can and probably do flow from each of the events in the proposed causal sequence, that is, at each point there is a divergence of causative influences, whereas the hypothesis is about one linear path through the sequence. For example, a critical item of support for the involvement of calpain as the agent that unmasks glutamate receptors is the rather dramatic effect of leupeptin in increasing glutamate receptors. The idea here is that leupeptin quite specifically inhibits calpain. However, other workers have pointed out that the action of leupeptin is not entirely specific to calpain. Other leupeptin-sensitive agents exist that may mediate the effects of leupeptin (see Routtenberg, 1984).

Because it participates in a variety of critical neural events, including the neuronal membrane channel openings and closings that govern the changes in electrical potentials assumed to represent information flow within and between neurons, alteration (e.g., phosphorylation and methylation) of brain proteins is a favored candidate for modification underlying learning and memory processes. Detailed considerations supporting this suggestion are provided by Routtenberg (1982). This idea, used in the hypothesis of Lynch and Baudry, has recently been encouraged by several studies in addition to their work, indicating that training of various sorts alters the phosphorylation state of certain proteins. For example, Routtenberg, Lovinger, and Steward (1985) have conducted an in vitro analysis of protein phosphorylation in samples of tissue taken from the dentate gyrus. Prior to hippocampal dissection, LTP in experimental animals was induced by stimulation of the axonal pathway (the perforant path) that carries information to the dentate gyrus from the entorhinal cortex. The results showed a significant increment in phosphorylation of a 47,000 dalton protein, F1, in the tissue previously manifesting LTP, compared to tissue from controls that received stimulation at a lower, nonpotentiating frequency. The correlation between the protein F1 phosphorylation state and the change in amplitude of the neuronal population spike used to index LTP was quite high ($r = .89$, $P < .01$), supporting the idea that the phosphorylation of protein F1 may represent an important link in the causal chain leading to the production of LTP.

Routtenberg (1984) describes evidence indicating that protein F1 is possibly acted upon by a protein enzyme, protein kinase C, suggesting that kinase C induces the phosphorylation of F1, which in turn induces LTP. In addition, Routtenberg points out that there are several neutral thiol proteinases, other than calpain, that are inhibited by leupeptin. One of these is protein kinase C. Thus, Routtenberg's alternative explanation of the data of Lynch and Baudry states that the action of leupeptin-sensitive neutral thiol proteinases is directed not at glutamate receptor function, but rather at kinase C and protein F1. At present, there is little information permitting a detailed assessment of these alternatives. However, it should be noted that an advantage of the account given by Lynch and Baudry is its completeness. Thus, increased glutamate receptors are proposed to be the proximate cause of LTP, whereas the link between the proximate cause, protein F1 phosphorylation, and LTP is not specified in Routtenberg's model.

Invertebrate models

Independent of its contribution to the understanding of learning and memory processes, the technical virtuosity that has been brought to bear upon the molecular substrates of LTP is of value in that it provides knowledge of the molecular mechanisms that govern the excitability of neurons. Yet, despite the evidence linking the hippocampus with memory in humans, and the correlations between training level and synaptic efficiency changes in the hippocampus in animals subjected to training experiences (e.g., Berger, 1984; Weisz, Clark, and Thompson, 1984), the assumption that LTP at particular synaptic junctions in the hippocampus represents storage of information during learning requires a substantial leap of faith. In this regard, models that have been developed recently employing invertebrate animals have the virtue that the plasticity observable in the nervous systems in these animals has demonstrable causal relevance to behavior.

Although there are several excellent invertebrate models (e.g., Alkon,

1982; Crow and Offenbach, 1983), we choose here to present in some detail the studies of short-term sensitization in the marine mollusc *Aplysia*, carried out by Kandel and his colleagues. An excellent review of this work is provided by Kandel and Schwartz (1982), and a theoretical paper that extends principles from the work on sensitization to more complex forms of learning in *Aplysia* is provided by Hawkins and Kandel (1984). Although much of what is presented by these workers is hypothetical, their project is without precedent in neuroscience as an example of the power of the "model system" approach, that is, the thoroughgoing investigation of the interrelations of behavior, neurophysiology, biochemistry, and biophysics, for a single instance of behavioral modification.

Sensitization is a form of behavioral plasticity defined as an increase in responsiveness following the presentation of a noxious (sensitizing) stimulus. For example, if a noxious stimulus is presented to the tail, the withdrawal response that *Aplysia* normally makes to a touch of the siphon shows an increased amplitude. The increase persists for minutes or hours depending on the amplitude of the sensitizing stimulus, and repeated applications of the sensitizing stimulus prolong the effect for weeks (Figure 11.2).

It had been well established that the critical changes underlying sensitization occur at the synapses of sensory neurons upon interneurons and upon motor neurons that drive the behavioral withdrawal (e.g., Castellucci et al., 1970; Byrne, Castellucci, and Kandel, 1974). The increased excitability of the interneurons and motorneurons following a sensitization stimulus and the progressive decrease of excitability that accompanies repeated elicitation of siphon withdrawal (habituation) have both been shown to result from changes in the amount of neurotransmitter released by the sensory neuron terminals. Recently, electron microscopy has suggested that the source of the sensitizing effect is a group of modulator neurons whose cell bodies are in the Aplysia abdominal ganglion, and whose axons terminate at presynaptic sites near the terminals of the sensory neurons. Several lines of evidence indicate that the modulator neurons release serotonin, and that it is the activation of cyclic adenosine monophosphate (AMP) in the sensory neuron terminals, triggered by serotonin, which is responsible for the enhancement produced by the sensitizing stimulus (Klein and Kandel, 1980). The effect of the serotonin-activated cyclic AMP is to enhance the voltage-dependent calcium ion influx into the sensory terminal (i.e., the influx that normally follows the invasion of the terminal by the action potential) and that is known to be critical for transmitter release. Further studies using the voltage clamping technique indicate that the increase in calcium influx is not a direct effect of cyclic AMP on calcium ion channels. Instead, cyclic AMP mediates a lessening of the flow of potassium ions through their channels. The potassium flow normally contributes to the repolarization of the membrane (i.e., the falling phase of

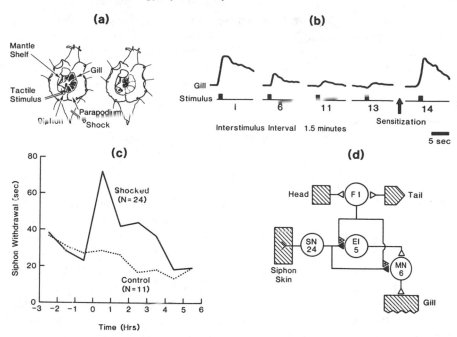

Figure 11.2. Sensitization elicited in the *Aplysia*. (a) Method of elicitation: A tactile stimulus (water jet) to siphon causes gill-withdrawal reflex. A shock to the tail serves as the sensitizing stimulus. (b) Photocell recording displaying reduction of response with repeated tactile stimulation (habituation) and increased response when the tactile stimulus is preceded by a noxious (sensitizing) stimulus (shock, represented by arrow). (c) Time course of sensitization: After a single, strong, electrical shock (at time zero), gill withdrawals were significantly longer than those of controls for up to 4 hrs. (d) Neural circuitry controlling gill-withdrawal reflex. SN, sensory neuron; MN, motor neuron; EI, excitatory interneuron; and FI, facilitating interneuron. (Adapted from Kandel and Schwartz, 1982.)

the action potential) following a spike. Thus, the restriction of the potassium channels is tantamount to the prolongation of the spike-induced depolarization of the sensory neuron terminal (i.e., the condition that normally induces calcium influx and transmitter release). Since it is widely believed that the action of cyclic AMP almost always involves protein phosphorylation, Klein and Kandel proposed that the serotonin-stimulated increase in cyclic AMP induces the dissociation of regulatory subunits from a cyclic AMP-dependent protein kinase in the sensory neuron terminals. The catalytic subunit of the protein kinase would then phosphorylate the potassium channel protein to inactivate the channel and therefore to slow membrane repolarization.

The serotonin-driven changes in cyclic AMP, and its effects on potassium and calcium currents provide an account of short-term sensitization. Kandel and his colleagues (e.g. Hawkins and Kandel, 1984) have invoked the same basic set of actors, with some additions, to account for other learninglike phenomena manifested by *Aplysia*, including long-term sensitization, habituation, as well as single-stimulus and differential classical conditioning. Again, the studies of Kandel and his colleagues are without precedent in providing a complete causal account of the biological substrates of plasticity. Yet, again, it is important to bear in mind the hypothetical nature of the account. At every link in the postulated causal chain, there are possible branches to alternative pathways not envisioned by the stated account, which nevertheless could be the "true" path to plasticity. Some specific alternatives to the proposed pathway in Aplysia are suggested by Routtenberg (1984).

Mammalian behavioral models

Resolution of the question of the precise causal path involved in LTP or in *Aplysia* sensitization may prove to be of little importance in relation to the substrates of learning and memory if the chief worry about these approaches, that the mechanisms revealed are unrelated to processes of human learning and memory, proves to be justified. In this regard, reductionist studies investigating well-defined learned behavior in vertebrate animals seem more likely to provide descriptions of substrates analogous to those that really operate in the human brain. Of course, the disadvantage of vertebrate models has already been confronted during the history of the reductionist studies – the overwhelming complexity and inaccessibility of the neural substrates in vertebrate and mammalian animals. Nevertheless, certain recent approaches, described next, appear to have avoided the obvious pitfalls discussed previously.

Experience-dependent changes in morphology of cells in vertebrate nervous systems. If, as the opening statement of this chapter suggests, learning mechanisms represent finetuning of behavior to meet environmental demands too unpredictable to be encoded genetically, it may be reasonable to speculate that changes in cellular morphology that represent substrates of learning and memory may be similar to the changes that occur during ontogenetic development. Perhaps learned behavior and memory are the products of continual elaborations and repeated rearrangements of the axonal and dendritic extensions that represent the very endpoints of neuronal ontogeny.

There have been many proposals in the history of neuroscience about possible morphological changes underlying information storage, including such things as changes in the numbers and shapes of dendritic spines and changes in the size, shapes, and numbers of synapses. Among the first empirical approaches to this issue was the classic work on the effects of exposure of young rats for varying periods of time (e.g., the third through the eighth week of postnatal development) to "enriched environments," that is, play areas containing brightly colored toys, objects of various sizes and shapes that could be contacted physically, moved, and so forth. The controls in these

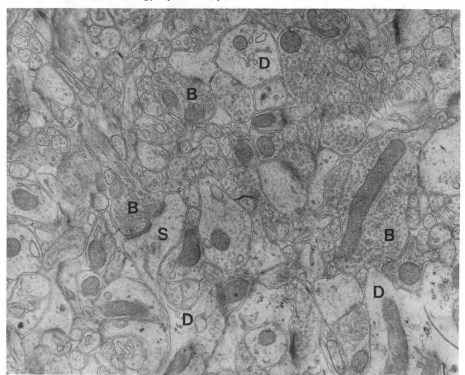

Figure 11.3. Sample electron micrograph used for examining changes in spines and synapses as a result of rearing in an enriched environment. S, spine; D, dendrite; B, bouton.

studies were kept in standard laboratory cages and received only handling experience that is comparable to that of the enriched rats. Studies of this kind have shown enrichment to be associated with several interesting changes in the nervous system, such as increases in the weight and thickness of the cerebral cortex (Bennett et al., 1964) and increases in the size of neuronal cell bodies and the number of glial cells (Diamond, 1967).

More recently, the magnification factor has been increased by use of a histological preparation of brain tissue, called the Golgi technique, and electron microscopy [recent reviews are provided by Greenough (1984) and by Greenough and Chang (1985)]. With the Golgi method, a dye injected into brain tissue is selectively taken into the intracellular space in a small proportion of the neurons occupying the site of injection. It is not clear why only some neurons take up the dye, but because that is true, it is possible to visualize under a standard microscope many if not all of the parts of the filled neuron including the cell body, dendrites, axon, and so forth (see Figure 11.3). This stark outline of the cell's shape then permits the measurement and quantification of its structural properties. Electron microscopy provides an even greater magnification, permitting measurement of such things as the size of synapses and the number of transmitter-containing vesicles in a synaptic terminal.

Several prior reports had indicated increased numbers of cortical dendritic spines and amounts of dendrites per neuron, using light microscopic assays of tissue stained using the Golgi technique. Assuming that the extra dendrite is actually innervated, Greenough postulated that there ought to be a greater number of synapses in the enriched animals. To test this idea, estimates were made of synapses per neuron, using a computer-based system for double-blind entry of data from electron photomicrographs of brain tissue and stereological corrections of estimates required when volumetric measures are based on two-dimensional (tissue section) samples. His results indicated that the animals given enriched experience had about 20% more synapses per neuron than controls in the upper layers of the occipital cortex. These results suggest that one way in which experience is encoded in the nervous system is by the formation of synapses.

In order to assess more specifically whether synapse numerosity changes during learning, Greenough and his co-workers have examined tissue from rats trained for 3–5 weeks in a variety of maze problems. In one of the experiments, the rats received "split brain" surgery (cutting the corpus callosum to impair interhemispheric communication) and eye occluders during maze training to localize the maze experience to one hemisphere. The results showed that some but not all of the changes that occur during the enrichment experience also occur during maze training. Specifically, the apical (upper) dendritic fields of layer IV and V occipital cortical pyramidal neurons were more extensive in trained than in merely handled rats. These effects occurred only in the trained hemispheres of the split brain rats trained with one eye occluded. Layer IV stellate neurons, which are affected by enrichment, were not affected by maze training.

Greenough contrasts synapse formation with protein synthesis, a process frequently invoked by investigators of the LTP effect and of invertebrate behavioral plasticity. He argues that synapse formation is likely to be involved in long-term storage of information. However, given the very short lifespan of proteins, protein synthesis is considered a reasonable candidate only for short-term storage and/or the kind of transient "learning" (e.g., sensitization) manifested by many invertebrates.

The specific changes brought on by maze training quite possibly represented engrams (i.e., they may have represented some facet of the information needed for the performance of the maze habits). Yet, there is no way to be sure that the changes revealed by these studies represented learned information, nor is proof claimed by the investigators. It is well recognized that the effects may reflect trophic changes produced by such things as increased utilization of sensory and motor processing mechanisms, altered activation levels, and so forth – effects that are indeed induced by training, but not directly relevant to the storage of learned information. Yet, perhaps at the present stage of the development of knowledge in this area, proof should not be the expected outcome. Instead, what is important about these data is their potential to become linked with the ever expanding body of developmental neurobiology, to provide a full description of the continuum of genetic and experience-driven changes in the structures of neurons. As this field matures, it will yield a menu of neuromorphological candidates for changes underlying learning. Thus, the approach will ultimately provide the empirical constraints needed to select reasonable candidates from the plethora of logically possible forms of change. This advantage does not accrue to artificially induced forms of plasticity such as LTP, since it is not known whether such forms actually function in behaving subjects.

Reductionist models of learning in birds. The discovery of hemispheric asymmetry of the neural control of human behavior was followed by a substantial period in which such asymmetry was regarded as a hallmark of the phylogenetically advanced human nervous system. However, in recent years, there has been an increasing recognition that functional asymmetry and lateralization is by no means the exclusive province of humanity or even of the primate order. A rather substantial body of literature now exists demonstrating asymmetry in subprimate mammalians and in birds (see reviews by Walker, 1980; Denenberg, 1981). Moreover, some of the most intriguing instances of asymmetry in birds have been uncovered by recent investigations of neural substrates of learning. For example, studies have indicated hemispheric lateralization of the neural system for the production of the acquired vocalization pattern in canaries (Nottebohm, 1977), visual discrimination learning in a task requiring chicks to search for food (Rogers and Anson, 1979), and the system involved in the acquisition of imprinting behavior (Horn, 1981; Bradley and Horn, 1981). Very recently, S. P. R. Rose and colleagues have demonstrated morphological changes in the medial hyperstriatum ventrale [(HV) a region of the chick forebrain] following the acquisition of passive avoidance behavior in newborn chicks (Stewart et al., 1984), that is, learning to avoid pecking at bright beads coated with unpleasant-tasting substances. Following training, the volume density of the presynaptic terminals, the number of synaptic vesicles per unit volume of neuropil, and the mean number of vesicles per terminal all increased. The increases were greater in the left than in the right hemisphere. An additional parameter, the length of the postsynaptic thickening, decreased with training, and this occurred preferentially in the right hemisphere, that is, in this instance, the hemispheres were asymmetrical before training, but they became equivalent after training.

Prior to the demonstrations of the lateralized morphological changes, rather extensive work had been done by Rose and others demonstrating several molecular changes in the chick HV consequent upon two varieties of learning, passive avoidance training and imprinting. The changes included transient alteration of cholinergic receptor binding (Rose, Gibbs, and Hambley, 1980; Alexsidze et al., 1981), increased incorporation of the amino acid leucine into the protein tubulin (Mileunsnic, Rose, and Tillson, 1980), and the incorporation of fucose into synaptic membrane glycoproteins (Rose and Harding, 1984). Moreover, the HV was the site of a dramatic increase in metabolic activity as indicated by the uptake of radioactive glucose (2-deoxyglucose) following passive avoidance training (see Stewart et al., 1984). Telencephalic ablations that included part of this region produced a deficit in avoidance learning (Benowitz, 1974). Several additional metabolic and electrophysiological alterations in this region were shown to accompany the acquisition of imprinting behavior (Horn, McCabe, and Bateson, 1979: Horn, 1981). Thus, an extensive amount of information has been obtained, at several biological levels, in regard to two forms of learning in chicks. With these data now in hand, it becomes feasible to attempt to integrate the observations made at the different levels. One possibility for doing this involves going back to examine which if any of the biochemical and electrophysiological changes is lateralized in parallel with the morphological changes. Any subset of the changes manifesting lateralization in parallel would, in combination with the morphological changes, quite possibly represent a unitary subsystem in the chick HV that subserves a specific functional component of the learning process being indexed in these studies. In general, these results illustrate the power of the "model system" approach to the study of the neural mechanisms of behavior, i.e., the extensive study of neural substrates at many levels of analysis, for a single form of learning in a single species.

Overview of the reductionist approach

The immediate objective of the reductionist studies is to contribute to the identification of the necessary and sufficient causal substrates of a particular form of plasticity, such as sensitization, habituation, classical conditioning, LTP, imprinting, passive avoidance learning, maze learning, and so forth. However, this statement fails to define the full rationale for these studies, for it is clear that the practitioners of the approach intend more than the identification of substrates for these highly specific forms of plasticity. In addition, they believe that the substrates identified will turn out to be substrates for many other, and perhaps all forms of plasticity and learning. For example, Kandel's rationale for the analysis of behavioral plasticity in *Aplysia* was at the outset based on the assumption of a strong likelihood that the invertebrate findings would generalize to the mammalian system (see Kandel, 1976). Similarly, Lynch and Baudry (1984) regard the explanation of LTP in hippocampus as applicable to "telencephalic" learning processes, such as are deficient in humans with damaged limbic systems, and such as are revealed by rats' inability to master radial mazes after hippocampectomy. Greenough views synapse formation in pyramidal neurons of the visual cortex as possibly essential to the formation of long-term memory, presumably in all mammalians It is clear from comments by Routtenberg that the phosphorylation of protein F1 is believed to be a general substrate of learning, and Alkon (1984) has argued that there exists a general neurobiological mechanism (involving calcium-induced lessening of potassium conductance across membrane channels) that subserves all forms of learning and plasticity. Thus, the ultimate as opposed to the immediate objective of the reductionist studies would appear in each case to approximate the historical objective (i.e., to find the magic bullet, the biological "essence" of learning).

One problem with the claims, explicit and implicit, that there is a single or just a few essential biological substrates for learning is that there is to date very little agreement as to what the substrate or substrates may be. Of course, variation of implicated substances is expected by virtue of the different levels of analysis addressed by different studies. Thus, one can hardly expect studies of morphological changes revealed by the Golgi staining method to draw the same conclusions as drawn by studies of protein synthesis. These studies measure different aspects of plasticity, but the view favored by the reductionist philosophy is that they reflect different aspects of a common mechanism. Nevertheless, if there does indeed exist only one or a limited set of substrates for plasticity, then certainly the conclusions from studies that measure identical or similar substrates should converge. Yet, although a case has recently been made for the operation of certain common elements as substrates of plasticity (Alkon, 1984), it is nevertheless true that each of the studies, reviewed in this section on reductionist studies, which attempt to describe the biochemical substrates of LTP and sensitization, has yielded its own substrate, distinct and separate from those yielded by the other studies. The absence of convergence for these relatively simple phenomena suggests that if one or a few critical substrates exist, none or only one of the projects has made contact with it. At the very least one would hope for agreement on relatively general issues such as whether plasticity is primarily a province of presynaptic mechanisms as indicated by the work of Kandel and his associates, or of postsynaptic mechanisms, as indicated by Lynch and Baudry.

The functionalist approach

Rationale

As an alternative to the reductionist position, one could deny the assumption of one or a limited set of general plastic processes, arguing instead that each category of learned behavior is a product of multiple interacting brain systems, each system functioning in the service of a particular component of the operative learning program. Indeed, the study of learning processes by experimental psychologists since the turn of the century has demonstrated the existence of various categories of learned behavior and memory, such as conditioned responses acquired through the use of classical and instrumental procedures, acquisition of skill, episodic "data-based" memory for events and situations, schemas, conditioned motor and emotional responses, and so forth. In addition, behavioral studies of learning indicate that particular categories of learning are made up of component processes that interact to yield the acquisition and expression of the learned behavior. For example, there are reasons for believing that the acquisition of instrumental avoidance behavior is based on two component processes, an emotional response of "fear," classically conditioned to environmental circumstances that predict aversive stimulation, and the instrumental learning of adaptive skeletal responses to avoid the aversive stimulus (Mowrer, 1960; Rescorla and Solomon, 1967; Solomon and Corbit, 1974). Other evidence supports the existence of sets of subprocesses such as memory encoding and retrieval, working memory and reference memory, automatic memory encoding, behavioral and memorial priming, motor learning, the associative linkage of contextual stimuli to the neural mechanisms for stimulus selection and response priming, the long-term storage of information, and so forth. Each of these subprocesses may be further decomposable. Ultimately, one would arrive at a level of subprocess that is definable in terms of the functioning of particular brain circuits, and some or all of the neurophysiological and biochemical actors in each circuit may manifest plasticity in the service of the general learning program that is in operation. Rather than conducting a search for the single substrate, it may be more useful to conduct searches for neural circuits and systems that correspond to the component processes that operate for a given instance of learned behavior. This it turns out is the goal of what we have referred to as the functionalist approach.

In fairness to the reductionist position, it must be noted that the existence of categories and components of learned behavior does not rule out the possibility that there are one or two fundamental biological substances that are essential for all instances of learning, regardless of the category or component. Indeed, the discovery of a general substance would be of enormous importance, particularly in relation to the clinical restoration of memory and cognitive functions. Yet, the functionalists would argue that such a discovery does not seem imminent, and even if it were to occur it would not take us very far toward knowing how learning works in the brain because it would not reveal the interactions of the neural circuits in which the component processes that complexly govern learning and learned behavior are instantiated.

Historical background

Present knowledge of the categories and components of learning and memory is largely a product of experimental psychology. Therefore, the entire tradition of this discipline in a sense represents a major part of the historical context relevant to functionalist studies of the psychobiology of learning and memory. Nevertheless, two relatively recent events mark the beginning of meaningful contact between experimental psychology and neuroscience in relation to research concerned with learning processes. One of these was Karl Lashley's historic 30-year search for the "engram" (Lashley, 1950), beginning in the early 1920s. This work involved extensive study of the effects on learned behavior in various paradigms of experimentally induced brain damage in rats and primates (Lashley, 1929). The second occurred in the mid-1950s when Scoville and Milner (1957) began to report on the profound disturbances of memory that resulted from bilateral temporal lobectomy in epileptic patients.

Bitemporal amnesia. These patients seemed as able as unoperated patients to recall information stored before the surgery, that is, they did not suffer retrograde amnesia. Moreover, they were able to store information for brief periods (i.e., short-term memory appeared to be intact). However, if distracted, they experienced difficulty in recalling information acquired just before the distracting event. Thus, for example, a trip to the drinking fountain would serve to prevent recall of the name of a physician to whom the patient had just been introduced. These observations and more formal assessment of the memory deficiencies led Milner to conclude that the brain damage had produced total anterograde amnesia (i.e., an inability to form new permanent memories). This conclusion led to the further hypothesis that the damage had disabled the process of consolidation (i.e., the ability to convert short-term labile memory traces, based on reverberatory patterns of nerve impulses, into long-term memory traces, based on permanent structural change in the brain). Analysis of the surgical records revealed the severity of anterograde amnesia to be correlated with the extent of damage to the hippocampal formation, a limbic system structure buried within the temporal lobe. This work seemed to indicate that the essential key to understanding the brain's memory process had been found. What was needed was a detailed analysis of the hippocampal mechanism for mediating memory consolidation. Moreover, the conclusion that the memory deficit was not a global memory loss, but rather a specific deficit of consolidation, implied that distinct neuroanatomical substrates would soon be found for the remaining components of memory (e.g., the repositories of short-term and long-term memory). These arguments provided a compelling rationale for the pursuit of a functionalist strategy.

Neuropsychology of the infrahuman hippocampus. In the wake of the findings of Scoville and Milner there followed a major research thrust toward a detailed analysis of the imputed hippocampal mechanism of consolidation (reviewed by Douglas, 1967; Kimble,1968; Altman, Bruner, and Bayer, 1973; Gabriel et al., 1980b). Much of this work followed the tradition established by Lashley wherein the acquisition and performance of learned behavior were observed in animals (rats, cats, dogs, etc.) with surgically induced experimental damage in the hippocampus and related limbic structures. The principal goal of these studies, initially, was to establish that hippocampal damage would impair the consolidation mechanism. Following this, it was assumed that a more detailed neurochemical and neurophysiological analysis of the mechanism could be carried out.

It soon became apparent that the simple demonstration of a hippocampal role in consolidation was not going to be immediately forthcoming. Performance of learned behavior was indeed impaired by damage to infrahuman limbic structures. For example, a deficiency in the acquisition of aversively motivated behavior has frequently been observed to follow damage to the cingulate cortex (Kimble and Gostnell, 1964; Lubar and Perachio, 1965; Ursin, Linck, and McCleary, 1969). Moreover, bilateral damage in the hippocampal formation has repeatedly yielded performance impairment in complex mazes (Winocur and Breckenridge, 1973; Kimble and Dannen, 1977), serial patterning tasks (Kimble and Pribram, 1963), alternation behavior (Means, Leander, and Isaacson, 1971), and extinction (Jarrard, Isaacson, and Wickelgren, 1964). Common threads that linked many of these studies were the findings that animals with hippocampal damage were hyperactive and hyper-responsive, and they manifested a peculiar form of behavioral rigidity when the task rules or contingencies were altered as, for example, in reversal training. In such circumstances, they appeared defective in the detection of the new contingencies and in the reshaping of new behavior, as if the already formulated coping strategy precluded the development of a new strategy.

These results are compatible with the idea of limbic system involvement in learning and memory processes, but they did not compel the research community to adopt this conclusion. For one thing, animals with lesions of the hippocampal formation were capable of normal learning and retention in a variety of experimental situations, such as instrumental approach behavior (Means, Walker, and Isaacson, 1970), classical conditioning of many forms (Schmaltz and Theios, 1972; Solomon and Moore, 1975), instrumental avoidance learning (reviewed by Black, Nadel, and O'Keefe, 1977), and simple discrimination learning, such as standard Y- or T- maze learning of discrimination based on brightness or pattern cues (Kimble and Kimble, 1970; Isaacson and Kimble, 1972).

Behavioral insufficiencies in dealing with altered task contingencies notwithstanding, the apparently normal behavioral acquisition manifested in many tasks by damaged subjects was perceived as incompatible with the idea that infrahuman hippocampal function is essential to the formation of permanent memory, as argued by Milner. Moreover, other authors put forward carefully reasoned arguments based on the neuropsychological results that the hippocampal formation and related limbic structures contribute to the mediation of functions, such as arousal reactions and orientation, that occur to novel and/or meaningful stimuli (Grastyan et al., 1959; Routtenberg, 1971; Bennett, 1975), or nonassociative modulation (facilitation and inhibition) of behavioral responses (McCleary, 1966; Altman et al., 1973). Theories of this kind, considered collectively, seemed to gain greater acceptance than theories that postulated a role for infrahuman limbic structures in learning and memory. No single theory clearly held favor, and the decade of the 1960s ended on a confused and pessimistic note in relation to brain organization relevant to learning and memory processes. Whereas the case that the hippocampal formation and related limbic structures were critically involved in human learning and memory seemed virtually unassailable, the case that one could devise an animal model for human limbic memory processes was devoid of empirical currency.

Current functionalist research

Continuing work during the 1970s and 1980s has produced some new perspectives on the functional relevance of the limbic system in learning and memory processes. Although there is still no commonly accepted animal model for human limbic amnesia, progress toward understanding the apparent discrepancies between the effects of human and animal limbic damage has been made, and there is now a strong consensus that limbic structures in animals and in humans are involved in processing relevant to learning and memory. Moreover, the new consensus includes the view that the limbic contribution represents one or more component memory processes important in both human and nonhuman mammalian systems. The remainder of this chapter considers representative research that has led to the current viewpoint.

Newer perspectives on human limbic amnesia. The fact, well-established by research in the 1960s, that animals with damaged hippocampi are capable of manifesting several forms of learning, and other data discussed below, have led to a lessening of enthusiasm for the idea that the hippocampal formation is the neural substrate of the consolidation process. This idea has been replaced by other interpretations of hippocampal functioning to be discussed next.

Retrieval, not storage, impaired. Milner's hypothesis stating that the amnesia produced by bilateral temporal lobectomy was a manifestation of failure of consolidation was based on her conclusions that the patients manifested total anterograde amnesia (i.e., inability to form new memories after surgery), along with a complete absence of retrograde amnesia (i.e., normal memory for events prior to surgery). Both of these conclusions have been questioned. Thus, Weiskrantz and associates have shown that, when special procedures are used, storage of information in learning situations can be demonstrated in amnesic patients who normally yield total anterograde amnesia. Patients show above chance recall of previously experienced words when they are given recall "hints" or cues, such as partially reconstructed printed versions of the words (Weiskrantz, 1970). These data have been interpreted to indicate that information is stored by amnesics and is available for recall but that the recall process is disturbed. When that process is given a "leg up" with the use of recall cues, stored items can be brought forth.

Additional studies have indicated that when amnesics do manifest recall, previously stored information tends to be recalled inappropriately in place of more recently stored information (Starr and Phillips, 1970; Warrington and Weiskrantz, 1973). For example, amnesic patients have a higher incidence of intrusion errors when recalling various lists acquired in sequence. Any procedure that improves the distinctiveness of the items to be recalled or the list membership of the items (e.g., providing cues for the appropriate items or rendering the various lists distinctive by composing them of items in the same semantic category) makes the performance of amnesics look more like the performance of normal subjects.

The idea that the patients lack retrograde amnesia (i.e., had normal memory for events that occurred a long time before surgery) was originally based on anecdotal reports and rather casual observation. This was not due so much to neglect as it was to the difficulties in finding a reasonable estimate of normal performance in attempting to compare remote memory performance in various patient populations. In order to deal with this issue, Warrington and Sanders (1971) administered a "retrospective questionnaire" to 300 normal subjects 40 to 80 years of age. The questionnaire contained items relating to particular years sampled over a 40-year period. A group of 5 amnesic patients showed a substantial impairment relative to the normal group, for all of the historical intervals tested. The amnesics did not perform at a better than chance level on the multiple choice items of the test. These results indicate that it is not only the ability to form new memories, but also the recall of information stored long before surgery that is impaired in patients with temporal lobe damage.

Since problematic storage predicts anterograde but not retrograde amnesia, Weiskrantz (1978) proposed that the memory problems of the patients do not result from deficient storage (i.e., consolidation), but rather from deficient retrieval, i.e., the ability to get information out of storage and to communicate it to the outside world.

Inadequacy of retrieval theory. A positive outcome of the debate over the storage/retrieval issue has been the direction of attention to the complex nature of the memory process and the need to address models that characterize the subprocesses of memory. In addition, the retrieval interpretation itself, and the data that give rise to it, have offered some glimmer of hope for bringing together the animal and human literatures relevant to the effects of brain damage on learning and memory: If the difficulty with damaged patients is one of using stored data, rather than one of storage itself, then it is not so disturbing to find certain forms of behavioral acquisition unimpaired in hippocampectomized animals. Moreover, the data indicating that errors made by amnesic patients are often of the intrusive variety is reminiscent of the difficulty that hippocampectomized animals have in changing already learned behavior in response to altered task contingencies. Nevertheless, serious conceptual problems arise when one attempts to account for the full spectrum of data pertaining to the amnesic syndrome using the retrieval theory alone. Thus, a pure retrieval theory predicts that all recall in amnesics will be impaired, including recall of information stored long before, just before, immediately after, or long after the amnesia-producing event. Yet it is clear by any measure that memory of pretraumatic experience in amnesics is substantially superior to that for postraumatic experience. Moreover, a pure retrieval theory cannot explain the occurrence of "gradients of remote memory" (i.e., impaired recall of events experienced 2 or 3 years before the amnesia-producing event) and relatively intact recall of events experienced long before (e.g., 5–10 years) the amnesia-producing event. Evidently, the recallability of stored information undergoes continuing improvement, perhaps in direct relation to the frequency of its use, following initial storage. In any event, it can be seen that neither pure retrieval nor pure storage theory adequately account for human limbic amnesia.

This impasse had led to the emergence of the idea that the storage–retrieval dichotomy may not be a useful conceptual basis to account for the amnesic syndrome. Instead, the focus of research has shifted from this distinction to a consideration of other component processes of the memory system.

Limbic and nonlimbic memory systems. The juxtaposition of abnormal and normal learning processes is one of the most striking features of the amnesic patients. Thus, whereas their performance in uncued recall of word lists is very poor, they show no impairment in learning of motor skills and solution of puzzles (Brooks and Baddeley, 1976) or reading reversed images of printed words (Cohen and Squire, 1980; Moscovitch, 1982). Each of these behaviors involves the acquisition of a kind of skill and each is relatively gradually acquired. The learning curves manifested by amnesics in these performances are indistinguishable from those of controls, even though recognition of the learning situation and the awareness that the skill has been acquired are absent. These data have led to the idea that the brain has distinct learning systems, one involving the hippocampus and related limbic structures and the other based on nonlimbic mechanisms. Moreover, various suggestions have been made as to the general nature of the processes carried out by each system. Thus, for example, Cohen and Squire (1981) have suggested that the limbic and nonlimbic systems, respectively, subserve declarative and procedural memory, where declarative memory denotes "data-based" memory (i.e., memory for such things as places, events, and facts about the world) and procedural memory is the repository of rules and procedures for accomplishing particular goals. Others have proposed analogous distinctions as, for example, the distinction between episodic (limbic) and semantic (nonlimbic) memory processes (Kinsbourne and Wood, 1975), and the distinction between working (limbic) memory and reference (nonlimbic) memory (Olton, Becker, and Handelmann, 1979). These ideas are useful in that they provide a conceptual framework that more readily embraces the data of human limbic amnesia. Rather than attributing amnesia to loss of storage or retrieval, the newer views account for the facts of amnesia by saying that an entire memory system, such as the working memory system or the declarative knowledge system, is lost in limbic amnesia. Presumably the loss of such a system involves the loss of both storage and retrieval processes that operate within the confines of the system, but these views also allow for the existence of other memory systems (with storage and retrieval) that remain intact following limbic system damage.

Primate models. It is ironic that the principles of mass action and equipotentiality that Lashley derived from his 30-year search for the engram have led some to doubt the relevance of cytoarchitectural and connectional definitions of neural systems to functional categories of learning. The irony is that the continuing pursuit of Lashley's research strategy has led to an ever increasing recognition that even subtle anatomical gradients in the brain are associated with functional differences, when appropriate searches for those differences are conducted. Indeed, one of the most important lines of this work, involving experimental cerebral damage in monkeys, was initiated in the early 1950s by students of Lashley (Blum, Chow, and Pribram, 1950), and carried through to the present day by numerous investigators. The early work provided a clear indication of cerebral cortical localization relevant to learning processes by demonstrating that damage to the inferotemporal region of the cortex is associated with a specific deficiency in the acquisition and performance of discrimination behavior in tasks involving visual stimuli. Typically, in such tasks, rhesus monkeys are given a fixed number of daily trials (e.g., 20) in which one of two simultaneously presented objects must be picked up in order to obtain a reward (e.g., a raisin), or in which one of two panels illuminated with distinctive visual patterns is pressed to obtain reward. The same stimuli are presented repeatedly and the cue–reward relationship is maintained until a high level of discriminative performance is achieved. Performance in comparable tasks involving auditory, tactile, and olfactory discriminanda was unaffected by inferotemporal damage. Subsequently, other nonprimary cortical zones were implicated in the mediation of learned behavior involving auditory (Neff, 1961) and tactile (M. Wilson, 1957) discriminanda, supporting the principle that in primates there exist separate cortical fields essential for learning involving separate stimulus modalities. Controls in these studies have indicated that the observed deficits are associative ones, not due to impaired sensory and/or motor functions. Analogous findings have been obtained in studies with cats and rats as subjects (e.g., Warren, Warren, and Akert, 1961; Thompson, Lesse, and Rich, 1963).

A major contributor to the work with primates, Mishkin, and his colleagues have presented a theoretical synthesis that states that learning involving stimuli in a particular sense modality requires unimpeded flow of information from the sensory cortical areas for the modality, through a series of associational areas, to the telencephalic areas of the limbic system. The synthesis is based on many findings indicating that performance of a particular discriminative habit remains intact, provided that the connections from the sensory area (e.g., the visual or auditory cortex) critical for the processing of task discriminanda, to various limbic areas (e.g., entorhinal cortex, hippocampus, amygdala) are intact. This remains true even if the intact sensory (or limbic) component of the system is in one hemisphere and the complementary sensory or limbic component is in the opposite hemisphere, provided that the callosal or other commisural pathways permitting communication between the two are undamaged (Mishkin, 1979).

Many of the most recent studies from Mishkin's laboratory have been concerned with the limbic side of the sensory limbic pathway. The point of departure for this work was the discovery (Mishkin, 1978) of a severe learning disability due to combined damage of the hippocampus and the amygdala. (The effects of damage in either limbic structure alone were consistent with many previous reports in primates and in subprimate mammalians, indicating only minimal impairment.)

The behavior blocked by the combined lesions was performance in what is called "delayed nonmatching to sample." In this task, presentation and removal of a sample set of objects is followed after some seconds (5–60 sec) by a new set containing items from the sample set and one novel item. The task is to select the novel item in the test set. Significant impairment occurred in this task only for the larger values (e.g., more than 20 sec) of the delay.

This task is very different from the standard primate discrimination task in that the information needed to perform successfully changes from trial to trial as new samples are presented. Thus, a premium is placed on the loading, use, and unloading of information into memory, with relatively brief intervals between exchanges. This flexible use of rapidly changing storage has been termed "working memory." Other forms of learned behavior, particularly those described above involving repeated and consistent association of certain cues with reward and others with nonreward, are almost entirely unaffected by the combined lesions.

The fact that two limbic structures must be damaged in order to impair performance of delayed nonmatching to sample suggests that there may exist at least two subsystems underlying the limbic contribution to learning processes and that the contribution of each subsystem is somewhat redundant with that of the other. In summary, these studies suggest a somewhat redundant duality within the limbic system for the performance of working memory tasks and a larger duality in which the working memory system is contrasted with yet another system, presumably involving the nonlimbic neocortical areas, for the processing of stimuli that repeatedly predict a given outcome.

The deficiency of performance in the nonmatching to sample task is reminiscent of the difficulties that human patients with temporal lobe damage have in maintaining good records of recently encountered information in the face of distraction. Moreover, the unimpaired ability in animals with combined lesions to master discriminative problems in which there is repeated exposure to consistent stimulus contingencies has been likened to the unimpaired ability of the human patients to acquire skills, such as reverse mirror tracing and learning to read reverse images of printed words efficiently. Thus, although the fit is less than precise, it is generally agreed that Mishkin's primate model provides the best approximation to date to an animal model for human limbic amnesia.

Neurophysiological models: classical and instrumental conditioning. An increasingly influential approach to the understanding of the neural mechanisms of learning has been the application of the model system strategy to the study of mammalian classical and instrumental conditioning. A particularly important aspect of this work has been the emphasis on study of the firing of neurons in the brain during training in the awake behaving subject. This method is the only one currently available that permits one to obtain a millisecond by millisecond record of activity in specific fields and pathways of the brain during a training experience.

Among the most advanced projects of this kind are included studies of the substrates of Pavlovian conditioning of the cardiac response (D. H. Cohen, 1982; Kapp et al., 1982), the conditioned nictitating membrane response (NMR) in rabbits (Thompson et al., 1983), conditioned facial movements in cats (Woody, 1982), and discriminative avoidance behavior in rabbits (Gabriel et al., 1980a,b, 1985).

The nictitating membrane is the "second eyelid" found in many mammals, a translucent membrane that sweeps across the corneal surface in response to mechanical stimulation. It rarely moves spontaneously in rabbits, but it can readily be made to move by mechanically or electrically stimulating the region surrounding the eye. Gormezano and his colleagues have shown that the NMR can be brought under control of originally "neutral" stimuli, that is, it can be conditioned using classical and instrumental procedures. Furthermore, the NMR manifests all of the standard properties that characterize other conditioned responses (Gormezano, Kehoe, and Marshall, 1983).

Thompson and his associates have used the procedure of Pavlovian classical conditioning, the repeated presentation of a neutral tone conditional stimulus (CS) followed after 750 msec by the unconditional stimulus (UCS) of an air puff to the eye. The work with this model, beginning in the early 1970s, has been directed at discovering the essential neural change, or "engram," for the production of the conditioned NMR.

In the mid-1970s, Berger and Thompson reported some rather remarkable neuronal correlates of NMR conditioning in the hippocampus and in related structures (Berger and Thompson, 1978). In particular, hippocampal pyramidal cell discharges elicited by the air puff UCS increased dramatically after just two or three pairings of the tone with the UCS. With just a few more pairings, cellular firing began to anticipate the NMR, occurring to the CS just before UCS onset. Even later in the training the discharges occurred just before the conditioned NMR. Most remarkably, the temporal course of neuronal firing frequency just before the NMR almost perfectly mimicked the temporal course of the NMR as registered on a polygraph. These findings led Thompson and Berger to suggest that the engram for the conditioned NMR was localized in the hippocampus.

Studies of the effects of lesions have supported the idea that the hippocampus contributes to the performance of the NMR, since lesions have been shown to produce such things as impaired reversal learning (Berger and Orr, 1982), flattened generalization gradients (Solomon and Moore, 1975), and loss of certain "higher-order" phenomena of conditioning such as the blocking effect (Solomon, 1977) discovered by Kamin (1969). These changes are analogous to the deficiencies, described above, that occur when other learned behaviors are measured in hippocampectomized animals. Yet, the suggestion by Berger and Thompson that the hippocampus is the site of the NMR engram has been difficult to sustain in the face of several studies, carried out before and after the electrophysiological data were obtained, indicating that acquisition of the conditioned NMR is completely unaffected by total bilateral hippocampectomy. Thus, the hippocampal contribution to the NMR, while important, is not essential for basic acquisition and performance of the behavior.

More recently, a discovery has been made that may turn out to be the key breakthrough in terms of understanding the neural locus essential for NMR acquisition (Thompson et al., 1983; Desmond and Moore, 1982). Lesions in the medial dentate and interpositus nuclei (i.e., the deep nuclei of the cerebellum) and lesions in related areas of the brainstem completely abolish the conditioned NMR. Such lesions do not at all affect the reflexive elicitation of the NMR by the UCS. Thus, impaired motor expression can be ruled out as a factor in accounting for the deficit. The lesions are effective if given before or after acquisition, and the effect appears to be irreversible. An important feature of the data is the apparent specificity of the lesion site. A very small lesion, appropriately located unilaterally in the critical region, will obliterate only the homolateral NMR. Finally, neurons in the critical region manifest dramatic increases in firing to the tone CS in conjunction with the NMR training procedure.

These results have led Thompson and his colleagues to propose that synaptic changes (i.e., engrams) in the cerebellar deep nuclei and/or in associated regions of the cerebellar cortex may enable CS-elicited auditory volleys in the brainstem to trigger neuronal firing in the cerebellar circuitry, and the firing of this circuitry in turn may ultimately excite the motor neuronal pool in the vicinity of the sixth cranial nerve nucleus, the abducens, to elicit the NMR. In other words, the engram for the NMR resides in the cerebellar circuitry. If this hypothesis is validated, it will be the first localization of the engram for a mammalian learned behavior. It will also open the way for a reductionistic analysis of the synaptic changes underlying NMR conditioning. This possibility underscores the potential that exists for combining the functionalist and reductionist approaches in the final stages of analysis of a mammalian model system.

It should be noted that the indication from the work of Thompson and his colleagues that the hippocampus influences the performance of the conditioned NMR in special circumstances, but is absolutely nonessential for the acquisition and performance of the behavior, provides an instance of the general distinction, discussed above, between limbic and nonlimbic functional systems relevant to learning and memory. Further elucidation of the brainstem circuitry involved in mediating the NMR should contribute to our understanding of at least part of what is meant by the "nonhippocampal" learning system, the existence of which seems to be required to account for learned behaviors controlled by consistent and repeating stimulus contingencies (i.e., the kinds of behavior that survive hippocampal and amygdalar damage in primates).

A possible complication, in relation to the future elaboration of the engram for the NMR, concerns the validity of the proposed circuitry that Thompson and his colleagues claim is operating in the service of the NMR. The proposed circuit diagram is a simple one, involving convergence of CS-driven sensory activity upon motor neurons. All that must be postulated to produce this effect is an increased efficiency at one synapse (purported to be cerebellar). The consequent enhanced output of this synaptic junction is proposed to excite the motor neuronal pool enough to trigger the NMR. Of course, the elegant simplicity of this area is attractive; however, there are certain fundamental properties of conditioned behavior that do not "fall out" when this model is adopted. One of these is the well-demonstrated finding, originally noted by Pavlov, that contextual stimuli of the experimental environment exert a considerable degree of control over conditioned behaviors (e.g. Gabriel, 1970; Winocur and Olds, 1978; Spear et al., 1980; Balaz et al., 1982). Although there are, to the authors' knowledge, no direct demonstrations, it seems very likely, based on these results, that conditioned responses in trained subjects will not occur, or they will occur only very weakly, unless the subjects are located in the training environment, ensconced in the full experimental regalia, including the devices attached to the rabbits for stimulus delivery and recording. If this is true, then associative changes (engrams) essential to NMR performance are called for to account for the control exerted by context, as well as for the specific efficacy of the CS in the conditioned subject. Yet, in the simple conditioning model described above, presentation of the CS should under any circumstances elicit the NMR in the trained animal, provided that obvious requirements, such as normal awake sensory and motor functioning, were met. Thus, it is difficult to see where, in the simple model, such control by contextual stimuli could be realized. The same question can be raised with regard to other uniprocesses, "single synapse" models of mammalian conditioning (e.g., D. H. Cohen, 1982). Unfortunately, the experiment needed to demonstrate the essential role of the experimental context for the NMR has not been done. Moreover, if contextual control was demonstrated, there are no doubt several ways to alter the existing model in order to incorporate this property. Such alterations would complicate the neurophysiological analysis of NMR conditioning, but if contextual control was present, then a more complicated model would be called for.

A unique opportunity provided by the neurophysiologically oriented mammalian model systems, not yet widely exploited, is the possibility of combining "in vivo" neuronal electrophysiology and discrete lesions in the same subjects during learning. With this approach, the electrophysiological data provide a window on the development of associative changes in various brain systems.

The window then permits the assessment, with lesions, of the causal roles that discrete anatomical projections have in relation to training-driven changes. The behavioral changes associated with the lesions provide information on the functional relevance not only of the damaged tissues but also of the downstream neural systems that are disrupted by the damage. This combination of techniques has been used by Gabriel and his colleagues to study interactions and behavioral relevance of the systems that mediate discriminative avoidance behavior in rabbits (e.g., Gabriel et al., 1980a,b, 1983; Foster et al., 1980).

The rabbits in these studies are trained to hop to avoid footshock whenever they hear a tone (CS+). They also learn to ignore a second tone (CS−) differing in auditory frequency from the CS+ and not predictive of the footshock. The CS+ and CS− are presented in a random order, and the rabbits receive 120 trials daily (60 with each stimulus) until a criterion of behavioral discrimination is attained. The rabbits complete criterial acquisition in three to five daily sessions, and at this time they avoid the footshock on an average of 85% of the trials, and they respond to the CS− on less than 8% of the trials.

These procedures establish the CS+ and CS− as signals for danger and safety, respectively. This acquisition of signal meaning occurs gradually, and is indicated generally by the acquisition of discriminative behavior. The basic experimental question that these studies poses is: In which brain systems does one observe the development of different (discriminative) neuronal firing patterns to the CS+, relative to the CS−? If discriminative firing does develop, it means that the brain systems manifesting it are in some way involved in processing the differential meanings of the signals acquired through learning. Thus, by studying the neuronal correlates of training, one obtains a preliminary "map" of the brain indicating the systems that participate in the acquisition of knowledge of stimulus significance. Following this, hypotheses regarding the causal antecedents of the discriminative activity in a particular system and the neural and behavioral consequences of the discriminative activity can be tested.

The initial studies with this model system concerned the neuronal activity in a triad of limbic structures [the posterior cingulate cortex (area 29), the anterior ventral nucleus (AVN) of thalamus, and the hippocampal formation]. Each of these areas is interconnected, monosynaptically and reciprocally, with the remaining two. Discriminative neuronal discharges [i.e., greater discharges to the positive CS (CS+) than to the negative CS (CS−)] developed in the deep cortical layers (5 and 6) in the first training session. The AVN and the upper cortical layers (1–4) developed discriminative firing later in training, when the discriminative behavior was first performed at its maximal (asymptotic) level (Gabriel et al, 1980a; Figure 11.4). At this late training

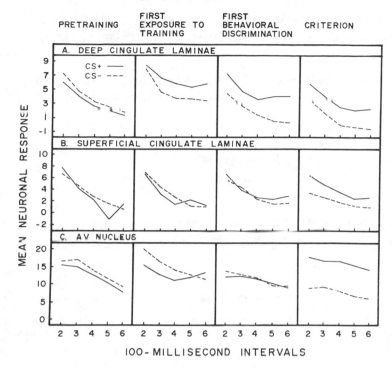

Figure 11.4. Mean neuronal response to CS+ and to CS− in successive stages of training recorded from deep and superficial laminae of the posterior cingulate cortex and anteroventral (AV) nucleus of the thalamus. (Adapted from Gabriel *et al.*, 1980a.)

stage, the cortical records that rapidly discriminated showed a diminution and/or loss of the effect (Foster et al., 1980).

A theoretical model based on these results provides a "sample product" of the functionalist neurophysiological approach (Gabriel et al., 1980b). It is at once a conceptual model, but stated in terms that include real neural entities, whose activities provide the empirical support for their inclusion in the theory. The model labels the discriminative activity as a neural code for the associative significance of the CS. The code, it is postulated, is a necessary but not a sufficient neural precursor of the behavior. The early-developing activity in the deep layers of area 29 is the *initial* code for significance. With continued training, the coding function is "relegated" to the AVN, an effect of the repetition of inputs to AVN via the corticothalamic pathway. Thus, thalamic neurons are "taught," via cortical "instructions," to produce their own discriminative code. Once acquired, the discriminative discharges of AVN neurons drive neuronal firing in the upper cortical layers, accounting for the late-developing activity therein. This input "informs" the cortex that

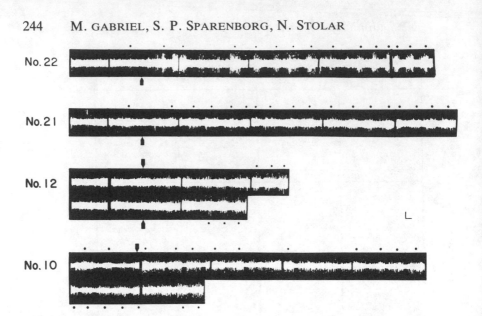

Figure 11.5. Oscilloscope photographs of hippocampal multiple unit activity from four rabbits during a training trial. Pointers indicate CS onset. A behavioral conditioned response occurred at the end of each trial; dots indicate neuronal bursts (defined as a cluster of at least three spikes). Horizontal calibration mark represents 50 msec and vertical mark represents 20 μV. (From Gabriel and Saltwick, 1980, reprinted with permission.)

the thalamus has acquired the code, promoting cortical *disengagement* from the task.

A second necessary condition for the behavior is the "priming" of the learned response. Priming, a tonically enduring acquired modification wherein the behavior to be performed is placed into a mode of readiness, is initiated upon introduction of a trained subject to the training environment. The stimuli that elicit priming are the background "contextual" stimuli of the training environment. The sites of response priming are assumed to be the neostriatum and/or the pons, projection targets of the limbic cortical and thalamic structures likely to be involved in response elicitation. If a response has been primed during training, CS-elicited volleys of code projected via the corticostriatal and/or the corticopontine pathway release the response.

Hippocampal neurons produce rhythmic, theta-like bursts increasing to 10/sec just prior to the avoidance response (Gabriel and Saltwick, 1980; Figure 11.5). Axonal inputs from the hippocampal formation bring about tonic suppression of the firing of neurons in area 29 and AVN. As this effect is being exerted, the hippocampus also receives polysensory information about the training environment (i.e., context), from the entorhinal cortex. When the CS + and the training environment are both present, the burst frequency increases, signifying a progressive removal of suppression by the hippocampus. This allows the cingulate cortical and/or AVN codes to be projected to the priming system, to release the behavior.

Hypothesis testing. Following the basic observations and model formulation, studies were undertaken, using a combination of electrophysiological recording and discrete CNS lesions, to test specific hypotheses derived from the model. The first tests concerned the function of the AVN.

A most intriguing finding seen in several studies is development in AVN of a large discriminative response, *after* the development of discriminative behavior. As the late-developing discriminative activity develops in AVN, the activity in the superficial layers of area 29, which receive axonal terminations of AVN neurons, also becomes discriminative for the first time. These results suggested that the late activity in the superficial layers is produced by direct driving exerted by the discriminating neurons of the AVN. Moreover, prima facie, the late developing activity cannot be viewed as a neural code for the early developing discriminative behavior of the rabbit. Instead, we proposed that it subserves maintenance and retention of the well-learned behavior. To assess these hypotheses, lesions of AVN were made after behavioral acquisition but prior to a test for retention given 8 days after the lesions (Gabriel et al., 1983). The lesions eliminated behavioral responding to the CS+ during the extinction phase of the retention test. Both control rabbits and rabbits with lesions were able to reacquire the original discrimination and the reverse discrimination. However, the performance of the rabbits with lesions was significantly impaired relative to that of controls during these task phases. These results supported the hypothesis that the late discriminative activity in AVN contributes to performance and retention of the well-learned behavioral response.

In addition, the AVN lesions abolished the CS-elicited neuronal discharges that normally occur in area 29. The loss of elicited discharges indicates that the thalamocortical projection is a major source of excitatory input to area 29.

Further work supported the idea of specificity of the functional relevance of the AVN to the well-learned behavior. Lesions in AVN given before training did not impair behavioral acquisition, although they did eliminate the excitatory CS-elicited activity in area 29. A similar behavioral effect – impaired performance of the well-learned behavior but not its original acquisition – followed lesions in area 29 directly (Ragsdale et al., 1984). Thus, the late-developing code in AVN and area 29 is not a trivial indication of function. One can eliminate its contribution either by eliminating area 29 or the AVN. In either case, the behavioral outcome is impaired performance of the well-learned behavior but not its original acquisition. A parsimonious account of these results is provided by the idea that the late discriminative code in area 29 *is* the AVN code relayed through area 29 enroute to systems for response production.

The role of the hippocampal formation. The literature referred to in the section on the neuropsychology of the mammalian hippocampus indicates that damage in area 29 is associated with deficient movement initiation, and a damaged hippocampal formation is associated with a hyper-responsive syndrome. Based on these data, several authors (Ursin et al., 1969; Altman et al., 1973) have postulated an antagonism between a cingulate cortical "go system" and a hippocampal "stop" system. These accounts represent the "nonassociative" explanations of the effects of limbic damage on behavior in nonhuman mammalians, mentioned above. Our model provided an associative, neurophysiological account of hippocampal functioning by virtue of its provision that the hippocampus tonically suppresses the activity of the neurons in area 29 and the AVN while monitoring contextual stimuli. The buildup of theta-like bursts manifested by hippocampal neurons after CS + onset, in anticipation of the behavioral response (Gabriel and Saltwick, 1980), reflects a progressive withdrawal of the suppressive effect, as "evidence" accumulates that the context is the appropriate one for the behavior. As hippocampal suppression is withdrawn, the cortical and thalamic codes projected to the neostriatal response priming system reach sufficient levels, and the primed response is released. Subjects with hippocampal damage are able to respond to the CS + . Indeed, because the code suppressive function of the entorhinal–hippocampal complex is lost, they respond too much, but they are lacking in capacity for a detailed assessment of the stimulus context and the attendant behavioral response inhibition mediated by the hippocampal system.

To test these hypotheses, the neuronal activity and behavior in animals with damage in the dorsal subicular complex and area 29 was studied (Gabriel et al., 1984a,b). The dorsal subicular complex is the region of the hippocampal formation that sends axonal projections to area 29 and to the AVN. The subicular lesions significantly increased the frequency of behavioral responses to the CS + and CS − . This effect was evident during training, and during extinction and reacquisition studied after the 8-day retention interval. At the neuronal level, the overall magnitude of the CS-elicited discharge in area 29 was significantly attenuated, and the expected early-developing discriminative activity was abolished in the subjects with subicular lesions (Figure 11.6). However, the overall level of firing in AVN was *enhanced* in the lesioned subjects (Figure 11.7). A similar enhancement of firing in AVN, as well as clear acceleration in the development of discriminative AVN firing, occurred following area 29 lesions (Figure 11.8). These results indicate that the early-developing discriminative activity and the normal magnitude of the CS-elicited discharge in area 29 depend critically upon information received from the

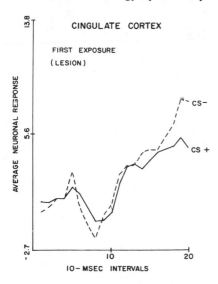

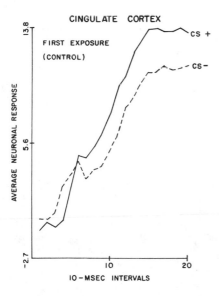

Figure 11.6. Average neuronal discharge to CS+ and to CS− during the first session of conditioning, recorded from the cingulate cortex in subjects with subicular lesions and in controls.

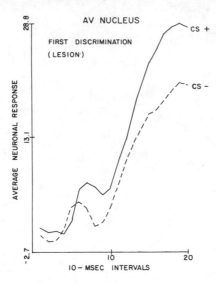

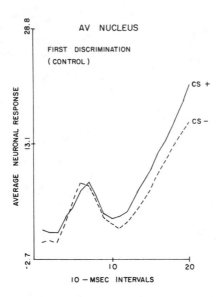

Figure 11.7. Average neuronal discharge to CS+ and to CS− during the session of the first significant behavioral discrimination. The records were obtained from the anteroventral (AV) nucleus of the thalamus in subjects with subicular lesions and in controls.

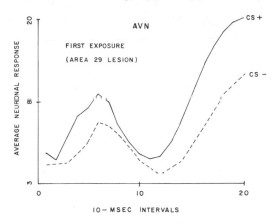

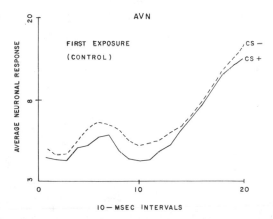

Figure 11.8. Average neuronal discharge to CS+ and to CS− during the session of the first exposure to conditioning; the records were obtained from the anteroventral nucleus of the thalamus in subjects with lesions in area 29 and in controls.

hippocampal formation. However, afferents to the AVN from the hippocampal formation and area 29 serve to suppress the CS-elicited firing of AVN neurons and to delay the onset of discriminative firing in that structure. Thus, the AVN does not develop discriminative firing in the late training stages because it is intrinsically slow to "learn." Moreover, the early-developing discriminative activity in the deep layers of area 29 does not "teach" the AVN how to discriminate between the CSs, as originally proposed. Rather, the late development of discriminative activity in AVN is the result of a suppression of firing and a retardation of discriminative encoding imposed upon AVN by hippocampal and cingulate cortical afferents.

Clarification of a past anomaly. The past studies of the sequence of discriminative coding in the limbic triad yielded a rather anomalous and not previously interpretable finding, that is, "reverse" discriminative activity in AVN in the very first conditioning session (second panel from left, bottom row, Figure 11.4). That is, CS− elicited a significantly greater discharge in AVN than the CS +. This effect was, of course, followed in the later training stages by no discriminative activity and, eventually, a large "appropriate" discriminative response. The anomalous finding becomes interpretable in light of the recent findings, as a direct consequence of the suppressive afferents from area 29. The fact that the suppressive input from the layer 6 projection neurons of area 29 is of a discriminative character in the early stages of training directly predicts the occurrence of "the reverse" neuronal discrimination in AVN.

Behavioral effects of the lesions: an account of a paradox. In addition to the unmasking of excitatory and discriminative activity in the AVN, damage in area 29 was associated with a significant reduction in the frequency of performance of the conditioned response. The impairment is of a similar magnitude to the impairment produced by AVN lesions, and it also occurs in the late training stages (Ragsdale et al., 1984). These results present a paradox. Subicular damage that increases CS-elicited excitatory activity in the AVN *facilitates* behavioral responding in the late stages. Why then should damage in area 29, which also increases the AVN response, reduce behavioral responding in the late training stages?

The answer proposed is inherent in the remarkable neuroanatomical isolation of the AVN. To our knowledge, the only efferents of AVN are those that project to the subiculum and to area 29. Area 29 neurons project in turn to the pons and the neostriatum, structures likely to be involved in the production of the behavioral response. Thus, area 29 lesions impair performance because they block the flow of the discriminative code of the AVN to response elicitation systems. Subicular lesions enhance responding because they facilitate the AVN response and do not block code outflow.

It is important to note that the lesions in limbic triad structures altered performance of the well-learned behavior. Yet, except for the increased responding in the first training stage in subjects with subicular damage, the lesions do not seem to affect the original acquisition of the behavior. The implication is that whereas the AVN outflow may exert an influence on behavior in the early training stages, that influence is redundant with the influence of some other neural system. This other system is able by itself to mediate behavioral acquisition, but it is not adequate alone to mediate performance of the well-learned behavior. Thus, the coding process of the AVN is specialized and essential for normal performance in the late stages in repetitive tasks with unchanging requirements. The advantage of such a contribution may be to free the first system to process novel problems, should the need arise.

The anterior cingulate cortical system and original acquisition. The unimpaired acquisition of the discriminative behavior in rabbits with lesions of area 29 and the AVN raises a question about the earlier data indicating severely impaired original acquisition in animals with cingulate cortical lesions (reviewed in Gabriel et al., 1980b), a finding that has been replicated for the discriminative avoidance behavior of the rabbit (Lambert and Gabriel, 1982). In all of these studies, the lesions involved the entire cingulate cortex, including area 29 as well as the anterior cingulate cortex (area 24). This implies that the anterior cingulate cortex, not area 29, may be the "other" limbic mesocortical component that must be intact if original acquisition is to occur in our paradigm. This interpretation is supported by our recent studies of activity during training in area 24 and the reciprocally interconnected n. medialis dorsalis (MDN) (Benjamin, Jackson, and Golden, 1978), which show that these structures develop discriminative activity during the course of training. As in area 29 and AVN, the discriminative activity in area 24 develops in advance of the activity in the MDN. However, the discriminative activity in area 24 develops more rapidly (after fewer trials and at shorter latencies) than the discriminative activity in area 29. Also, the discriminative activity in the MDN develops more rapidly than in the AVN (Gabriel and Orona, 1982; Orona and Gabriel, 1983). Thus, the prefrontal system is the "leading edge" system for stimulus significance coding during discriminative avoidance learning. It is left for future work to establish whether corticothalamic interactions and effects of the hippocampal afferents occur in area 24 and the MDN similar to those observed in relation to area 29 and the AVN.

Revised working model: integration of recent results. The data give rise to the idea that two distinct functional systems operate in mediating discriminative avoidance learning. The first is a diencephalic system represented by the AVN, MDN, and perhaps other midline and dorsal thalamic regions, wherein the principal codes for stimulus significance are formed. The codes are discriminative, excitatory, CS+ elicited discharges that, once acquired are projected through areas 24 and 29 to pontine and neostriatal sites, where they initiate release of the primed behavior. The code in the MDN mediates the original acquisition of the behavior, whereas the code in the AVN, perhaps in tandem with the MDN code and/or codes formed in other midline and dorsal thalamic nuclei, mediate the well-learned behavioral response.

The second system is telencephalic, represented by activity in portions of the cingulate cortex (deep laminae of area 29 and area 24) and the hippocampal formation. Cingulate cortical neurons are driven primarily by the excitatory input from their relevant thalamic nuclei, but the fate of the flow of information from the thalamic nuclei is determined by afferents to the cingulate cortex from the subicular complex. Thus, excitatory volleys from AVN to area 29 in the early training stages are returned (echoed) to the AVN, via the layer 6 corticothalamic neurons as negative feedback to suppress continued activity of AVN neurons. The echoing is initiated by the subicular afferent whenever novel task contingencies are detected by the hippocampal formation. The detection and initiation are indexed by the appearance of discriminative responding in the deep laminae of area 29, activity that is presumed to represent the firing of layer 6 neurons. It is very likely that discriminative firing of layer 6 neurons will be detected in layers 4 and 5, since these layers contain large proximal dendrites of the layer 6 cells (Vogt and Peters, 1981). These phenomena are particularly prominent during the early stages of training and the early stages of reversal training (i.e., transitional training stages where novel task contingencies evoke deep laminar discriminative activity).

When task contingencies are no longer novel, as in the late stages of training, the hippocampus acting through the subicular afferent to area 29 redirects the discriminative excitatory code arising from AVN. Rather than echoing the code back to the AVN via layer 6 corticothalamic fibers, the code is routed through layer 5 neurons, which project to the pons and neostriatum (Domesick, 1969; Wiesendanger and Wiesendanger, 1982; Wyss and Sripanidkulchai, 1984), that is, the centers postulated to subserve behavioral elicitation. The firing of these large pyramidal neurons is indexed by discriminative activity in layers 1–4, that is, layers occupied by their dendritic aborizations (Vogt and Peters, 1981). This mode of the system's operation is reflected by the late-developing discriminative discharges in the upper layers of area 29. As training progresses, an increasing proportion of the thalamic code is routed to the response elicitation systems.

It should be noted that the patterns of afferent terminations in area 29 are quite compatible with these assumptions. Thus, terminations of AVN and subicular fibers interdigitate in area 29. Anteroventral nucleus terminals occur in layers 1 and 4 (Domesick, 1972; Berger et al. 1980; Vogt, Rosene, and Peters, 1981), whereas subicular fibers terminate in layers 2–3 (Finch, Derian, and Babb, 1984). One could assume that tonic inputs from the subiculum to layer 5 cell dendrites in layer 2–3 could lessen layer 5 cell excitability to AVN inputs in layer 1. The same subicular activity arriving at layer 6 cell synapses could *increase* the excitability of layer 6 cells to AVN inputs in layer 4. Both effects in concert would account for the echoing function that operates to suppress AVN firing when novel task contingencies are encountered. In the later training stages, tonic inputs from the subicular afferents would subside, lessening the excitability of layer 6 neurons and allowing excitatory pulses of AVN origin to drive layer 5 neurons, thus to trigger behavioral responding. Presumably, analogous relationships hold for the area 24 and the MDN during the early training stages, an hypothesis to be tested in future studies.

On this view, the temporal course of behavioral acquisition is controlled largely by the rate at which the excitatory thalamic code is redirected by the hippocampus. When the hippocampus is damaged, all of the thalamic code is sent to the response systems, and the rate of thalamic code acquisition becomes the limiting factor controlling behavioral acquisition. This accounts for the hyper-responsiveness and the increased acquisition rate seen in several paradigms in subjects with hippocampal damage.

The redirection of the code may also occur within single trials during the acquisition process. Thus, in the early portions of the CS–UCS interval, the code is fed back to the AVN, but as milliseconds elapse the subicular afferent may reduce its tonic impulse flow to area 29, so that the code is routed to the response elicitation system. This transition is reflected by the buildup of theta-like neuronal bursts (and of the theta rhythm itself) seen in the subiculum and hippocampus in anticipation of behavioral response performance (Gabriel and Saltwick, 1980).

The suppressive influence of the early-forming cingulate cortical discriminative activity stands in contradiction to our past working hypothesis, which viewed the early developing discriminative cortical activity as instrumental in bringing about the later-developing discriminative activity in the AVN. A most important consequence of the newer view is the resultant simplification of the analytic problem. Telencephalic afferents do not seem to support coding processes in AVN directly. Thus, given the limited number of subcortical afferents that reach AVN, a rather complete neurobiological causal description of the thalamic significance coding process seems forseeable. It is the thalamic activity that would seem to be the essential substrate of the learned behavior.

Although these results and the model that they suggest are new, they seem to reveal again a relatively long-standing principle of brain function – the idea that the phylogenetically newer structures of the cerebral cortex are involved in suppressive control of the older "protoreptilian" structures (see MacLean, 1978). The recent results suggest that the protoreptilian system is the basic source of excitatory drive that is used by the telencephalic system either to trigger behavioral outputs at appropriate moments or as negative feedback to suppress continuing thalamic excitation.

It is interesting to compare the conclusions derived from Mishkin's studies, to those suggested by the studies of the substrates of discriminative avoidance behavior in rabbits reviewed in this chapter. In both instances, damage to two major limbic subsystems appeared to be a prequisite of behavioral impairment. Moreover, the two subsystems implicated in each case appear to have properties in common, one involving the hippocampus and related structures of the Papez circuit (the anterior thalamus and posterior cingulate cortex) the other involving the amygdala and related structures (the anterior cingulate cortex and the MDN). However, there is a major difference between the conclusions of the two projects. One project involves sample nonmatching behavior controlled by unrepeating stimuli with changing predictiveness. The other involves behavior controlled by repeating stimuli with consistent predictive relationships. In order to test the possibility that the two projects are tapping analogous learning systems, future studies will be needed to establish whether combined anterior and medial thalamic damage and cingulate cortical damage in primates is disruptive of standard discrimination learning with consistent and repeated cue–outcome relationships. In addition, studies are needed to test the hypothesis that combined amygdalar and hippocampal damage in rabbits is especially destructive to working memory performances, but not to discriminative avoidance using repeating and established cue–outcome relationships.

Epilogue: reduction and function

This chapter has maintained a rather strict dichotomy between reductionist and functionalist approaches to the study of the neural substrates of learning. However, it is our final recommendation that the future of studies in this field lies in the merging of the two approaches. It can be hoped that this development will occur as the completeness and specificity of the functionalist models increases. As this happens, it may make sense to begin to ask reductionist questions in relation to the functionalist systems. When these questions are ultimately pursued, they will not concern the magic learning substance. Rather, they will describe the molecular underpinnings of specific neural circuits that have identifiable and limited functional relevance to one of several categories or components of learned behavior. Many of the presently known but functionally undefined varieties of molecular change will in all likelihood find a place of residence within neural circuits having known functional relevance.

12 A neurobiological view of the psychology of memory

Michael Gabriel, Steven P. Sparenborg, and Neal Stolar

General approach

A central theme of Chapter 11 is that a prerequisite of progress in understanding how the brain learns and remembers is an active behavioral science of memory that specifies fully, at both the conceptual and empirical levels, the properties and limits of memory systems. The relatively new cognitive psychology of memory as exemplified by the material presented by Schacter in Chapter 10 has added not just a series of new phenomena but a new outlook to the collective wisdom of traditional experimental psychology. The new outlook, it seems to us, has added considerably to the "friendliness" and adaptability of the behavioral science of memory to the goals of physiological psychology.

An inevitable boundary condition for this discussion concerns the very likely prognosis that progress in understanding the neural mediation of learning and memory processes is going to involve the continuing use of animal studies. With this in mind, the establishment of a meaningful dialogue will have to involve those aspects of cognitive psychology that seem applicable to existing and future animal models.

Now, at this point it seems appropriate to caution the cognitive psychologist readership not to expect too much. The animal models in physiological psychology are not likely, at least for some time, to yield full-blown explanations for the phenomena described by Schacter. Instead, what can be hoped for are what may be termed animal analogue phenomena, that is, behavioral phenomena in animals that seem analogous to phenomena of interest in humans. Some of the existing and to be developed analogues may seem appallingly remote from their counterparts in humans. Nevertheless, if their implementation in the brain in animals can be worked out by physiological psychologists, the results may hopefully stimulate theoretical models in cognitive psychology that have as their terms structures, pathways, and activities of the brain, rather than the black boxes and distributed processing networks constituting present day neural models in cognitive psychology. In any case,

255

the comments to follow present our thoughts regarding areas in which physiological psychology may converge upon some of the ideas and topic considered in Chapter 10 by Schacter. In some cases, we will mention convergences that already exist, and in others, possibilities for future convergences. In addition, we will attempt to comment on the prospects for understanding the physiological underpinnings of phenomena in animals that are analogous to the phenomena in cognitive psychology.

Areas of convergence

Context effects

The learning context is the full set of relatively static, tonically acting stimulus conditions present when a to be remembered stimulus is presented. It seems clear from a variety of findings that the learning context is importantly involved in many forms of learned behavior.

To illustrate, it is instructive to try to imagine what would happen if the theoretical models that do not acknowledge the role of contextual control were actually true. Suppose, for example, that classically conditioned hopping responses in rabbits to a 4 kHz pure tone conditional stimulus (CS) were purely the result of a hardwired engram, that is, an enhanced synaptic transmission between 4 kHz neurons in the auditory pathway and the motor neurons for the conditioned response. The ungainly aspect of these assumptions is that any 4 kHz stimulus would trigger the behavior, regardless of the situation occupied by the rabbit. Thus, should one of the conditioned rabbits escape to the woods, the whine of a jet passing overhead, or the call of a bluejay, would trigger hopping behavior that would appear inexplicable and maladaptive. Similar maladaptive behavior might be observed in the home cage of a nonescaped rabbit, in response, for example, to sounds made by cagemates, air conditioners, or cage cleaning equipment. An example closer to home might be the case of the man who "stepped on the brake" in the presence of any red stimulus. Fortunately for rabbits and other species, conditioned responses are not absolutely wired to physical stimuli, but instead they are wired in a relative sense in that they tend to occur preferentially, in circumstances in which the specific CS is accompanied by the contextual stimuli present when the conditioned response (CR) was formed. All the pharmacological and behavioral literature on state dependence of learned responses (Overton, 1968, 1972) would seem relevant to the idea of contextual control of behavior. In this instance, the context that is manipulated is an internal one, generated by the particular physiological state of the subject during acquisition. In addition, there are many excellent examples of external contextual control of conditioned responses in the current literature of the psychology of animal learning (see Chapter 11 for references), and, in fact, this topic is probably the single most active one currently in that discipline. Thus, there is no dearth of potential animal models that could conceivably permit the study of the neural mechanisms of the contextual effects.

The topic of encoding–retrieval interactions discussed by Schacter refers to the linkages between contextual circumstances and specific item encoding during acquisition and the consequences that such linkages have for later retrieval. It might be suggested that this topic in cognitive psychology intersects with issues of contextual control, as studied in physiological psychology and animal learning. What may be helpful at this point would seem to be the development of a plausible animal model specifying the neural circuitry and plastic changes that bring about the establishment of contextual control, and the mechanisms whereby its effects are exerted when learned responses are retrieved. Thus, on this issue, the ball would seem to be in the court of physiological psychology.

Unfortunately, although the issue of contextual control has long been recognized (see Pavlov, 1927; Konorski, 1967), and specific neurological proposals have been made suggesting ways in which the problem of the neural implementation of contextual control might be approached (Konorski, 1967; Hirsh, 1974; Gabriel, 1980b), there are few explicit empirical attacks on this issue.

One of the exceptions to this lack is the substantial body of work that was started in the early 1970s indicating that hippocampal neurons tend to show a remarkable sensitivity to static properties of the experimental environment (O'Keefe and Nadel, 1978; Best and Best, 1976; Kubie and Ranck, 1982). In addition, the effects of hippocampal lesions on radial maze performance have led to the idea that hippocampal neurons are involved in the encoding of stimuli of the extramaze environment, so crucial to adequate maze solution (see Olton, Walker, and Gage, 1978). These results seem quite compatible with the idea that hippocampal neurons encode the properties of environmental contexts. It is not far-fetched to consider the additional possibility that the hippocampus is importantly involved in such things as the detection of familiar contexts and deviations from the familiar and the use of contextual information for retrieval of learned responses.

Our own work has yielded some progress in relation to this issue. Thus, neurons of the subicular complex of the hippocampal formation that project to the cingulate cortex would appear to affect the cingulate cortex in a tonic rather than in a phasic fashion. (A tonic rather than a phasic influence is generally a property to be expected of the neural manifestations of contextual control, as contextual stimuli that affect conditioned responses are usually conceived to be continuously present environmental features, to be distinguished from the phasically occurring events, i.e., "cues" that trigger behavioral responses.) Moreover, the results may be interpreted to indicate that when the tonic influence is operating, neural excitation flowing through the cingulate cortex is routed back to its origin (the thalamus) to suppress further excitation. This mode of activity occurs when the existing context is being violated by the occurrence of novel stimulation, and it is essential to suppress behavioral responding until the consequences of the new information can be evaluated. Alternatively, when the context is familiar and the behavior well learned, the subicular influence is withdrawn and the neural excitation is routed to centers for response elicitation. Our account of behavioral hyper-reactivity in animals with hippocampal lesions concerns the failure of the suppressive mode of functioning. Specifically, it is presumed that the switching circuitry of the cingulate cortex is permanently left in the "on" position in lesioned animals, resulting in the flow of all thalamically generated significance codes to centers for elicitation of primed behavior.

This model refers to the consequences of contextual governance for neural circuitry that is "close to" the behavioral output. In this regard, it is interesting to note that any one of the links in the hypothetical causal sequence, such as the detection of the expected context, the modulation of subicular neurons based on that detection, or the setting of cingulate neurons to communicate with the behavioral centers, are part of what has been called retrieval. This thought brings to mind a possible contribution of functionalist physiological models to the cognitive/physiological dialogue (i.e., the specification of global concepts such as retrieval through provision of specific and hopefully precise concepts denoting the set of neural processes leading to the production of learned behavior).

Our empirical work still leaves completely unbreached the question of how inputs to the hippocampal formation from the polysensory entorhinal cortex are recognized by hippocampal circuitry as signaling the expected context, and what events are involved in the detection of deviations from the expected configuration of contextual inputs. In order to deal with these issues, we will have to start studies that involve explicit manipulation of contextual variables, with observation of correlated events in the entorhinal and hippocampal circuitry. Nevertheless, it is interesting to wonder whether the explicit circuitry that these studies indicate to be involved in the performance of active avoidance behavior in rabbits may have any explanatory value in relation to fundamental processes of contextual control and retrieval of the contents of human memory.

A basic premise of our account is that CS-elicited discharges in the anterior ventral (AV) and medial dorsal (MD) thalamic nuclei routed through cingulate cortex serve as an internal "go" signal to trigger the elicitation of the learned response. We have assumed that the behavioral response, locomotion by the rabbit, is in some sense "primed" and held in an output mode, awaiting a signal for its elicitation, in the trained animal.

The mechanism being proposed is analogous in some respects to the "timing fence" mechanism found on many computer printers of the "dot matrix" variety. The timing fence is an elongated plastic strip, an inch or so in width, tied across the front of the printer inside the case. The strip contains alternating translucent and opaque vertical bars, each about 1–2 mm in width. As the print head proceeds from left to right, a small lamp that moves with the head projects its beam through the timing fence upon a detector that is also moving with the head on the opposite side of the timing fence. Because the light only penetrates the strip at the translucent bars, the detector receives pulses of light at regular intervals. The duration of the intervals and the pulses depends upon the speed of the print head. Each time a pulse enters the detector a character is printed. That is, the light pulse triggers electrical circuitry that causes a digital letter code, already loaded into a hardware buffer, to be converted into a pattern of solenoid movements, which cause a particular subset of stiff wires to be injected through the inked ribbon, to impress a character in a particular column of the paper. This arrangement has the effect of coordinating the printing of letters with the velocity of print head movement, which is notoriously variable depending on lubrication and adjustment of the clutch and bearing surfaces. Thus, the characters always end up in the intended column despite variations of the print head's velocity. The several points of analogy between the printer and the neural model include the light beam, which may be likened to the general source of relatively nonspecific CS-driven afferent activity arising from the AV and MD thalamic nuclei to trigger primed behavior; the timing fence which, like the cingulate cortex, is the site at which the afferent beam is gated on and off, depending on appropriateness of circumstances; and the printer's character buffer, which contains the digital letter code (the primed response), awaiting the signal that converts it into the forward thrust (the overt response) of the specific subset of dot matrix solenoids.

A problem arises in applying these ideas to human learning situations because the tests that appear sensitively to reflect amnesia resulting from limbic system damage are typically such things as free recall tests or paired associate learning, each test involving many items. Thus, these tasks tap stimulus and response selection processes not tapped by the conditioning paradigm. The question is, how may limbic significance coding processes operate in situations involving multiple stimuli and multiple responses?

This problem could be approached, for example, with regard to paired associate learning, by assuming that all selection occurs at the level of the thalamic significance code itself. In other words, we could argue that a qualitatively unique form of neuronal response develops in the AV and MD nuclei and that each unique code through stimulus–response (S–R) association elicits the appropriate response. This view places all of the burden of response selection upon the stimulus encoding properties of the thalamic nuclei.

We do not believe that this idea represents a tenable approach to the problem because it does not allow the properties of the responses themselves to contribute to recall, a contribution well-established in experimental psychology (e.g., Noble and McNeely, 1957; Underwood, Runquist, and Schwartz, 1959; Paivio et al., 1968). In addition, our own data do not lend themselves to the idea because of the requirement that encoding occurs in terms of qualitative differences in neuronal firing patterns, one for each stimulus on the list. Our data indicate instead that it is not the qualitative properties of the neuronal response that encode significance but rather the quantity of firing. Neuronal discharges of substantial magnitude in AV and MD predict behavioral output, whereas smaller or absent responses predict no output. The suggestion is that a primed response will be triggered when a certain threshold of excitation is exceeded.

In order to incorporate these ideas into both the animal and human situations, it could be assumed that there is response priming in both cases, and that the primed response is sensed by the limbic significance coding system. The sensing must take the form of a neural code identifying the unique properties of each response. In the case of the rabbit's performance, there is only one primed response, and its identity code is a constant contextual input to the hippocampal system. When the CS+ input and other environmental and interoceptive contextual inputs are added to the response identity code, the hippocampal formation detects a complete context, and it allows thalamic code to be gated through cingulate cortex to the response elicitation systems in the pons and/or the neostriatum. The progressive buildup during the CS –UCS interval, of theta-like neuronal bursts in the hippocampus (Gabriel and Saltwick, 1980) just prior to behavioral response elicitation reflects hippocampal detection of "context completion." The progressive nature of this buildup is owed to the fact that the context grows progressively more appropriate for the behavior as time elapses during the CS–UCS interval. Thus, like the printer's timing fence, the hippocampus is responsible for the coordination and timing of outputs in relation to subtle momentary changes in the stimulus context.

In the case of human paired associate retrieval, one could assume that all responses on the list are primed, but that only one at any given time can occupy the "chute" (i.e., a state of immediate readiness for elicitation). Perhaps each response item gets a turn to occupy the chute periodically, for example, once every 2 sec, this interval depending on many factors such as list length, item meaningfulness, and so forth. Occupation of the chute is the condition that enables the response's identity code to be projected to the hippocampus, to act as a contextual input. Perhaps chute occupation is also associated with the feeling of knowing. The particular condition that will trigger response output is context completion in the hippocampus [i.e., joint input of codes for a list stimulus item and the proper (paired) response item] This brings about the routing of thalamic significance code, which is elicited in a fairly generalized way by all items, through the cingulate cortex to the response elicitation system. When the code is projected, the response that happens to occupy the chute will be elicited.

It is recognized that this rough, qualitative anecdote will probably not lead to the conduct of critical experiments elucidating the nature of paired associate retrieval. The intention, however, is not to solve the problem of retrieval, but instead to reveal the kinds of terms and parameters that might be found in a model of human information processing that is based on functionalist physiological models and data.

Qualitative nature of encoding operations

If as has been suggested (Greenough, 1984; Rose, 1981) the alterations observed in reductionist studies of neuronal ultramorphological changes accompanying learning reflect storage (i.e., encoding) of task-relevant information, it should be the case that one can manipulate the character of the morphological changes by manipulating circumstances such as expectancy or amount of prior knowledge, factors that affect the qualitative nature of the encoding operations, as indicated by Schacter in Chapter 10. Such manipulations would provide an opportunity to employ a strategy of converging operations to test hypotheses about the functional relevance of the morphological changes. Thus, the study of encoding operations in cognitive psychology may have potential to stimulate interaction with physiological psychologists of the reductionist persuasion.

One caution in relation to this strategy arises from consideration of possible alternatives to the interpretation that states that the various preconditions said to engage qualitative levels of encoding may instead engage quantitative levels. For example, the chess master may remember board positions better than a novice not because he or she encodes the board in a different way than the novice, but because he or she has acquired certain perceptual skills – skills that confer an increased board-scanning efficiency. Thus, for a given amount of viewing time the board position is encoded more thoroughly (a condition for superior recall) by the master than by the novice. If this is true, one could, at least in principle, teach a novice how to chunk chessboard configurations so he or she could do as well as the master. A similar issue can be raised with regard to the other manipulations that have been interpreted to engage various qualitative levels of encoding. If the quantitative interpretation is correct, the strategy of manipulating factors, such as prior knowledge, expectancy, and semantic set, may well prove useful in reductionist studies of storage mechanisms as means for generating converging operations for hypothesis evaluation. However, the various manipulations applied to animals prior to, for example, dendritic measurement might be expected to affect the frequency or magnitude of dendritic alterations, rather than the qualitative form of the alterations, or the particular places in which they occur.

Priming and implicit memory

Priming and implicit memory refer to such things as enhanced word completion and perceptual identification performance that occur when the subjects have had some prior experience with the items to be completed or identified. Because they are conceptually similar to the ideas of priming discussed by physiological psychologists (e.g., Lashley, 1951), and because a hypothetical process bearing the same name is used in our own models, these phenomena in cognitive psychology seem to be good candidates for stimulating interaction between cognitive and physiological psychology.

Particularly intriguing is the fact that performance enhancements occur even when the prior exposure to the material does not enhance performance in more conventional tests such as free recall or recognition. Thus, items encoded by the amnesic brain will be revealed by a perceptual identification task, but not in free recall or recognition.

At first consideration, the paradigms used to demonstrate the priming and implicit memory phenomena would appear to be without counterparts in physiological psychology or in animal learning. This is understandable in the case of word completion due to the nonverbal nature of animal subjects. On the other hand, it could be argued that the behavioral tests of learning in animals are all priming studies, because animals never tell us explicitly what they do and do not know. Rather, they are always "tricked" into revealing knowledge by observing them while they perform what are to them sensible instrumental behaviors to obtain desirable results. This leads to the following thought. The feeling of familiarity discussed by Schacter in Chapter 10 should be irrelevant to the tasks that are used to demonstrate priming and implicit memory, because the subjects performing these tasks do not realize that memory is being tested and they are not called upon to extract contents from memory. Likewise, in all of the animal studies, the subjects are not asked to extract memory contents, but rather they are engaged in a kind of game-playing or prize-seeking behavior. Perhaps the reason for the difficulty in finding a convincing animal model for human limbic amnesia is that the demonstrations of human amnesia involve tasks that depend upon knowledge that one knows or the "feeling of knowing," whereas the animal performances and the priming and implicit memory phenomena in humans do not depend on the feeling of knowing. The dependence on this function may be an exclusively human one that is tied in some way with the linguistic system. Thus, when amnesic patients are asked to produce a correct answer they do not feel knowledgeable, they do not try to produce an answer, and, consequently, they perform poorly or not at all. These speculations suggest that the memory performances of humans that do not depend on the feeling of knowing may be more directly comparable to standard animal performances. This line of reasoning may prove helpful in trying to resolve the empirical and conceptual gap that has for some time separated studies of human and animal neuropathologies of memory.

A psychologist's reply

Daniel L. Schacter

The commentary of Gabriel, Sparenborg, and Stolar in Chapter 11 contains the kind of constructive suggestions and questions that will need to be posed and pursued if a useful dialogue between the cognitive psychology and neuroscience of memory is going to develop. In Chapter 13, I distinguish among collateral, complementary, and convergent relations between the two disciplines. It may be therefore of some interest to consider briefly the comments of Gabriel et al. from this perspective. Clearly, their discussion takes us out of the domain of collateral relations. However, does it draw us into the domain of complementary or convergent relations? The answer, I think, is a little bit of both.

The discussion by Gabriel et al. of how neuroscientists might contribute to the analysis of encoding–retrieval interactions is along the lines of what I describe as a complementary relation (see Chapter 13). For example, they suggest that it would be desirable to develop "a plausible animal model specifying the neural circuitry and plastic changes that bring about the establishment of contextual control, and the mechanisms whereby its effects are exerted when learned responses are retrieved" (Chapter 11). The emphasis here is on specifying the neural concomitants of a cognitive phenomenon and hence expanding our understanding of it. Their subsequent description of the role of hippocampal projections and other neural pathways is one that at some point might provide a very useful *complement* to a well-articulated cognitive model that is concerned with encoding–retrieval interactions. However, it is not clear whether studies of this kind would facilitate understanding of the phenomenon *at the cognitive level*, that is, although there can be little doubt that a model of the neural circuitry involved in encoding–retrieval interactions would broaden our understanding by providing a description of the phenomenon at a previously uninvestigated level of analysis, it is less certain that such a model would have direct implications for cognitive theories and hence deepen our understanding at that level. Similar comments would apply to the interesting suggestions by Gabriel et al. concerning possible neural underpinnings of differences among encoding operations.

In contrast to the foregoing, I think that the suggestions regarding priming effects and implicit memory phenomena come closer to the sort of discussion that can occur when a convergent relation exists between the two fields with respect to a particular issue. Gabriel et al. consider the properties of a phenomenon observed in cognitive psychology, discuss whether it might occur in animals, and then present an hypothesis that concerns both the neural and cognitive aspects of the phenomenon. Now, it happens that I am not entirely in agreement with the specifics of their hypothesis for various reasons that need not be discussed here (Schacter, 1985b). Suffice it to say that there is experimental evidence indicating that monkeys succeed on memory tasks that reveal impairments in human amnesics (e.g., delayed matching to sample), suggesting that amnesics' deficiencies are not attributable to a failure of mechanisms that are bound up with linguistic functions (see also O'Keefe and Nadel, 1978; Olton, 1984a,b; Schacter and Moscovitch, 1984; Squire and Zola-Morgan, 1983). More important, however, is the fact that understanding of implicit memory phenomena at the neural level can and should have direct consequences for theories at the cognitive level. If, for example, it is possible to provide a precise delineation of the functional neural system that is necessary for explicit memory and to distinguish it from the neural system that mediates implicit memory, cognitive theorists will have to recognize that a nonunitary mechanism underlies the memory phenomena that they observe. Hence my suggestion in the previous commentary that this issue represents an emerging example of convergent relations between the two fields.

Gabriel et al. are correct to caution against expecting too much from the interdisciplinary approach at this early stage of the game. It seems clear that we are just beginning to scratch the surface of the numerous problems that need to be addressed, and it seems equally clear that differences in terminology, methodology, and fundamental assumptions should not be underestimated. However, the fact that the dialogue has been initiated is itself a noteworthy development, one that bodes well for the future prospects of a cognitive neuroscience of memory.

13 A psychological view of the neurobiology of memory

Daniel L. Schacter

In 1860, the Scottish neurologist and philosopher Thomas Laycock published his two-volume treatise *Mind and Brain*. Although neither the author nor the book are well known today, both exerted a good deal of influence on mid-nineteenth century psychologists and neurologists (see Amacher, 1964; Greenblatt, 1965; Schacter, 1982, Chapter 7). In *Mind and Brain*, Laycock argued that it is not possible to gain a scientific understanding of memory and other psychological phenomena without careful study of the pertinent neurophysiological mechanisms. Noting that both clinical and experimental facts demonstrated that changes in the brain affect memory (Laycock, 1960, Vol. I, p. 149), he provided a detailed critique of the opinions offered by philosophers who "refused to expressly recognise the validity of these facts" (p. 150). Laycock argued forcefully that theories of memory should encompass both the psychological and neurophysiological levels of analysis.

Times have changed since the publication of Laycock's tome, and there are no longer any serious students of the subject who would argue that memory exists independently of the brain. However, a reading of Chapter 11 by Gabriel, Sparenborg, and Stolar and Chapter 10 by Schacter reveals that much contemporary research concerning memory focuses on one level of analysis – either cognitive or neurophysiological – and largely ignores the other. The task of the present discussion is to consider whether an interaction between cognitive and neurobiological approaches to memory can be achieved, or whether the two enterprises represent separate pursuits that neither require nor enrich each other and hence should proceed independently. After considering the ground covered in Chapter 11, it seems to me that there is no absolute, yes-or-no resolution of this issue and that a response must be couched in somewhat more conditional terms: It depends. The kind of relation that is possible between cognitive psychology and neuroscience depends upon the types of evidence and issues that are at stake. I will suggest that three relatively distinct kinds of relations can and do exist between the cognitive psychology and neuroscience of memory: collateral, complementary, and convergent relations. I will define and illustrate each of these relations with respect to the issues that were discussed in Chapter 11.

Collateral relations between cognitive psychology and neuroscience exist when an issue is pursued in one of the fields at a level that cannot be mapped on to the other; research in one discipline thus proceeds independently of developments in the other. Consider, for example, the question discussed by Gabriel, Sparenborg, and Stolar in Chapter 11 concerning whether plasticity is a presynaptic or postsynaptic phenomenon. Clearly, this is a fundamental question for the neurobiology of learning. Yet, it would be difficult to find many cognitive psychologists whose thinking about memory is constrained by current knowledge pertinent to this question or, for that matter, many who believe that their thinking could or should be constrained by whatever understanding ultimately is reached concerning this issue. After all, knowledge of whether memory occurs presynaptically or postsynaptically tells us little about the principles that govern the complex mnemonic functions that interest most cognitive psychologists. I am not in disagreement with such sentiments. However, I believe that one reason why resolution of the presynaptic versus postsynaptic issue has little bearing on contemporary cognitive psychology has to do with our currently impoverished knowledge of how neural events are related to complex cognitive phenomena. It seems entirely conceivable that at some point in the future – probably the distant future – understanding the presynaptic versus postsynaptic issue could impose constraints on a cognitive theory, if the theory were to specify the properties of memory mechanisms at a sufficiently fine level of detail. Thus, I would suggest that an issue such as the presynaptic versus postsynaptic one is not, in principle, irrelevant to the cognitive psychology of memory; it is simply beyond our current level of understanding even to attempt to relate this issue to cognitive psychology in any meaningful way. The same thoughts apply to a number of other issues discussed in Chapter 11, particularly to those that are classified under the reductionist umbrella. For example, questions such as whether leupeptin acts specifically on calpain or whether alternation of the phosphorylation state of brain proteins is crucial for memory currently have no direct bearing on cognitive theories, even though they might at some point in the future. In these cases, then, there is a collateral relation between the two disciplines: Cognitive research is not yet in a position to take account of the neurophysiological data and, hence, it cannot be influenced by them; neurobiological research operates at a molecular level at which the more global concepts of cognitive psychology are not applicable.

In contrast to the foregoing, *complementary* relations between cognitive psychology and neuroscience are observed when a description of a phenomenon in one discipline can meaningfully supplement the description of a similar phenomenon in the other discipline. Here, the gap between the cognitive and neurophysiological levels of analysis is not nearly so broad as in the afore-mentioned examples. Much of the classical work on localization of memory processes, described by Gabriel, Sparenborg, and Stolar in Chapter 11, in the context of the functionalist approach, provides an example of a line of research that can have a complementary relation to cognitive models. A cognitive model attempts to delineate the nature of and relations between the structures and processes that comprise memory or a specified subdomain of memory. A complementary model in neuroscience may attempt to do much the same thing, except that the focus is on the component neuroanatomical structures and their relations rather than on hypothesized mental mechanisms. Gabriel's own model of discriminative avoidance learning provides an ex-ample of the kind of research that may bear a complementary relation to cognitive psychology. Consider the following extracts from Gabriel, Spar-enborg, and Stolar's description of the dynamics of avoidance learning in Chapter 11:

Cingulate cortical neurons are driven primarily by the excitatory input from their relevant thalamic nuclei, but the rate of the flow of information from the thalamic nuclei is determined by afferents to the cingulate cortex from the subicular complex. Thus excitatory volleys from AVN to area 29 in the early training stages are returned to the AVN via the layer 6 corticothalamic neurons as negative feedback to suppress continued activity of AVN neurons ... On this view the temporal course of behavioral acquisition is controlled largely by the rate at which the excitatory thalamic code is redirected by the hippocampus. When the hippocampus is damaged, all of the thalamic code is sent to the response systems, and the rate of the thalamic code acquisition becomes the limiting factor controlling behavioral acquisition.

Clearly, there is strong emphasis in this description on specifying *where* the activities that are critical for learning occur. A cognitive psychologist who was attempting to model performance on the same task would not be partic-ularly concerned with the "where" of processing; he or she would focus exclusively on the "how." However, any mental mechanisms proposed by a cognitive psychologist to describe task performance would gain a great deal of credibility if they could be instantiated in a neural model such as the one proposed by Gabriel, Sparenborg, and Stolar. Thus, neural modeling that attempts to specify the brain structures involved in learning can have a com-plementary relation to cognitive theorizing, insofar as it describes a phenom-enon of memory at a level that is distinct from, but can be related to, the level addressed in a cognitive theory.

Complementary relations between cognitive psychology and neuroscience may also exist at the methodological level. A task developed in the cognitive psychology of memory may prove useful for investigating an issue in neu-roscience, such as the effects of a particular drug on a specific aspect of memory. Similarly, an investigative tool that is used primarily by neuro-scientists, such as the averaged evoked potential, could provide useful insights into the temporal dynamics of mnemonic processing on tasks that are of interest to cognitive psychologists (cf. Posner, Pea, and Volpe, 1982).

The third type of relation between cognitive psychology and neuroscience – a convergent relation – is in many ways similar to the complementary relation just described. I distinguish between complementary and convergent relations, however, because I think that there is a subtle yet important difference between them. A complementary relation exists when the separate agendas of neuroscience and cognitive psychology produce methods, data, or theory that can be exchanged profitably; a cognitive psychologist or a neuroscientist may be able to use knowledge from the other discipline to help make sense of a phenomenon within his or her own discipline. By contrast, I would suggest that a convergent relation exists when cognitive psychologists and neuroscientists coordinate their agendas in such a manner that they address a common issue to bring to bear the various conceptual and experimental tools of their respective disciplines to analyze it. When a convergent relation exists between the two fields, experimental findings and theoretical ideas from cognitive psychology may help a neuroscientist to understand a phenomenon *at the physiological level* and, conversely, findings and ideas from neuroscience may help a cognitive psychologist to understand a phenomenon *at the psychological level*.

Convergent relations represent a potentially exciting and productive form of interaction between cognitive psychology and neuroscience that may ultimately lead to a unified theory of memory function that bridges cognitive and neurophysiological levels of analysis. However, it seems clear that convergent relations will not be achieved easily because they require investigators to overcome differences in methodology, terminology, and theoretical assumptions. As suggested earlier, there are some levels of analysis at which these differences are presently so large as to preclude the establishment of convergent relations.

Are there currently any possibilities for convergent relations between the cognitive psychology and neuroscience of memory? I think that Chapter 11 touches on one of the emerging candidates: the question of whether distinct and dissociable "forms of memory" or "memory systems" can be identified. In spite of the many obvious differences in the content of the two chapters, both Chapters 10 and 11 devoted some attention to evidence suggesting that distinct neural and cognitive systems underlie various forms of learning and memory. The evidence for dissociable forms of memory cited by Gabriel, Sparenborg, and Stolar in Chapter 11 derived partly from their work on avoidance learning in the rabbit and partly from work by Mishkin on one-trial recognition and discrimination learning in monkeys; the evidence cited in Chapter 10 derived from various studies of college students and amnesic patients. One reason why it seems appropriate to describe this line of research as representing a convergent, rather than a complementary relationship between cognitive psychology and neuroscience is that investigators in each discipline are increasingly taking account of developments in the other when formulating hypotheses and discussing them. Thus, discussions of multiple forms of memory by cognitive psychologists have begun to reveal the influence of developments in neuroscience (e.g., Wickelgren, 1979; Tulving, 1983; Johnson, 1983; Schacter, 1984, 1985b; Schacter and Moscovitch, 1984; Tulving, 1984), and recent formulations by neuroscientists have been similarly affected by developments in cognitive psychology (e.g., Oakley, 1981; Moscovitch, 1982; Squire, 1982; Oakley, 1983; N. J. Cohen, 1984; Mishkin, Malamut, and Bachevalier, 1984; Olton, 1984a, b; Squire and Cohen, 1984). Several reasons why understanding of this issue may be facilitated by interaction between cognitive psychologists and neuroscientists have been outlined elsewhere (Shallice, 1979; Tulving, 1983; Schacter, 1984). These recent developments encourage speculation that we may soon reach the point at which a satisfactory theory will have to account systematically for relevant data in both disciplines. Such a state of affairs has not been common in memory research, and its attainment would confirm that a convergent relation exists between the two fields with respect to the issue of dissociable forms of memory.

In summary, the diverse territory covered in Chapter 11 indicates that the prospects for interaction between the cognitive psychology and neuroscience of memory depend upon the specific issues that are at stake. I have suggested that collateral, complementary, or convergent relations exist between the two fields with respect to different issues and levels of analysis. We do not yet know whether convergent relations can be established with respect to a variety of issues, nor do we know whether the more molecular levels of analysis in neuroscience will ultimately prove refractory to this approach. An important challenge for those who believe that a cognitive neuroscience of memory is possible will be to delineate the issues that are amenable to cross-disciplinary analysis and to discuss the form that theoretical explanations should take (cf. Shallice, 1979; Kean and Nadel, 1982; Posner, Pea, and Volpe, 1982). Perhaps it is not entirely unrealistic to hope that a future discussion of a similar chapter will be able to conclude that convergent relations between the two fields have been established at many levels of analysis.

A reply from neurobiologists

Michael Gabriel, Steven P. Sparenborg, and
Neal Stolar

Schacter's three types of relations between cognitive psychology and neuro-science provide a useful organizational scheme for various forms of interaction between cognitive psychology and neuroscience. Also, the tenor of both Schacter's (Chapter 10) and our own (Chapter 11) chapters suggests that there is a general feeling shared by cognitive psychology and neuroscientists that some degree of active interaction between disciplines is to be encouraged. Beyond this general aura of agreement though, there is little in the present dialogue, or in other writings, that attempts to assess the degree of importance that interaction may have for the advancement of the disciplines or the extent to which it is reasonable to formalize the interdisciplinary relationship. As in the case of the evaluation of progress in cognitive psychology discussed by Schacter in Chapter 10, one could take either an optimistic or a pessimistic view of this issue.

On the optimistic side it could be argued that we are presently witnessing an optimal interaction and that a continuation of the status quo will provide the best possible climate for nurturing exploration of the convergent relations between the two disciplines, and an ever-deepening, more meaningful inter-action. On the pessimistic side, a case could be made that the surface has merely been scratched and that a much more active and formal relationship between the disciplines will have to emerge if the benefits of interaction are to be realized.

Our own view falls between these two extremes, but it leans to the pessi-mistic side. It is the idea that an interaction between the disciplines more formal and deliberate than that currently operating should be sought, without infringing in any way on the continuing independent pursuit of current goals within each discipline. The new model should perhaps take the form of a separate interdisciplinary science. This more formal relationship is necessary, we believe, even approximately to realize the benefits that interaction seems to promise.

Below are mentioned some general reasons for taking this position. The arguments to be presented have been made elsewhere in this dialogue and in other places. However, it may be useful to restate them here, succinctly, in support of our position.

Much work in contemporary neuroscience is directed toward providing a description of such things as molecular and biophysical substrates of neural activity, the brain's biochemical contests, neurophysiological events, cytoar-chitectures, and pathways. These descriptions are made with no reference at all to their behavioral or psychological relevance. Thus, these aspects of neuroscience are in a sense "mindless."

We are not casting aspersions on this work. It has provided the bulk of the discipline to date, and it is the very foundation on which the understanding of functional relevance is based. It is also the most mature component of neuroscience in general, and it can be argued for some areas (e.g., the connectional neuroanatomy of the retina and many other systems) that the essential work has been completed, that is, the important stories have been fully played out and what remains to be done is largely in the nature of a "mop up" operation.

Yet, one must consider where we want to be when we have had a completely exhaustive neuroscience. What is the end product that is envisioned? It is to understand fully brain function. The vast bulk of what is left to be done is still undone, whereas descriptive neuroscience is quite far along. We have already made the case in our main chapter that what is organizing and stimulating the most vital and exciting work on the neural substrates of learning is existing knowledge and practice in cognitive and experimental psychology. It seems reasonably clear that this interaction is essential to the future vitality of neuroscience and to the realization of its ultimate objective. This is especially true now that a significant portion of the work of descriptive neuroscience has been completed.

Cognitive psychology seeks to describe mental structures that are needed to account for observed facts of behavior. This approach is in a sense "brainless" (i.e., the cognitive structures postulated make no reference to the brain) in the great majority of cases. Again, as in the case of descriptive neuroscience, it is not necessary to belabor the obvious and enormous value of cognitive psychology as the principal source of knowledge of human capacity, and as a "complementary" stimulant to the study of brain function. Yet, there has been a malaise in the field as indicated by Schacter in Chapter 10. There is some question about cyclic reinvention of the wheel, and some of the products once believed permanent have proved ephemeral. Finally, mature cognitive theories in their final stages do not always create a feeling of intellectual compulsion, normally expected of and manifested by mature theories in the physical and biological sciences.

It is the case that the brain is an alternative set of structures to the ones more commonly invoked in cognitive psychology to explain behavioral facts. Moreover, the brain is known a priori to perform the functions for which cognitive structures are postulated. It would thus seem reasonable that some effort be directed toward the construction of models couched in realistic neurological terms. This approach might provide a way to devise independent criteria and constraints upon theory that could screen out theoretical ephemera. Perhaps one could proceed with business as usual but when a candidate structure is invented, with reference to the nervous system, manipulation of the brain could be the ultimate reality test.

It is obvious that the pursuit of cognitive structure based on the brain is going to require a considerably varied set of skills and background knowledge on the part of many practitioners, if the kind of interaction being advocated is to occur. This is the reason for suggesting the need for a new interdisciplinary science. It is not likely that a body of adequately trained scientists is going to emerge fortuitously.

Finally, it seems clear that the new discipline will have to encompass more than neuroscience and cognitive psychology as presently constituted. The information and strategies employed in the psychology of animal learning and cognitive psychophysiology will also be essential. Because ethical constraints prohibit routine invasive studies of the human nervous system, animal models are needed to to provide the critical bridge between cognitive and behavioral studies, on the one hand, and neuroscience, on the other. Cognitive psychophysiology, and especially the use of event-related potentials recorded from the scalp during human information-processing tasks, provide an additional important link between cognitive processes and activities in the nervous system.

Part IV

Emotion

14   The psychology of emotion

Ross Buck

The relationship of cognition and emotion, and the nature of their interaction, is a topic of great current interest and controversy that bears upon the very essence of human nature. Questions about whether human behavior is governed by reason or passion, whether human nature is fundamentally good or evil, and whether human beings are no different from other animals or somehow separate from the animal kingdom have appeared perennially in western philosophy. The answers to these questions generated in the course of history have been important, for they have determined the sense of whether human beings are above nature, with the ability to control its forces, or a part of nature, participating in its processes and ultimately subject to its whims.

This chapter presents a view of emotion that sees human behavior as simultaneously governed by biology, knowledge, and reason – upon structures that are built into the organism, upon the organism's learning experiences in its environment, and upon linguistic structures that are unique to humans and to some extent independent of both biology and individual experience. In doing so, it suggests a view of the interaction of cognition and emotion that attempts to accommodate the approaches not only of cognitive psychologists and neuroscientists, but also of ethologists and personality and social psychologists.

One of the major assumptions of the present approach is that a major distinction must be made between: (1) "special purpose" mechanisms of behavior control that are hard-wired into the nervous system, (2) "general purpose" cognitive mechanisms that are programmed by the learning experiences of the individual in the environment, and (3) mechanisms of behavior control that are based upon language and are unique to humans. Emotion is seen to be an aspect of the special purpose systems that interacts with the cognitive systems. The details of how this interaction occurs is the subject of much contemporary controversy, with some arguing that emotion is the primary driving force in this interaction (Zajonc, 1980, 1984) and others arguing for the primacy of cognition (Lazarus, 1982, 1984). This chapter discusses this controversy, arguing that the former position is the more plausible. However, in human beings there is another system of behavior control that is so powerful and has become so important that the role of the biologically based, special purpose systems is often ignored. This system involves language. This chapter argues that it is language that truly distinguishes human and animal behavior, freeing the individual from his or her personal experience and allowing the consideration of events that have never been, and indeed could never be, experienced.

This chapter first considers the nature of emotion and its relationship with motivation. It argues that emotion is fundamentally a readout of motivational potential, with three kinds of readout corresponding to three functions of emotion. It then discusses three basic functions of emotion and the methods for studying them. These include bodily adaptation, social communication, and subjective experience. Next, the relationship of emotion and cognition is discussed, considering both traditional views and contemporary controversies, and culminating in a general model of emotion that attempts to reconcile these conflicting views. The final section considers the implications of language for the integration of cognitive and neuropsychological approaches to emotion.

The nature of emotion

Emotion and motivation

One of the first requirements for a coherent and integrated view of cognition and emotion is to clarify the relationship of emotion and motivation. Most theorists of emotion tend to avoid the mention of motivation, and the converse is also the case. However, the two concepts have much in common (cf. Buck, 1976). This section defines motivation and emotion and discusses how they are related to one another.

Defining motivation. Motivation is responsible for the control of behavior, that is, the *activation* and *direction* of behavior. We shall define motivation as a *potential* that is inherent in the structure of systems that control behavior. This conception of motivation is analogous to the notion of potential energy in physics. For example, a piece of matter may have potential energy by virtue of its position (i.e., a weight raised to a height) or arrangement of parts (i.e., a coiled spring). The energy is actualized by an appropriate stimulus: dropping the weight or releasing the spring. In the same way, systems of behavior control may be seen as having potential by virtue of their particular qualities, which are actualized by appropriate stimuli. This potential is motivation.

The qualities of biological systems of behavior control have been dictated by the process of evolution. Thus, biological motives are potentials for the activation and direction of behavior that have been hard-wired into the arrangement of the nervous system. For example, the knee-jerk reflex is a potential inherent in the arrangement of the sensory and motor nerves running between the knee and spinal cord; it is actualized by a tap of the knee. Similarly, hunger is a potential inherent in a complex neurochemical system that is actualized by a lack of nourishment; and Tomkins' (1962, 1963) primary affects (happiness, sadness, fear, surprise, disgust, and anger) are potentials inherent in the structure of the brain that are set off by appropriate stimuli [a baby's smile, a threatening stare (cf. Buck, 1984)].

As these examples imply, these systems of behavior control exist at different levels in the organism. At a relatively simple level, *reflexes* such as the knee jerk serve quite specific (although vital) functions and are based upon relatively simple neurochemical structures. In contrast, the need for food sets off much more complex homeostatic regulatory mechanisms within the body resulting in the *drive* of hunger. In the brain, this involves processing at the level of the hypothalamus. Tomkins' *primary affects* are potentials inherent in higher structures of the brain (i.e., limbic system structures).

Cognitive motives. The foregoing examples all concern special purpose processing systems in which the behavior tendencies in question are built into the biological structure of the organism. It is also possible to consider the motivation underlying the general purpose processes of learning and cognition – where the behavior is structured by the experience of the individual organism in the environment – in these terms. Thus, the motivational force behind the construction of the cognitive system – White's (1959) effectance motivation – can be seen as a potential inherent in the developing cognitive structure.

Piaget (1971) has described how the cognitive system "constructs its own structure" in the course of the individual's interactions with, and adaptation to, the environment. Through the assimilation of new experiences and the accommodation of those experiences into the developing cognitive system, a more and more complex internal representation of reality is acquired by the organism. The result is a cognitive growth cycle in which exploration leads to learning and cognitive integration, followed by further exploration (Elkind, 1971).

The motivational force behind this process involves the organism's intrinsic attraction to "aliments" that are incompletely assimilated and thus act as "food" to the developing structure. Thus, a child might show intense interest in a puzzle or game while it serves as an aliment for the developing structure, only to lose interest once that activity is accommodated and gain interest in a more complex activity. Here, the motivational force can be seen as a potential inherent in the processes of assimilation and accommodation, and the nervous system structures underlying those processes, that is set off by the presence of the aliment (i.e., the system is constructed so that incompletely assimilated activities are intrinsically rewarding; when an aliment is encountered, it is analogous to a hungry organism encountering food).

Language and motivation. All of the examples of motivation considered thus far are biologically based and can be seen to function in animals as well as humans. However, language constitutes a system of behavior control that occurs in all groups of humans and which is not seen in animals. Language is composed of sounds – phonemes – that number about 100 in all human languages. These are combined according to grammatical rules into morphemes, which are *symbols* with arbitrarily defined meanings. Human languages typically have a few hundred thousand morphemes. The magic of language is that these can be combined, again according to grammatical rules, into a virtually infinite variety of sentences that can be decoded by any human with a knowledge of the morphemes and the rules of combination. This quality of language is unique to humans, and it, in effect, frees humans from the restraints of individual experience. Humans can communicate and think about things that have never been and could never be experienced, such as the number of angels that can stand on the head of a pin, or the nature of the interior of a black hole.

Language contains its own motivational implications that are quite different from the motives based upon biological systems. However, the notion that motivation is a potential inherent in systems of behavior control still fits. Language is a system of behavior control – a very powerful one – and inherent in that system are *rules* that must be followed if that system is to operate. These rules involve principles of reason and logic (i.e., if something is black it cannot simultaneously be white). Following or otherwise manipulating these rules is intrinsically rewarding to humans. There is deep satisfaction for the logician in constructing a clear and reasoned argument; for the poet or novelist in evoking strong and clear imagery; or for the propagandist in constructing a persuasive message. Likewise, playing with the rules of language is the source of much humor and amusement, as in puns.

Many analyses of complex human motives involve tendencies toward cognitive completeness and consistency. Thus, many social psychological theories are based upon needs for understanding (as in attribution theory) or cognitive consistency (as in cognitive congruity, balance, or dissonance theories). These "needs" are characteristics of linguistic systems. The classification and categorization of experience into complete or consistent systems makes sense only if that experience is organized linguistically. Experience *qua* experience in its presentational immediacy in the phenomenology of the organism is unique – one's immediate experience of a rose is *always* unique and constantly changing. As Piaget (1971) has shown, it takes many years and a long period of cognitive development before a child begins to organize its experience by principles of reason and logic: to learn that "a rose is a rose is a rose."

It might be noted that an analogous argument can be made for mathematics (e.g., it is a system with its own internal rules that must be followed if the system is to operate). As with language, following the rules of mathematics can generate very strong motivational forces that appear to be unique to humans.

Defining emotion. We have defined motivation as potential inherent in systems of behavior control, ranging from simple reflexes and drives to systems of language and mathematics. However, when something is a "potential," it is never actually experienced in itself. For example, in physics, energy is a potential that is never experienced in itself but rather is actualized or manifested in matter (i.e., in heat, light, and force). Similarly, motivation may be viewed as a potential that is manifested in emotion: with emotion being, in effect, a readout of motivational potential (Buck, 1984, 1985). The *source* of the readout process is in all cases the motivational potential built into the system of behavior control, but the *target* of the readout varies according to its function.

The functions of emotion

There are three targets of the readout process that correspond to three functions of emotion: these involve bodily adaptation, social communication, and subjective experience.

Bodily adaptation

Adaptation and homeostasis. In the first and most basic readout process, termed Emotion I, the target of the readout is the body, and the function served involves the regulation of the internal environment of the body in ways that support life. This involves both *homeostasis*, the maintenance of a relatively constant internal environment that supports life, and *adaptation*, the response of the body to changes in the external environment. These involve constant steady states that must be regulated and maintained at a relatively uniform level. Thus, we must ingest food to maintain optimal blood sugar levels, body temperature, water balance, and so forth within narrow limits despite changes in the external environment.

Three systems are particularly important in bodily adaptation and homeostasis, always in interaction with one another. One is the *autonomic nervous system*, which has two antagonistic branches, the sympathetic nervous system, active during emotional excitement, and the parasympathetic nervous system, active during rest and quiet sleep. The second system is the *endocrine system* of ductless glands (including the pituitary, the adrenal glands, gonads, etc.) that discharge their contents (hormones) into the bloodstream. The third system is the *immune system*, which includes cells formed in the bone marrow (B cells) and thymus gland (T cells) that function to attack and destroy foreign substances interfering with bodily functioning. These cells are particularly active in the lymph nodes, where they cleanse the body's lymphatic fluid, and in the spleen, where they cleanse the blood. The autonomic nervous system and endocrine and immune systems interact with one another in complex ways that are just beginning to be understood, and the well-established relationship between stress and disease is one of the results of this interaction.

Stress and disease. The study of the role of emotion in adaptation and homeostasis has been greatly affected by the work of two pioneering scientists: W. B. Cannon and Hans Selye. Cannon based his "emergency theory of emotion" on his extensive investigations into the functioning of the autonomic nervous system. He argued that whereas the parasympathetic branch servers vegetative functions, such as digestion and the buildup of energy reserves, the sympathetic branch prepares the defense mechanisms of the body for stress. Cannon (1915) termed the latter the *fight or flight response*.

The work of Selye (19750, 1978) greatly clarified the coordinated roles of the autonomic nervous system, endocrine system, and immune system in the response to stress and the problems that can result from the disruption or long-term activation of these systems. Selye defines *stress* as occurring to the extent that the normal homeostatic regulatory mechanisms of the body fail to adapt to the situation: thus, any situation may be stressful if the organism cannot adapt to it.

Selye distinguishes three stages in the response to stress. The first is the *alarm reaction*, which includes an initial shock phase in which the body's resistance to stress actually diminishes, and a countershock phase in which bodily resistance is mobilized. This involves Cannon's fight or flight reaction, which includes the discharge of epinephrine and norepinephrine into the bloodstream from the adrenal medulla. It also involves endocrine responding, including the release of ACTH (adrenocorticotropic hormone) into the blood from the pituitary gland, which among other things signals the adrenal cortex to secrete a number of steroid hormones that are collectively termed corticoids. Selye suggests that these responses function to maintain life while more specific local defenses are organized for dealing with the stress. The latter include immune system responses. The second stage in the stress response is the *stage of resistance*, in which the most appropriate local defense against the stress has been organized. The general stress response is no longer necessary, and autonomic and endocrine system arousal falls to normal levels.

These local responses may eliminate the stress. However, if the stress is too strong or persists for too long, or if new stresses occur, the ability of these local responses to contain the stress break down. In this case, the final *stage of exhaustion* occurs. The autonomic and endocrine arousal recurs, and may prolong survival for a time, but eventually bodily resistance declines and death eventually follows. The body's attempts to deal with stress can themselves be stressful to the body, resulting in what Selye calls *diseases of adaptation*. For example, high levels of ACTH and corticoids can lead to ulceration, cardiac disorder, and high blood pressure.

The Emotion I process involves the most basic function of emotion, but it does not operate independently of the other functions. As we shall see in the next section, the overt expression of emotion is intimately involved in bodily adaptation and the response to stress.

Social communication

In the second readout process, termed Emotion II, the target of the readout involves responses that are accessible to other animals via sensory cues. These accessible responses include certain chemical releases (i.e., pheromones); facial expressions; bodily movements, postures, and gestures; vocalizations – indeed any behavior that can be seen, heard, smelled, felt, or even tasted by other animals. The function served by this readout process is social communication.

The evolution of communication. The importance of this function of emotion has been emphasized in work of the ethologists, who study the role of innate or instinctive patterns in behavior – the "biology of behavior" (Eibl-Eibesfeldt, 1970). Their studies, together with investigations of insect societies and population biology, have formed the basis of *sociobiology*, which approaches social behavior and structures from the point of view of evolutionary theory (E.O. Wilson, 1975). These studies have demonstrated that much social behavior is based upon innate patterns of communication that have evolved within a given species. We shall define communication as the process by which the behavior of one animal (the sender) influences the behavior of another (the receiver).

In 1872, Darwin argued that animals signal their readiness to fight, flee, court, and care for one another through a variety of calls, postures, and facial expressions, and that such "nonverbal communication" is necessary for the regulation of social behavior (cf. Andrew, 1963, 1965). Several general types of such communication may be distinguished, each of which addresses a common social problem. Courting behavior, for example, addresses problems of identifying a suitable mate, overcoming fears and spatial barriers, and coordinating behavior so that fertilization occurs. These problems have existed throughout the evolution of sexual reproduction, and instinctive communication mechanisms involving sex can be seen in the simplest of creatures. Another common social problem that is dealt with by innate communication systems involves aggression. Lorenz (1966) and others have noted that fights between animals of the same species rarely lead to serious injury and death, apparently because instinctive gestures have evolved that regulate aggression. Threat gestures stand in for physical attack, while submissive gestures appear to inhibit aggression in others. In many species, a *dominance order* or "pecking order" is established in which the individuals in a group come to know each other's capabilities, often as they are growing up together. This order can be maintained by gesture of threat and submission, rather than costly and injurious fights.

Many have argued that instinctive behavior does not occur in humans and that learning is of overwhelming importance (e.g., Montagu, 1968). While it is true that learning increasingly controls behavior as one goes up the phylogenetic scale, it also seems likely that instinctive mechanisms are still at work in hidden ways, and that they have more influence than we often realize. We learn much about a person through their nonverbal communication for example, their posture (erect or slumped), eye behavior (do they "look you in the eye"), and tone of voice (loud or soft, high or low-pitched, fast or slow).

Spontaneous versus symbolic communication. This kind of communication differs substantially from the kind of communication process in which a sender consciously constructs and relays a message, or proposition, to a receiver. The latter is clearly *intentional* on the part of the sender, and also *symbolic*, in that the elements of the message are symbols that bear arbitrary relationships with their referents (i.e., the English word "tree" as opposed to the referent object). Such a communication process is also *culturally structured*, in that it requires that both sender and receiver know the meanings of the particular symbols and rules for their combination (i.e., they must "know the language"). Finally, such communication is *propositional* in that it can be false. Bertrand Russell (1903) has defined a proposition as a statement that is capable of being logically analyzed, and the most elementary kind of logical analysis is the ability to be true or false (cf. Buck, 1983, 1984).

In contrast, emotional expressions based upon instinctive communication mechanisms are *spontaneous* as opposed to intentional, that is, they must involve a relatively "automatic" expression of the internal motivational/emotional state that is not under the sender's conscious control. Second, such expressions are *biologically structured* as opposed to culturally structured, that is, they must be based upon sending and receiving tendencies that have evolved within the species, so that the communication process is literally innate. This requires that the expressions in question appear universally in the species, and we shall see that the expressions of certain emotional states do appear to be universal or at least widely generalized in the human species. Third, spontaneous expressions are not "symbols" in that the relationship between the expression and its referent (the emotional state) is not arbitrary. Rather, the expression is a part of the referent – an externally accessible *sign* of the referent. Because a truly spontaneous expression is a part of the referent, it cannot occur in the absence of the referent. Thus spontaneous communication is *nonpropositional* in that it cannot be false: if a spontaneous expression occurs, the relevant emotional state must be present by definition (cf. Buck, 1983, 1984).

Measuring spontaneous communication. If one is to evoke spontaneous emotional expressions from people, one must first evoke emotion, and do it in a setting where the person is not trying to put on a socially appropriate display. Adapting a procedure developed by Robert E. Miller to study the communication of emotion in monkeys, Buck et al. (1972) presented subjects with a series of emotionally loaded color slides. The slides included sexual pictures of nudes, scenic pictures of pleasant landscapes, pleasant pictures of happy people, unpleasant pictures of severe injuries and burns, and unusual pictures showing strange photographic effects. An observer watched the subjects' facial expressions via a hidden television camera (without audio) while they watched and talked about the slides. The observer tried to guess what kind of slide the subject saw on each trial, and how the subject felt about each slide. Results indicated that observers were quite accurate in their overall ability to make judgments about the slides. However, there were large individual differences in the overall expressiveness of different senders. Women were better senders than men; the observers were much more correct when judging the slides viewed by women. Later studies suggested that this sex difference is learned: there is not a large sex difference in expressiveness in young children, but as boys get older (between 3½ and 6 years) they become *less* expressive to the slides, while girls did not change with age. Another interesting result was that persons who had little facial expression also showed a large autonomic (heart rate and skin conductance) response to the slides. Thus "blank-faced" persons who appeared to be feeling nothing, actually showed large internal reactions. In general, males tended to show this *internalizing* pattern of small overt but large physiological responses, while females tended to show the opposite, *externalizing* pattern of response (cf. Buck, 1979).

Emotional expression and health. These results suggest that males and females learn different "display rules" in our culture concerning emotional expression and that this learning may have physiological consequences. Potentially, these could contribute to physical illness: the increased level of physiological responding involved in the internalizing response mode could well become stressful if it were sufficiently strong or carried out over a sufficiently long period of time. Interestingly, Buck, Miller, and Caul (1974) found that persons showing an internalizing mode of response also had apparent difficulty in describing their emotional experience to the slides, describing the content of the slide rather than their own feelings. This is consistent with the concept of *alexithymia* (literally, "no words for mood") that has been found to be common among persons with psychosomatic illness (Lesser, 1981). Nemiah and Sifneos (1970) state that such patients recount events that happen to them and their own actions, but there is a striking absence of the verbal description and expression of feelings: "the way to the patient's inner life seems to be blocked by an impregnable wall" (Nemiah and Sifneos, 1970, p. 156). It may be that the free expression of feelings, both in facial expression and in words, is important in moderating the bodily effects of stress. More research is needed before we fully understand the complex interrelationships between emotion, emotional expression and inhibition, stress, and disease, and the possible factors that moderate these relationships.

Subjective experience

In the third readout process, termed Emotion III, the target of the readout is the cognitive system, and the function is to inform the cognitive system about the state of the motivational system in question. This allows the cognitive system fast and easy access to motivational states in which cognitive processes may be instrumental in the process of satisfaction. For example, individuals appear to be directly aware of their needs for food, drink, temperature regulation, and air; the state of tendencies toward fight, flight, and courtship; and feelings of happiness, sadness, surprise, and so forth. This subjective experience of motivational and emotional states has often been ignored by behaviorally oriented researchers because of the difficulty of observing it objectively. However, new techniques of brain recording have made it possible to become more objective about the reality of subjective experience. Also, the logic of the ethologists about the evolution of the social communication of emotion can also be applied to the analysis of subjective experience, suggesting that such experience is not a mere fluke or epiphenomenon, but has a real function in the evolution of behavior control.

The argument goes as follows: we have seen that the cognitive system is a general purpose processing system that is structured by the organism's experience in the environment. It can be argued that, given that a cognitive system sensitive to the external environment has evolved in a given species, it would make sense that that system also be sensitive to the internal, bodily environment. Such internal sensitivity would make possible the detection of deficits in food or water *before* dangerous bodily imbalances occur, the sensing of fight or flight tendencies *before* overt actions actually become necessary, and the feelings of affects *before* they are so strong that they must be expressed uncontrollably. Such a feed forward mechanism would facilitate the *planned* response to motivational and emotional states. Just as an external readout of emotion has evolved because it fosters the coordination of behavior in social species, it can be argued that an internal, subjective readout has evolved in all species with significant cognitive capacities because it fosters the cognitive handling of those states (cf. Buck, 1984).

Emotion and cognition

The notion of Emotion III bring us directly to the question of the relationship of emotion and cognition. This section examines a number of theoretical viewpoints relevant to this relationship. On first glance, these theories appear to be incompatible with one another, and, in fact, they are generally considered to be competing theories. However, each theory can be considered to be correct as far as it goes but incomplete. Each may be correct in its treatment of some aspects of emotional responding, but incorrect if generalized to other aspects. If the differences in essential subject matter are recognized, the theories can be viewed as compatible with one another.

The James–Lange versus Cannon debate

The James–Lange theory. The central notion of the James–Lange theory is that an emotional stimulus produces both visceral and skeletal muscle changes and that these are the source of the subjective experience of emotion: "Bodily changes follow directly the perception of the exciting fact, and the feeling of these same changes as they occur *is* the emotion" (James, 1884, reprinted in 1968, p. 19). It is noteworthy that for James there is no intervening cognition of emotion that precedes the bodily response – the bodily response is the emotion. We see a bear, our heart rate rises, we tremble, and run. Our feeling of fear is our feeling of the heart rate change, trembling, and running (see Figure 14.1). This reverses the usual notion that we run because we are afraid, or cry because we are sad, or laugh because we are happy. For James, we feel sad because we cry, and happy because we laugh.

Cannon published a review and criticism of the James–Lange theory in 1927. Among his major criticisms was that bodily sensation is too slow, diffuse, and insensitive to account for the speed and wide range of human emotional

James - Lange

Thalamic

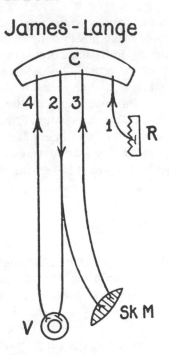
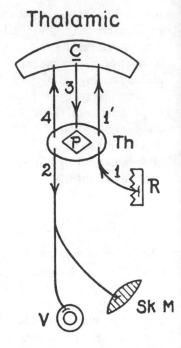

Figure 14.1. Left: Diagram of the James–Lange theory of emotion. R, receptor; C, cerebral cortex; V, viscera; Sk M, skeletal muscle. The connecting lines represent nerve pathways; direction of impulses is indicated by arrows. Right: Diagram of Cannon's thalamic theory of emotion. R, receptor; C, cerebral cortex; V, viscera; Sk M, skeletal muscle; Th, thalamus; P, pattern. The connecting lines represent nerve pathways; the directions of impulses are indicated by arrows. Corticothalamic pathway 3 is inhibitory in function. (From Cannon, 1931. Reprinted by permission.)

experience (Cannon, 1927, reprinted in 1968). In recent years, two kinds of responses have been made to Cannon's critique, which we shall review below. First, we shall examine Cannon's own theory of emotion. It has two aspects: one is concerned with the genesis of emotional experience; the other involves the analysis of the true functions of the bodily response.

Cannon's theory. The "thalamic" first proposed by Cannon (1931), served as a starting point for later models that stress the relationship of higher cortical mechanisms and emotional systems in the subcortical and limbic structures of the brain (see Figure 14.1). Essentially, the theory notes that stimuli reach the cortex via subcortical systems and suggests that if the stimuli are of an emotional nature, the subcortical systems simultaneously and independently inform the cortex (resulting in emotional experience) and the relevant peripheral bodily systems (resulting in bodily responses). Cannon thought, incorrectly, that the critical subcortical mechanism is located in the thalamus.

In 1937, Papez suggested that instead the bodily *expression* of emotion involves the hypothalamus, which in turn controls the autonomic and endocrine systems, while the *experience* of emotion involves the limbic system. This notion is now widely accepted and is consistent with many studies of brain functioning in both animals and humans (cf. Buck, 1976). The fundamental notion also appears in Malmo's (1975) notion of an archaic brain interacting with more newly evolved structures, and MacLean's (1970, 1973) theory of a triune brain consisting of hierarchical control systems associated with reptilian (subcortical), old mammalian (limbic system), and new mammalian (neocortical) levels of functioning.

Cannon concluded his critique of the James–Lange theory by arguing that the effects of peripheral bodily changes on emotional experience are, in fact, quite small.

The processes going on in the (body) are truly remarkable and various; their value to the organism, however, is not to add richness and flavor to experience, but rather to adapt the internal economy so that in spite of shifts of outer circumstance the even tenor of the inner life will not be profoundly disturbed (Cannon, 1927, reprinted in 1968, p. 51).

In short, the peripheral changes serve the functions of adaptation and homeostasis. This emergency theory formed the basis of *arousal theory*, which has an important place in theories of the interaction of emotion and cognition.

The cognitive–physiological interaction

Self-attribution theory. As noted above, two major kinds of attempts have been made to answer Cannon's critique of the James–Lange theory. One is that there are two sorts of emotional experience that interact with one another. The first is produced by cognitions arising from the individual's understanding of the situation that aroused the emotion – it is fast and well differentiated. The second is produced by the bodily sensations – it is slower and more diffuse. This notion is at the heart of Schachter's *self-attribution theory* of emotion (Schachter, 1964), which holds that emotion is a joint function of bodily arousal and cognitive interpretations or *labels* for that arousal. If either the bodily arousal or the cognitive interpretation is absent, Schachter suggests that the emotion is incomplete.

This notion allowed Schachter (1964) to integrate much of the discussion concerning the James–Lange versus the Cannon theories of emotion. In the first place, there is considerable evidence that visceral feedback from autonomic nervous system arousal often contributes to emotional experience, while, on the other hand, such feedback does not appear to be either necessary or sufficient for the occurrence of all kinds of emotional experience. For example, Hohmann (1966) interviewed patients who had suffered spinal injuries and asked them to compare their emotional experience before and after the injury. He found that the reported range, duration, and intensity of emotional experience is reduced in such patients, and that the extent of reduction is greater the higher the spinal damage (and thus the greater the loss of visceral sensation). For example, one described feeling a cognitive kind of emotional experience when he was some distance from shore in a leaking boat in a storm: "I knew I was sinking and I was afraid all right, but somehow I didn't have that feeling of trapped panic that I know I would have had before." Others reported acting angry without really feeling angry:

it's sort of a cold anger. Sometimes I act angry . . . yell and cuss and raise hell, because if you don't do it sometimes, I've learned people will take advantage of you, but it just doesn't have the heat to it that it used to. It's a mental kind of anger (Schachter, 1964, pp. 74–75).

In a related observation, a patient who had one side of the sympathetic nervous system removed due to cancer reported that "his previous and customary sensation of shivering while listening to a stirring passage of music occurred in only one side and he could not be thrilled on the sympathecto-mized half of his body" (Delgado, 1969, pp. 134–135).

One of the major effects of sympathetic activation is the increased secretion of epinephrine into the bloodstream from the adrenal medulla. The circulating epinephrine has bodily effects similar to those of sympathetic activation, and it has been suggested that this provides a chemical "backup" to the faster but more transient sympathetic arousal (cf. Buck, 1976). An injection of epinephrine thus artificially causes many of the bodily changes felt in strong emotion: the heart races, the hands tremor, the face feels flushed, and so forth. Bertrand Russell (1927, reprinted in 1961) noted that an injection of epinephrine given him by a dentist made him feel *as if* he was angry, or excited, or frightened, but he knew he was not because there was no *reason* to be. This is the opposite of the situation experienced by the patients with spinal injuries: with them the cognitive element of emotion was present, but the physiological aspect absent, with Russell the physiological aspect was present but the cognitions absent. Schachter and Singer (1962) suggested that if a person was in a state such as Russell's but had no ready explanation for the state, he or she would tend to search in the environment for clues about the reasons for their feelings, that is, people would be motivated to try to explain or label their feelings. The emotion that resulted would depend on the explanation: thus the same arousal state might be labeled as joy or anger, depending on the situation (see Figure 14.2).

To test this theory, Schachter and Singer (1962) studied the effects of different cognitive labels on the reaction to an injection of epinephrine. Some subjects were injected with epinephrine: others were injected with a saline placebo that produced no physiological effects. The epinephrine-in-

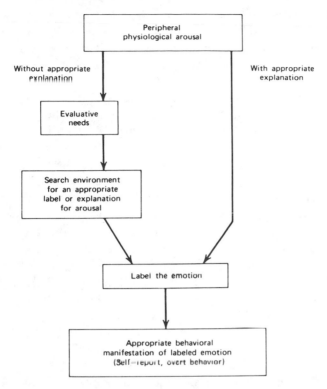

Figure 14.2. Schachter's self-attribution theory of emotion. (From Figure 2.6 in Buck, 1976. Copyright © 1976. Reprinted by permission of John Wiley & Sons, Inc.)

jected subjects were either *informed* correctly about the drug's effects, *misinformed* (i.e., told that it might cause numbness and a slight headache), or were *uninformed* (told nothing). All subjects waited with another person who had supposedly also received an injection, but who actually served as a model for the subject to follow in labeling the arousal. In one condition, the model behaved in a boisterous and euphoric manner; in the other, he behaved in an angry manner. The experiment examined the overt behaviors and self-reported emotional experience of the subjects to determine whether they were influenced by the model. As expected, the uninformed and misinformed subjects were more influenced than the informed subjects, acting and reportedly feeling more euphoric or angry, depending on the condition. Schachter and Singer argue that they were in a state of arousal for which they had no explanation. By providing an explanation by way of the model, different "emotions" were produced.

Appraisal and coping. Another research program that has investigated the interrelationships of emotion and cognition is that of R. S. Lazarus (1966).

He and his colleagues began by investigating the effects of how a threatening event is interpreted or appraised by the subject. Lazarus defines appraisal as the "evaluation by the individual of the harmful significance of some event" (Lazarus and Opton, 1966, p. 244). He has demonstrated that a given stimulus may or may not be stressful, depending upon the subject's cognitive appraisal.

Lazarus and his colleagues have studied the effects of gruesome films – one showing a circumcision-like ritual performed on adolescent boys in an Australian aboriginal culture, another depicting gory industrial accidents – upon autonomic (skin conductance) arousal. They showed that the arousal induced by the films could be lessened by commentaries denying the harmful significance (i.e., asserting that the ritual was actually a happy occasion), or encouraging intellectualization by describing the scene technically. On the other hand, the arousal to the films could be increased by a commentary emphasizing the horror and pain in the films (Spiesman et al., 1964). Other studies demonstrated that arousal to the films could be short circuited by denial information presented before the film showing. This was more effective, in fact, than denial information presented with the film (Lazarus and Alfert, 1964).

It is clear from these studies that "emotions are highly sensitive to changes in the person–environment relationship and the way these changes are appraised" (Lazarus, 1984, p. 128). However, does this mean that emotions are based upon appraisal? Lazarus argues that it does. Averill, Opton, and Lazarus (1969) suggest that potentially threatening stimuli are first evaluated by the cognitive process of appraisal. If the stimuli are appraised as threatening, coping strategies are cognitively selected to deal with the threat. Thus cognition is considered to be central to emotion. Recently, Lazarus (1984) has expanded on this theme, arguing that "cognitive activity is a necessary precondition of emotion because to experience an emotion, people must comprehend . . . that their well being is implicated in a transaction, for better or worse" (p. 124). This need not be a conscious, rational sort of cognition, however. Lazarus states that appraisal can involve a conscious, rational, symbolic process, but that it can also be a "primitive evaluative perception" (Lazarus 1984, p. 124) that is "global or spherical" (Lazarus, 1982, p. 1020). Lazarus appears to be making a distinction here between two different types of cognition, but this is not explicitly acknowledged. We shall return to this point.

Limitations of the interactionist viewpoint. The interactionist viewpoint is powerful and able to integrate much of the research on emotion, and many current theorists, including Averill (1980), Mandler (1984), and Scherer (1984), accept it in one form or another. However, there are instances where emotional responses can occur without apparent cognitive involvement. The theories of Cannon, Papez, and MacLean imply that emotion can occur prior to, and independently of, cognition, through the activation of subcortical and limbic system structures. Zajonc (1980) has reported evidence consistent with this view, suggesting that preferences between stimuli affect their evaluation even before the stimuli can be recognized (Kunst-Wilson and Zajonc, 1980; W.R. Wilson, 1979).

Another area of difficulty for a strong interactionist approach involves observations of the effects of brain stimulation and drugs, which can produce apparently uncontrollable and irresistible feeling states that bear little relationship to the person's understanding of the external situation (Buck, 1976). In humans, brain stimulation delivered via telemetry without the patient's knowledge has been shown to produce sudden rage (Mark and Ervin, 1970) or unexpected pleasure (Heath, 1964). Similarly, psychoactive drugs can produce feeling states ranging from euphoria to terror to rage that are largely, if not completely, independent of cognition.

These and other observations suggest that apparently complete emotional states, with appropriate expressive behavior, goal-directed behavior, and self-reported experience, can be elicited in humans from brain stimulation and psychoactive drugs and also from diseases that affect certain brain centers (Buck, 1976). This suggests in turn that emotional experience must to some extent be a direct function of activity in specific neurochemical systems in the brain. This is consistent with Cannon's view of emotion, and also Tomkins' theory, to which we shall now turn.

Tomkins' theory of emotion

The second answer to Cannon's criticisms of the James–Lange theory is based on the contention that the skeletal muscle activity associated with emotion, particularly facial expression, is responsible in part for emotional experience. Ekman, Friesen, and Ellsworth (1972) argue that the facial muscles respond sufficiently quickly and are sufficiently differentiated to account for emotional experience: "The face might . . . fill the informational gap left by a solely visceral theory of emotion, distinguishing one emotion from another, changing rapidly and providing feedback about what is occurring to the person" (p. 173). This hypothesis that facial expressions provide feedback to the responder that determines emotional experience and behavior is termed the *facial feedback hypothesis*.

One of the most influential exponents of the facial feedback hypothesis is Tomkins (1962, 1963, 1983). Tomkins posited that there are innate neural programs, probably in the subcortical and limbic regions of the brain, associated with the primary affects: happiness, sadness, fear, anger, surprise, interest, shame, and disgust/contempt. He argued that each of these is associated with a specific facial display that is universal in the human species.

The notion of universal facial expressions has received strong support, particularly from the work of Ekman and Izard. Ekman found that seven of Tomkins' primary affects (all except shame) consistently emerge when photographs of posed facial expressions are judged. In addition, Ekman and his colleagues found that these expressions could be correctly judged all over the world, even in remote preliterate cultures in Borneo and New Guinea (Ekman, Sorenson, and Friesen, 1969). Izard (1971) likewise found photographs of posed expressions associated with the primary affects to be meaningful in widely different cultures.

The facial feedback hypothesis, suggesting that these expressions are directly involved in the subjective experience of emotion, is more controversial and remains unproven (Buck, 1980). For example, it implies that if emotion were not expressed it would be "softened." This is not consistent with evidence noted above that the autonomic responses of overtly expressive persons are often smaller than the responses of less expressive persons. It seems clear that emotion does not "go away" if the face is unexpressive. It is noteworthy in this regard that Ekman has been recently quoted as acknowledging that "there is emotion without facial expression and facial expression without emotion" (Mandler, 1984, p. 170).

The various views of emotion and cognition that we have considered appear at first to be incompatible with one another. However, as we have seen, there may be differences in the ways that different theorists are using terms such as "cognition." These differences are relevant to the emerging evidence that the right and left hemispheres of the brain may be associated with different sorts of cognition.

Cerebral lateralization

Cerebral lateralization is a relatively new, but extremely promising area of study that has important implications for the analysis of the relationship of cognition and emotion. First, the study of the functions of the left and right hemispheres suggests that there are, in fact, different sorts of cognition – different way of knowing reality. Second, much research and clinical experience indicates that the right and left hemispheres play markedly different roles in the expression and control of emotion.

Cerebral lateralization and cognition. There is considerable evidence that the left and right cerebral hemispheres are associated with different kinds of cognition. The left hemisphere is associated with cognition in the usual sense of the term; it involves the *interpretation* of sense data. It is analytic, breaking sense data into meaningful pieces that are equivalent to one another, for example, a rose is a rose is a rose. It is also linear, tending to organize these pieces into sequences, and symbolic, in that it is adept at attaching sense data to learned shapes and sounds. These abilities explain the great importance of the left hemisphere in the control of language. In most persons, when the left hemisphere is damaged in certain places, crippling disorders of language (aphasias) result.

The right hemisphere, in contrast, is associated with a wholistic, synthetic sort of cognition that has been termed syncretic cognition (Tucker, 1981). Tucker suggests that this provides an integration of sensory information from different channels (visceral, visual, auditory, tactual, etc.) into a "superordinate conceptualization." He also suggests that such a conceptualization is particularly important in emotion, providing the "kind of perception and behavioral organization necessary for adequate emotional functioning" (p. 22).

The distinction between syncretic and analytic cognition may be the key to the resolution of the controversy between Lazarus and Zajonc. We saw above that Lazarus includes "primitive evaluative perceptions" in his definition of cognition, while Zajonc in contrast requires that sensory input be transformed in some way in order for something to qualify as cognition: " 'pure' sensory input, untransformed according to a more or less fixed code, is not cognition. It is just 'pure' sensation" (Zajonc, 1984, p. 118). Both accounts are reasonable descriptions of syncretic cognition, in Tucker's terms, although Lazarus does not distinguish it from analytic cognition, while Zajonc does not consider it to *be* cognition.

The distinction between syncretic cognition and analytic cognition is analogous to a long-standing distinction in epistemology between knowledge by acquaintance and knowledge by description. Bertrand Russell defines knowledge by acquaintance as "direct sensory awareness without the intermediary of any process of inference or any knowledge of truths" (Russell, 1912, p. 73). It involves the presentational immediacy of experience and is completely self-evident. Russell notes that the experience of feelings and intentions (emotional and motivational states) is knowledge by acquaintance. William James describes knowledge by acquaintance as follows:

I know the color blue when I see it, and the flavor of a pear when I taste it . . . but *about* the inner nature of these facts or what makes them what they are I can say nothing at all. I cannot impart acquaintance with them to anyone who has not made it himself (James, 1890, p. 221).

In contrast, knowledge by description involves the *interpretation* of sense data, which would seem to require processes of transformation and analysis (e.g., analytic cognition). It is noteworthy that both Russell and James appear to agree with Zajonc's view of the primacy of affect in that they argue that knowledge by description depends upon knowledge by acquaintance and follows it in time: "Feelings are the germ and starting point of cognition, thoughts the developed tree" (James, 1890, p. 222).

The evidence that syncretic knowledge by acquaintance is particularly associated with the right cerebral hemisphere functioning and analytic knowledge by description is particularly associated with the left puts this epistemological theory in an entirely new light, suggesting that we are approaching the point where a biological theory of human knowledge may become a powerful tool for understanding human nature. The details of how the right and left hemispheres function in this regard, and how they interact with one another, is of course very poorly understood at this time. However, there is evidence that emotion is intimately involved in these processes.

Cerebral lateralization and emotion. Recent evidence suggests that the input of the subcortical and limbic system mechanisms associated with emotion to higher brain structures is lateralized, even in lower animals. Thus, important fiber systems associated with motivation and emotion, such as the medial forebrain bundle, are right-lateralized, and Denenberg and his colleagues have shown that right hemisphere damage causes decreased emotionality even in rats (Denenberg et al., 1978; Denenberg, 1981). In humans, there are striking anatomical differences between the right and left hemispheres that have been interpreted as reflecting the right hemisphere's specialization for emotion and the left hemisphere's specialization for language in most right-handed persons (Geshwind, 1979).

There is much recent evidence suggesting that the right hemisphere plays a special role in both the expression and recognition of emotion. In normal right-handed persons, right hemisphere activation occurs when answering emotion-related questions (Schwartz, Davidson, and Maer, 1975) and during stress (Tucker et al., 1977). Also emotional information is processed more efficiently when presented to the right hemisphere via the left visual field (cf. Safer, 1981) or the left ear (Haggard and Parkinson, 1971; Carmon and Nachshon, 1972; Safer and Leventhal, 1977). Moreover, there is significant left-facedness during the posing of emotional expressions in right-handed subjects, suggesting either relative right hemisphere activation or left hemisphere inhibition of the display (Campbell, 1978; Sackeim, Gur, and Saucy, 1978; Borod and Caron, 1980; Ekman, 1980; Borod, Caron, and Koff, 1981; Ekman, Hagar, and Friesen, 1981; Buck, 1982).

Patients with right hemisphere damage show a marked lack of emotional responsiveness that has been termed "emotional indifference." This is also the case with normal volunteers whose right hemisphere has been temporarily inactivated by an injection of the drug amobarbitol into the carotid artery (Terzian and Ceccotto, 1959; Terzian, 1964; Rossi and Rossadini, 1967; Gainotti, 1969, 1972; Galin, 1974). Right hemisphere-damaged patients also have particular difficulties with emotional speech as opposed to propositional speech (Heilman, Scholes & Watson, 1975; Tucker, Watson, and Heilman, 1977) and with the recognition and discrimination of emotional faces and pictures (Cicone, Wapner, and Gardner, 1980; DeKooky et al., 1980; Benowitz et al., 1983). Also, such patients show reduced skin conductance responding to pain (Heilman, Schwartz, and Watson, 1978) and emotionally loaded color slides (Morrow et al., 1981).

There is also evidence that the left hemisphere may be involved in the control of the emotionality associated with the right. Buck and Duffy (1980) demonstrated that left hemisphere-damaged patients are more expressive on the slide-viewing measure of spontaneous communication than are right hemisphere-damaged patients, even though the former are much less able to communicate using intentional gestures and pantomime (cf. Duffy, Duffy, and Pearson, 1975). Data in this study suggested that the left hemisphere is associated with the control of the spontaneous display – altering, modulating, and masking the spontaneous expression of emotion with "facial management strategies" (Ekman and Friesen, 1975). Tucker (1981) has argued that the left hemisphere is associated with anxiety and the verbal modulation of the emotionality of the right hemisphere, and he and his colleagues have demonstrated that subjects use verbal techniques to control emotion (Shearer and Tucker, 1981; Tucker and Newman, 1981). In effect, the sequential, analytic, and verbal sort of cognition associated with the left hemisphere appears to control the syncretic cognition of the right hemisphere, which is connected more closely to the subcortical and limbic system mechanisms of emotion.

Additional evidence comes from a fascinating and important study by Sperry, Zaidel, and Zaidel (1979) on two patients with "split brains," in whom the right and left hemispheres have been surgically separated by cutting the corpus callosum. This is done to combat certain kinds of epilepsy. Previous research has shown that this creates two largely independent centers of consciousness, each exhibiting a pattern of abilities and qualities peculiar to that hemisphere (cf. Sperry, Gazzaniga, and Bogen, 1969). The patients were tested with a specially designed contact lens which restricted vision to one half the visual field. Since the left visual field projects exclusively to the right hemisphere and the right visual field to the left, this allowed the presentation of visual stimuli to the desired hemisphere, permitting prolonged examination and free scanning of the stimulus by the patient. The stimuli were pictures, photographs, drawings, and so forth in arrays of 4–9 items arranged on 25 × 25 cm cards. Emotionally loaded material, such as pictures of the patient; the patient's family, pets, friends, and belongings; familiar scenes; and familiar public/historical figures (Hitler, Churchill, Richard Nixon), were interspersed among neutral material. Patients were asked to evaluate the items by giving a thumbs up or thumbs down gesture, and they were asked to point to pictures they liked, disliked, or recognized.

When presented to the left hemisphere, the affect-laden stimuli produced prompt recognition and evaluation. When presented to the right hemisphere, they produced marked emotional reactions, and the patient could point to the critical item, but could not verbally explain the response. For example, in a case where pictures of the patient were unexpectedly included in an array, the patient examined the material for abut 7 sec, and then exclaimed: "Oh no!... Where'd you g... What *are* they?... (laugh)... Oh God!" (Sperry et al., 1979, p. 158). When the array was removed and the examiner asked what was in the picture, the patient responded in a loud emphatic voice: " 'Something nice whatever it was... Something I wouldn't mind having probably.' This was followed closely by another loud laugh" (p. 158).

Sperry et al. (1979) interpret this reaction as follows:

The emotional components of the reaction triggered in the right hemisphere crossed rapidly to the left hemisphere through brainstem mechanisms and colored the tone of speech in the vocal hemisphere. However, the content of the subject's remarks shows that the left hemisphere remained unaware of the exact stimulus material that had triggered the emotional reaction in the other hemisphere (p. 159).

They also note that the emotional responses from the right hemisphere were "more intense and less restrained" than those from the left (Sperry et al., 1979, p. 156). These observations seem consistent with the notion that the right hemisphere is responsible for emotional responses that are controlled by the left.

We have considered a number of theories relevant to cognition and emotion, arguing at the outset that they may be considered to be compatible views that are directed toward different aspects of emotional functioning. A model of emotion that contains aspects of all ·of these views is presented in the section on the readout model of emotion.

The readout model of emotion

This model is illustrated in Figure 14.3. It assumes that emotional stimuli impinge on the neural systems underlying motivation and emotion directly and without cognitive mediation. The impact of a particular emotional stimulus for a given person is determined by (1) the state of arousal and arousability of the neural system in question, and (2) the individual's relevant learning experiences associated with that stimulus, or the conditioned emotional response (CER) to the stimulus. Thus, stimuli are in effect "filtered" by the arousal/arousability of the relevant neural systems and the relevant

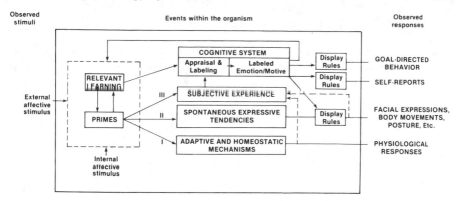

Figure 14.3. The readout model of emotion. (From Buck, 1984.)

learning experiences with the stimulus, and this filtering determines the impact of the stimulus for the given individual.

This impact is registered on both cognitive and emotional levels. On the emotional level, homeostatic and adaptive processes are activated (Emotion I), tendencies toward spontaneous expression occur (Emotion II), and subjective experience occurs, both directly as in the Cannon–Bard theory (Emotion III) and indirectly via feedback from the body as in the James–Lange theory. The latter includes both feedback from the adaptive/homeostatic responses and from expressive behavior (i.e. facial feedback) as is shown by the dashed lines.

On the cognitive level, the cognitive–physiological interaction occurs in which the individual appraises or labels the affective stimulus on the basis of past experience, the subjective emotional experience, and the present situation. Once this interpretation has taken place, the individual has a basis for coping with the stimulus, that is, making appropriate goal-directed responses. One aspect of these goal-directed responses involves following the learned display rules about what kind of emotion is appropriate in a given situation. These rules influence the self-reports of emotional experience and the expressive behavior.

The more emotional aspects of this process (i.e., on the lower right side of Figure 14.3) appear to particularly involve right hemisphere functioning, while the more cognitive aspects on the upper right appear to involve the left hemisphere. The whole process as described thus far can be seen in animals as well as humans. The difference between animals and humans, I suggest, lies in the fact that in humans the phenomenon of language introduces a new system of behavior control that is functionally independent of this readout process.

Language and human emotion

This section suggests how the control of behavior in humans and animals is in some ways fundamentally different. It first suggests that the fundamental difference between the control of behavior in humans versus animals involves language and then discusses the power of linguistically structured social rules in the motivation of human behavior.

Language and the control of behavior

Human motivation and emotion are based upon biological systems, as they are in all animals. In both humans and animals, learning and cognitive factors build up an internal representation of reality that influences these motives and emotions. Clearly this kind of cognitive–emotional interaction is not unique to humans. What *is* unique to humans is language, which has created a culturally patterned system of behavior control that is functionally independent of biology and fundamentally different from anything seen in animals. "Only in humans does behavior come so completely under the control of factors that are mediated by language, including logic, reason, and social rules" (Buck, 1985, p. 406).

This does not mean that there is a fundamental biological gap between humans and animals – that is clearly not the case. It is not necessarily the case that animals cannot learn language. There is evidence in fact that the higher apes (gorillas and chimpanzees) may be capable of significant linguistic skills if taught sign language. It *does* mean that if an animal did learn language, it would cause a fundamental change in the animal's behavior, because the behaviors would now be controlled by a different sort of structure, a different principle of organization, *on top of the old* (Buck, 1985).

The principles of organization inherent in language are functionally independent of the principles of organization in the motivational systems we have considered thus far. Language involves more than a simple elaboration of those systems: It involves new principles for the organization and control of behavior. Piaget (1971) has shown how the process of cognitive development culminates in the ability to use formal logical operations independent of experience to reason about things that have never been and never could be actually experienced. With language also comes the conformity and obedience to culturally patterned social rules.

The power of social rules

In 1963, Milgram published the results of a study that continues to stir considerable controversy. Essentially, he demonstrated that most subjects would continue to administer what they thought were increasingly painful and potentially dangerous electric shocks to a victim on the orders of an experimenter, even though the victim was screaming in pain, complaining of heart trouble, and begging to be freed. The subjects in the study were extremely hesitant to shock the victim, but continued to comply with the experimenter's orders nevertheless. The extent of the compliance was not expected. In fact, Milgram described the procedures of one condition to 40 psychiatrists, and asked them to predict how many subjects would continue to give increasingly severe shocks to the end of the scale. The psychiatrists predicted that less than 2% would give the strongest shock possible. Actually in that condition, over 60% went to the end.

Milgram argues that these studies demonstrate the enormous and largely unrecognized power of social norms and rules. The experimenter is an established authority figure, and the situation is structured so that the subject must break the rules and openly disobey such a figure. Many subjects could not do this, despite their personal reluctance to hurt the victim. Milgram suggests that there is a great reservoir of potential anxiety that prevents us from openly breaching social rules, but that this anxiety is unnoticed until we actually begin to break the rule (Tavris, 1974).

The system of behavior control represented by linguistically structured logic, reasoning, and social rules is responsible for human civilization. However, its effects are not universally benign. We noted above that many animal species have innate controls that regulate aggression between species members. Lorenz (1966) has suggested that such innate controls are relatively weak in humans because we are not biologically equipped with natural weapons that posed a threat during evolution. He posits an equilibrium between natural killing power and inhibitions of aggression, so that animals with natural weapons (e.g., the teeth of a wolf or the beak of a raven) have evolved strong inhibitions to prevent them from killing their own kind. Humans do not have such inhibitions, but our linguistic powers have led to the invention of artificial weapons with awesome potential. We cannot rely on natural inhibitions built in to human nature to prevent us from using these weapons on ourselves, just as the victims in Milgram's study (and many other human victims) could not rely on the very real concern of their tormentors to prevent them from being injured. We can only hope that the same ability to reason that led to the invention of these weapons will prevent their use.

The power of linguistically structured social rules in the control of human behavior explains the ability of purely cognitive approaches to social behavior and personality to account for so much without considering innate emotional expression and communication. However, it has become clear that behavior is simultaneously affected by phylogenetically ancient systems of control, and that an ultimate understanding of human nature will depend upon the consideration of both sorts of systems and how they interact with one another. This will depend, in turn, upon increased communication and collaboration between specialists in the biological, neural, cognitive, and social sciences.

Summary

I have argued that reflexes, drives, and affects such as happiness and sadness can all be considered to be motivational systems at different levels of organization and that emotion can be regarded as involving a kind of running progress report, or readout, about the state of those motivational systems. There are three sorts of readout corresponding to the three functions of emotion. The most basic kind of readout involves homeostasis and adaptation and takes place via peripheral bodily responses, such as endocrine and sympathetic nervous system arousal (Emotion I). The second kind of readout is the external display of the state of the motivational systems via responses which are accessible to others: odors, postures, color changes, facial expressions, and so forth (Emotion II). The third kind of readout involves the direct subjective experience of the state of the motivational systems (Emotion III). Different kinds of emotional behaviors are differentially involved in these systems. Emotion I is measured most directly by responses that reflect homeostasis and adaptation, such as autonomic nervous system responses, measures of immune system functioning, hormonal assays, and the like. Emotion II is measured most directly by expressive behavior, such as facial expressions and postures. It is important that these behaviors be *spontaneous*, that is, it is important that the responder be unaware that his or her expressive behavior is of interest, because expressive behaviors can be altered to follow learned and linguistically structured "display rules." Self reports of emotional experience are the most practical access to Emotion III in the absence of direct brain measures, although they may also be influenced by display rules.

15   The neurobiology of emotion

Joseph E. LeDoux

This chapter is about the biology, and in particular, the neurobiology, of emotion. The aim is to relate concepts concerning emotion to the structure and function of the nervous system. The availability of new knowledge about and ways of studying the nervous system make it at least conceivable that the last frontier of subjective existence, our emotions, will someday be knowable in terms of neurocellular and molecular events.

Several assumptions underlie this discussion of the neurobiology of emotion. First, the various processes related to emotion can be viewed as information-processing functions of the brain. Second, by tracking the flow of information from point to point in the nervous system, we can uncover the neuroanatomical substrates of the emotional processes. Third, neuroanatomical circuit specification is a crucial first step in the arduous task of relating functional concepts of emotion to cellular and molecular mechanisms.

In keeping with these assumptions, the focus of this chapter will be on the neuroanatomical pathways involved in transmitting emotional information between regions within the brain and between the brain and the peripheral nervous system. This selective review will not cover other important areas, such as pharmacological and biochemical studies of emotion and emotional disorders (for reviews, see Gray, 1982; Sourkes, 1981; Berger and Barchas, 1981). We begin with a brief survey of the historical development of our contemporary views of emotional biology and then turn to current issues and research.

The biology of emotion: some historical highlights

The history of ideas concerning emotion dates back to ancient Greece, if not before. However, the focus here will be on select historical developments that have shaped our current views of brain and emotion.

301

Descartes' passions and the brain

Contemporary thoughts about mental processes can be traced to the seventeenth century and, in particular, to Descartes. He codified the concept of mind as an internal system for representing events occurring in and knowledge about the external world. His view that mind is a nonphysical substance (i.e., consciousness), distinct from the body and other physical objects, set up the classic debate that has come to be known as the mind–body problem. Our interest here is not in the mind–body problem per se but instead is in one facet of Descartian consciousness, namely, the passions (emotions) of the soul (mind).

Descartes' view of the passions incorporates what is perhaps the first neurological model of emotion (Figure 15.1). For him, bodily actions common to man and animal (such as breathing, the circulation of the blood through the body, flight from danger) are purely mechanical responses managed without the aid of the mind. When asked by a skeptical colleague whether a mental state of fear does not intervene between seeing a wolf and running, Descartes replied that the brain regulates flight from danger automatically. The danger signal is carried to the brain as a visual message by way of the optic nerves and upon arrival the brain drives the "vital spirits" in the descending nerves to produce muscular and glandular action appropriate to the stimulus. These actions are accomplished in much the same way that the motions of a clock are produced by the force of its springs and the shapes of its wheels. There is a difference, though, between a human and an animal fleeing from danger: only the human experiences fear. Even in the human, the experience of fear does not cause flight. Fear and other passions reflect passive awareness by human consciousness of bodily actions. Passions are thus mental consequences rather than causes of muscular and glandular action. Because animals are without souls (conscious minds), they are without passions.

For Descartes, then, mind was not a function of the brain. The brain belonged to the body; the mind was of a completely different nonphysical substance. When the mind was influenced by the body (by way of the brain's pineal gland), passions were the result. The mind, of course, could will bodily actions (also by way of the pineal gland). Such actions, though, were the consequence of reason rather than of passion. And one of the important functions of the soul's capacity to reason was to bring the passions under control.

As a neurological model of emotion, Descartes' views were tremendously insightful, but incomplete. He recognized that bodily actions must be based on the evaluation of stimulus significance by the brain, but he failed to specify how significance was "mechanically" determined. Descartes' primary objective was not to explain emotion. He was first and foremost concerned with

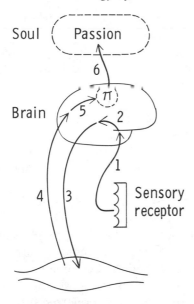

Figure 15.1. Descartes' model of the passions. In the seventeenth century, Descartes proposed that emotional experiences involve a chain of events starting with the reception of sensory input and its transfer to the brain (1). Within the brain, the sensory signal is "mechanically" (without conscious awareness) evaluated (2). Descending nerves (3) to the muscles and internal glands produce physical responses appropriate to the emotion-provoking stimulus. The brain is informed of the consequences of these actions by way of additional sensory nerves (4) and this information is relayed (5) to the pineal gland (π). When the pineal gland communicates (6) with the soul, emotion (passion) is experienced. This model anticipated the well-known James–Lange hypothesis, which also argues that emotional experiences are consequences rather than causes of overt emotional responses.

finding a philosophical justification for the existence of a human soul (in the theological sense of soul). His model of emotion was thus dictated by other concerns. Nevertheless, his view of the passions anticipated a tradition in emotional theory that persists today.

Nineteenth-century biology and emotion

Some 200 years after Descartes, the idea that conscious emotional experience is a consequence rather than a cause of emotional expression appeared anew in the writings of William James and Carl Lange. It is not generally acknowledged that their well-known hypothesis was essentially the view proposed by Descartes, albeit with some important additions.

In the time between Descartes and the James-Lange hypothesis, knowledge about biology had advanced considerably. By the late 1800s Darwin had made his mark with his theory of evolution and had gone on to propose that mind and behavior, like bodily parts, might also be subject to natural selection. Moreover, physiologists had begun to study the nervous system directly using experimental methods. Bell, Magendie, and Muller had found that sensory and motor nerves travel separately in the periphery. Studies by Munk, Ferrier, and others then demonstrated that sensory and motor functions were represented in different parts of the cerebral cortex. Information of this type made it possible to begin to construct theories of brain–mind relationships on the basis of empirical data.

William James (1884), being a physiologist, was particularly impressed with the observations of sensory and motor centers in the cerebral cortex, but noted that "the aesthetic sphere of the mind, its longings, its pleasures and pains, and its emotions, have been so neglected in all these researches." He then posed the question of whether these aesthetic aspects of mind correspond to processes occurring in sensory and motor centers or whether they have their own special centers yet to be discovered. James preferred the view that aesthetic processes are functions of sensory centers and based his theory of emotion on this choice.

In introducing his theory, James noted that the natural way to think is that the perception of a sensory stimulus gives rise to a conscious emotional experience which, in turn, leads to bodily changes (i.e., behavioral responses and alterations in the internal organs). However, in James' view, the bodily changes follow directly from the perception of the stimulus and our subsequent awareness of those changes as they occur *is* the emotion. We do not run from the wolf because we are afraid; we are afraid because we run.

In neurological terms, James argued that an environmental event received by the sensory areas of the cortex excites the motor areas and through descending nerves skeletal and visceral responses are produced. Sensory nerves in the skeletal muscles and visceral organs then bring back to sensory cortex a report of the modifications that have occurred. The arrival of this information at sensory cortex is the condition of emotional experience.

How does stimulus selection leading to bodily changes occur? Not all environmental stimuli impinging on peripheral receptors elicit emotional reactions, and when a reaction occurs, it bears some specificity to the stimulus. We approach a pet cat but avoid a mountain lion. Emotional expression, in other words, must be based on stimulus evaluation.

In James' view the nervous system is a bundle of predispositions to react in particular ways to particular features of the environment. Some of these reactions involve "conative adaptations" of the Darwinian type, while others, particularly in humans, must be "attached by education and association."

The reactions are thus dictated by either innate or acquired predispositions to respond to stimulus features, and these predispositions reflect physiological relations between sensory and motor cortex.

Thus, like Descartes, James removed conscious emotional experience from the sequence leading to emotional expression and argued that emotional experience is a consequence of emotional expression. In addition, James also attributed emotional expression to mechanical (automatic or unconscious) brain functions that analyze the meaning of sensory events. However, James went beyond Descartes in two important ways. First, he attempted to specify the brain mechanisms that evaluate the significance of sensory events and thereby mechanically control emotional expression. Second, he described conscious emotional experiences as functions of the brain (for Descartes, the brain was just an intermediary between mind and body). By attributing emotional experience to sensory cortex, James took an important step away from Descartian dualism and toward a scientific, biological analysis of emotion and other mental processes as functions of the brain.

One year after publication of James' hypothesis, Carl Lange (1885) proposed a similar idea. Historians say this was a serendipitous rather than suspicious coincidence, and the hypothesis now often bears both of their names. The similarity of their views is indicated by this passage from Lange:

If I start to tremble when I am threatened with a loaded pistol, does a purely mental process arise, fear, which is what causes my trembling, palpitation of the heart, and confusion; or are these bodily phenomena aroused immediately by the frightening cause, so that the emotion consists exclusively of these functional disturbances of the body?

For Lange the main bodily disturbances underlying emotional experience was alteration of cardiovascular activity. He proposed that these changes were effected by activation of the "vasomotor center" in the brainstem by either a simple sense impression (e.g., a loud noise or beautiful color combination) or a mental process (i.e., some memory or association of ideas). Since "most sense impressions are without emotional effects" Lange was mainly concerned with the mental elicitation of cardiovascular changes. He described in detail how ideas become associated by neural changes in sensory cortex and thereby allow sensory impressions to come to control vasomotor changes on the basis of acquired symbolic meaning. Lange's model, though similar in many respects to James', failed to provide a mechanism whereby vasomotor changes relate back to the brain to produce emotional experiences.

Emotional biology in the early twentieth century

The James–Lange hypothesis enjoyed considerable popularity for a number of years. In 1927, Perry, for example, wrote, "This famous doctrine is so strongly fortified by proof and so repeatedly confirmed by experience that it cannot be denied substantial truth." However, Perry's comment notwithstanding, there was little attempt to test the hypothesis directly through experiments designed for that purpose. The evidence was circumstantial and experiential, rather than experimental.

As more and more information about brain organization was accumulated, it became apparent that, contrary to James' opinion, there were areas of the brain specifically concerned with emotional functions. Studies by Sherrington, Cannon, and Bard on the consequences of brain lesions in animals and clinical observations of brain injured humans by Head and Holmes were beginning to point toward the involvement of subcortical regions in emotional functions. Moreover, a number of findings questioned the viability of the view that the backflow of impulses from the periphery to the brain could account for emotional experience. These latter findings were persuasively presented by Cannon (1927) as evidence against peripheral models of emotion.

Cannon's critique consisted of five points:

1. Total separation of the viscera from the central nervous system does not alter emotional behavior.
2. The same visceral changes occur in very different emotional states and in nonemotional states.
3. The viscera are relatively insensitive structures.
4. Visceral changes are too slow to be a source of emotional feeling.
5. Artificial induction of visceral changes typical of strong emotions does not produce these emotions.

Cannon's critique is less convincing today. We recognize some of his points as simply irrelevant. For example, the effects of visceral separation on emotional expression have little importance for resolving the issue of whether visceral factors determine the quality of emotional experience. Other points, as we shall see, have been negated by newer, more sophisticated techniques for studying visceral functioning. Nevertheless, the significance of Cannon's critique was not that it buried the peripheral model of emotional experience. Instead, its two major consequences were to stimulate empirical studies aimed at directly evaluating the role of visceral factors in emotion and to focus a great deal of work on the search for special emotional systems in the brain. These trends constitute the bulk of contemporary research in emotion and will be the subject of much of this chapter.

Neurobiology and emotion: current issues and research

The historical message is that emotion is not a unitary concept. At least three emotional processes can be identified: stimulus evaluation, bodily expression, and conscious experience. Any model of emotion must account for each of these. In the following, we will, therefore, examine how contemporary research on the brain and peripheral nervous system sheds light on emotional evaluation, expression, and experience. We begin with an overview of our contemporary understanding of the peripheral nervous system and its control over visceral functioning. We shall see that in spite of a great deal of progress in identifying the mechanisms through which visceral changes are expressed in emotional and nonemotional situations, the issue of whether visceral factors can determine the quality of emotional experience is still unresolved today. We then turn to studies of brain and emotion and consider what is known of the circuitry underlying emotional evaluation, expression, and experience. These considerations will lead us to a model that integrates the role of the brain and periphery in emotion.

Role of the peripheral nervous system in emotion

The peripheral nervous system consists of those nervous elements that exist outside of the brain and spinal cord. Through the peripheral nerves, the central nervous system (CNS) receives information from the external environment and from the tissues of the body and controls the muscular contractions that constitute overt behavior and adjustments in internal physiology.

With respect to emotion, the peripheral nervous system therefore functions in two ways: it is the afferent channel through which sensory information reaches the emotional systems of the brain and is the efferent channel through which the motor output of the central emotional systems is expressed (Table 15.1).

Afferent input to the central emotional systems originates in sensory receptors directed toward either the external environment (i.e., receptors in the eyes, ears, nose, skin) or the muscles, organs, and glands of the body. Nerves transmitting information from the external world are called exteroceptive afferents, while nerves arising in receptors tuned to the body itself are called interoceptive afferents. Interoceptive afferents can be further distinguished depending on where they originate. Interoceptive somatic afferents arise from receptors located in skeletal muscles and subserve the muscle sense, that is, proprioception. In contrast, visceral afferents have their origins in the internal organs and glands and carry impulses to the brain concerning visceral functioning.

Emotional expression is mediated through the efferent output of the central emotional systems. Emotional behavior is expressed through somatic efferent nerves leaving the CNS and terminating on striated muscles (i.e., skeletal muscles). In contrast, visceral efferent nerves innervate the viscera and mediate the internal expression of emotion.

Most work concerned with the peripheral nervous system and emotion has

Table 15.1. *Functional classification of peripheral nerves*

Peripheral nerve group	Origin	Termination	Example
Exteroceptive afferent	Peripheral receptor tuned to external environment	CNS	Nerves transmitting visual impulses to brain
Interoceptive afferent			
Somatic afferent	Receptors located in striated muscle	CNS	Nerves transmitting proprioceptive input from muscles
Visceral afferent	Receptors located in smooth muscle and internal organs and glands	CNS	Nerves transmitting input from viscera to brain (i.e. afferent limb of vagus nerve)
Somatic efferent	CNS	Striated muscles	Nerves carrying impulses to muscles from brain
Visceral efferent	CNS	Visceral tissues	Nerves innervating heart and blood vessels

centered on the visceral input–output channels. Consequently, we will focus on these in this chapter, beginning with the visceral efferent system, which is also known as the autonomic nervous system, and its role in emotional expression.

1. The autonomic nervous system and emotional expression. The autonomic nervous system (ANS) is the output network leaving the CNS and terminating in the viscera. The ANS controls the routine physiology of the body as well as physiological adjustments, such as changes in blood pressure, heart rate, respiration, and so forth, associated with emotional arousal.

a. Organization of the autonomic nervous system: the classic view. The organization and function of the autonomic nervous system was characterized in early studies by Langley (1921). His work, together with much that has come since, has given rise to the classic picture of autonomic organization and function.

Involuntary action: The ANS is so named because its functions are, for the most part, outside of conscious direction. While we can willfully control skeletal muscles, voluntary control over smooth and cardiac muscles is difficult to achieve. The involuntary nature of autonomic function allows for the internal housekeeping chores of the body to be performed without necessitating any conscious decision making. Thus, for example, the rate of the heart's pumping action, the constriction and dilation of arterial beds, the secretion of digestive fluids, pupilary adjustments, and so forth, are automatically controlled by the ANS.

Tonic and phasic control: The ANS maintains the basal activity level of internal organs as well as phasic adjustments in activity in response to changes in environmental, behavioral or other demands. The heart must continuously pump blood throughout the body in order for life to continue, and in order to pump the heart must be continuously excited by neural impulses. However, in addition to maintaining this tonic or life-sustaining level of activity, the autonomic nervous system also differentially modulates the heart rate to meet the specific needs of the moment. Thus, the heart rate will be less during relaxation than during exercise, will be greater during a fist fight than during sleep. It should be noted, however, that some smooth muscles do not require tonic neural input. These maintain their basal functioning through myogenic (i.e., of the muscle) tone. While the heart and blood vessels exhibit neurogenic tone, the gut has myogenic tone.

Divisions of the autonomic nervous system: The nerves of the ANS are generally classified as belonging to the sympathetic or parasympathetic divisions (Figure 15.2). These two divisions are often described as acting reciprocally so that if sympathetic activation excites an organ parasympathetic activation will inhibit it. In general, the sympathetic division responds as a whole, resulting in all or most of the tissue innervated by sympathetic nerves being simultaneously influenced, a phenomenon referred to as "mass discharge." In contrast, the parasympathetic nerves innervating specific organs can be activated individually.

For both sympathetic and parasympathetic nerves, the cell body of origin is in the central nervous system. This axon then terminates upon a peripheral neuron (a ganglion cell), which in turn makes contact with smooth or cardiac muscle. Nerves arising in the CNS and terminating on ganglion cells are referred to as "preganglionic" sympathetic or parasympathetic nerves, and nerves originating in ganglion cells and contacting target organs are called "postganglionic" nerves.

Certain important differences exist between sympathetic and parasympathetic nerves. While sympathetic preganglionic nerves arise in the thoracic and lumbar segments of the spinal cord, parasympathetic nerves arise in either the medulla of the brainstem or in the sacral region of the spinal cord. Sympathetic ganglion cells are located between the spinal cord and the target organ, whereas parasympathetic ganglia are usually located in the target organ itself (see Figure 15.2).

Autonomic neurotransmission: Preganglionic transmission (transmission at ganglion cells) is mediated by the release of acetylcholine (ACh) by preganglionic sympathetic and parasympathetic nerve terminals. Excitation of gan-

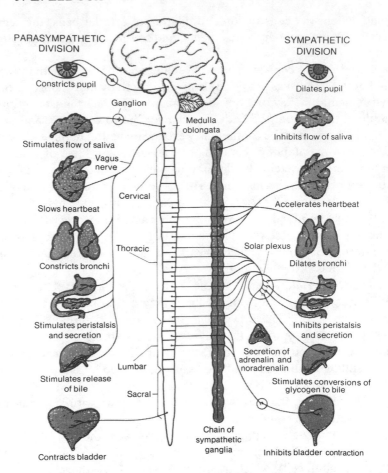

Figure 15.2. Organization of the sympathetic and parasympathetic components of the autonomic nervous system. Sympathetic preganglionic nerves originate in the thoracic segments of the spinal cord while the parasympathetic nerves have their cell bodies in the brain stem or in the sacral spinal segments. (From *Introduction to psychology*, 7th ed., by Ernest R. Hilgard, Rita L. Atkinson, and Richard L. Atkinson. Copyright © 1979. Harcourt Brace Jovanovich, Inc. Reprinted by permission of the publisher.)

glion cells by ACh then results in the release of norepinephrine (NE) by postganglionic sympathetic nerves and of ACh by postganglionic parasympathetic nerves. These substances then act directly on the target smooth or cardiac muscle to either excite or inhibit activity (Figure 15.3).

Studies of autonomic function have been greatly facilitated by the fact that neurotransmission along parasympathetic and sympathetic pathways can be selectively interrupted by appropriate pharmacological intervention. This is possible, in part, because of the specialized nature of postsynaptic receptors

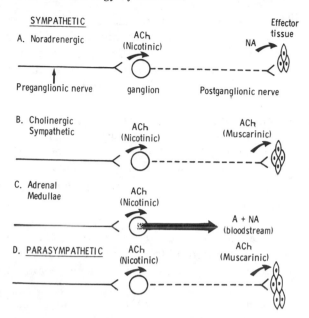

Figure 15.3. Autonomic neurotransmission. Preganglionic transmission in both sympathetic and parasympathetic nerves involves the release of acetylcholine (ACh), which acts on postsynaptic nicotinic receptors located on ganglion cells. Postganglionic sympathetic nerves release noradrenaline [(NA), also called norepinephrine] or ACh at target organs. The adrenal medulla directly releases adrenaline [(A) also called epinephrine] and NA into the circulation, and these substances are thereby carried to distal target organs. Postganglionic parasympathetic nerves release ACh, which interacts with postsynaptic muscarinic cholinergic receptors on target organs. (From Day, 1979, reproduced with permission.)

excited by the transmitters released at various synaptic junctions. Release of ACh by preganglionic nerves (sympathetic or parasympathetic) activates cholinergic "nicotinic" receptors, while ACh released by postganglionic parasympathetic nerves activates "muscarinic" receptors. Muscarinic transmission can be blocked by drugs, such as atropine, which selectively antagonize muscarinic receptors, while nicotinic transmission at preganglionic sympathetic and parasympathetic nerves is interrupted by drugs such as hexamethonium, which are appropriately referred to as ganglionic blockers. Adrenergic transmission by postganglionic sympathetic nerves can also be blocked. Propranolol blocks β-adrenergic receptors and phentolamine blocks α-adrenergic receptors located on target organs such as the heart (β receptors) or the blood vessels (α receptors). Both α and β transmission can be interrupted by drugs that interfere with the synthesis and storage of NE in the postganglionic nerve terminals (reserpine), which prevent the release of NE (guanethedine), or which "kill" the nerve terminals through toxic effects (6-hydroxydopamine).

The adrenal medulla: A special case exists for the sympathetic innervation of the adrenal medulla, where the sympathetic ganglion cells are located in the gland (Figure 15.2). The release of ACh by preganglionic sympathetic nerves results in the release of NE and epinephrine [(EPI) a metabolite of NE] from the ganglion cells in the adrenal medulla into the circulation. The NE and EPI then reach the target organs indirectly through the bloodstream. While considerably more EPI than NE is released from the adrenal medulla, these substances have similar, though not identical, effects. Thus, sympathetic activation doubly excites many organs: directly through postganglionic nerves and indirectly through circulating EPI and NE. Moreover, the half-life of the circulating substances is considerably longer than that of the same substance when released by postganglionic terminals, explaining in part the prolonged nature of sympathetic excitation.

b. Changing concepts of the autonomic nervous system. The classic concepts of the ANS for the most part have stood the test of time. However, studies in recent years have led us to alter the classic picture in a number of important ways.

Voluntary control of the autonomic nervous system. In spite of the fact that the ANS operates largely outside of voluntary control, it is possible to learn to control autonomic functions to a limited degree through special techniques. For example, some practitioners of yoga meditation can regulate their internal functions with some success. Similar results have been achieved through use of biofeedback techniques (Benson et al., 1969). Moreover, animals have been conditioned to alter their blood pressure or heart rate using procedures that reinforce spontaneous fluctuations in cardiovascular activity in the same manner that bar pressing is shaped by reinforcing spontaneous changes in behavior (Miller, 1969).

Nonreciprocal action of sympathetic and parasympathetic division. It is true that for some organs sympathetic and parasympathetic innervation functions in opposing ways. However, as shown in Table 15.2, many if not most organs are dominated by either sympathetic or parasympathetic innervation (Guyton, 1981), and, for those organs receiving dual innervation, sympathetic and parasympathetic excitation can have similar effects or completely unrelated effects (Day, 1979). The reciprocal action hypothesis is of limited value as a general characterization of ANS function.

Autonomic patterning versus mass discharge. While it is true that the sympathetic division is usually activated as a whole, it is not true that the sympathetic division has a single unitary response to any situation. We now know that although all or most sympathetically innervated organs will be activated, the extent of activation of each organ can be systematically varied by autonomic control systems. The sympathetic nervous system can thus respond in diverse ways by patterning the amount of outflow to different organs. Lacey's

Table 15.2. *Effects of sympathetic and parasympathetic excitation on visceral tissues*

Organ	Sympathetic stimulation	Parasympathetic stimulation
Pupil	Dilation	Constriction
Ciliary muscle	Relaxation	Contraction
Nasal gland	Vasoconstriction	Secretion
Lacrimal Glands	Vasoconstriction	Secretion
Sweat Glands	Copious sweating	None
Cardiac Muscle	Increased rate	Slowed rate
Bronchi	Dilation	Constriction
Liver	Glucose release	Slight glycogen synthesis
Kidney	Decreased output	None
Bladder:		
Destrusor	Relaxed	Excited
Trigone	Excited	Relaxed
Penis	Ejaculation	Erection
Blood vessels		
Abdominal	Constriction	None
Muscle	Constriction	None
	Dilation	None
Skin	Constriction	None

Source: Modified from Table 57–2 (Guyton, 1981).

work has been instrumental in illuminating this concept (Lacey, 1950; Lacey, Bateman, and van Lehno, 1953; Lacey and Lacey, 1970).

The enteric nervous system: a third division of the ANS. One of the pioneers in autonomic research, Langley (1921), proposed that the "enteric nervous system" (the neural elements controlling the gut) be treated as a third division. Most discussions of the ANS since have instead viewed the enteric system as a component of the parasympathetic division. However, Gershon, Dreyfus, and Rothman (1979) have recently called for a reinstatement of Langley's original divisions. Their argument is based on several points: (1) Many intrinsic ganglion cells of the gut do not receive preganglionic innervation from sympathetic or parasympathetic neurons. (2) The enteric system, unlike the other divisions, has a morphology that resembles the central nervous system. (3) The neurotransmitter profile present is different from that in the other divisions during development.

Multiplicity of ganglionic receptors. The view that ganglionic transmission is always cholinergic is being challenged by observations indicating the existence of many possible ganglionic receptors. Since the existence of a postsynaptic receptor for a substance is one of the essential requirements for that substance being considered a transmitter at the synapse in question, results indicating that noncholinergic receptors are present on ganglion cells raise the possibility that noncholinergic transmitters might modulate cholinergic transmission in the ganglia or that ganglionic transmission might in some cases be noncholinergic. Studies have, for example, shown the presence of catecholaminergic, seretonergic, and peptidergic receptors. The exact role of these receptors in ganglionic function, however, remains to be determined.

Autonomic neuropeptides. Recent work has shown that a number of neuropeptides are present in autonomically innervated regions (see Schultzberg, 1983). These include substance P, vasoactive intestinal polypeptide, enkephalins, somatostatin, cholecystokinin, neurotensin, and bombesin. Many of these have shown physiological effects, but their role as part of the ANS is still unclear. Studies of at least adrenal medullary-located enkephalins (which are released along with EPI and NE) have related peripheral peptides to emotional functions (Martinez, 1982), as described below.

c. Normal modulation of ANS during routine daily activities. It is generally accepted that emotional arousal is usually if not always accompanied by autonomic changes. However, autonomic changes occur in association with every movement of the body, whether willful, forced, or automatic, whether affectively charged or emotionally flat. Moreover, autonomic changes occur in anticipation of movements (Mancia and Zanchetti, 1981) and vary systematically with changes in mental states (Lacey and Lacey, 1970). These changes represent physiological adjustments imposed by the situational demands facing the organism on a moment to moment basis. The exact pattern of autonomic activity will, as a result, depend upon the specific requirements of the situation.

For example, during vigorous exercise there is an increase in heart rate and cardiac output and a redistribution of blood flow throughout the body such that flow is increased in the vascular beds of the active skeletal muscles. This is accomplished by the constriction of vessels in the nonactive striated muscles and in the viscera and by the dilation of the beds in active muscles. All of these activities (i.e., increased heart rate and cardiac output and vasoconstriction and vasodilation) are mediated by the autonomic nervous system, specifically by the CNS patterning of the neural impulses in the sympathetic and parasympathetic nerves innervating the heart and blood vessels so that the metabolic needs of the various active muscles can be efficiently satisfied.

Even relatively simple movements such as postural shifts are accompanied by dramatic and sometimes life-sustaining autonomic responses. During orthostasis (assumption of an upright posture), blood pressure and heart rate both increase, and blood flow is reduced in arterial beds throughout much of the body, but is increased in the carotid artery, the source of circulatory input to the brain (D. M. Cohen and Obrist, 1975). Such changes are necessary to overcome gravitational forces that if unresisted would draw blood away from the rising head and brain. Without the redirection of the circulation to the brain, we would lose consciousness and ultimately life by simply standing up.

Autonomic changes also occur in situations where little movement is involved. For example, autonomic activity shifts in predictable ways in association with EEG changes indicative of different stages of sleep. When cats enter into the desynchronized stage (the stage corresponding to dreaming in humans) blood pressure, heart rate, and cardiac output fall. The redistribution of the cardiac output is such that visceral arterial beds undergo vasodilation while vasoconstriction occurs in skeletal muscle (Zanchetti and Bartorelli, 1977). Vasodilation in visceral vessels is brought about by the inhibition of sympathetic activity in nerves supplying the internal organs and the constriction of vessels in skeletal muscle by a corresponding increase in sympathetic discharge.

Finally, it should be noted that autonomic adjustments can be brought about simply by changes in mental state. Lacey's work has illustrated how autonomic patterning is influenced by levels of arousal, attention to environmental events, intentions, and the like (Lacey and Lacey, 1970).

d. Characteristic patterns of autonomic activity during emotional arousal. It is often said that there is a unitary pattern of autonomic activity that is typical of emotional arousal (Woodworth and Schlosberg, 1954; Schachter, 1975); Mandler, 1975). This view, which dates back to Cannon's emergency theory (if not before) and which is the backbone of contemporary cognitive/arousal models of emotion, has some validity for states of intense aversive emotion but is incomplete as a general description of emotional physiology. Different patterns of autonomic activity accompany different emotions, as dictated by the unique demands imposed on the organism by the emotional state itself and by the somatomotor activities that constitute the behavioral expression of emotion. Let us first examine the so-called typical pattern of autonomic activation associated with emotion.

The emergency reaction. One of Cannon's most influential proposals was that the arousal of emotion induces an emergency reaction that prepares the organism for struggle, for fight or flight (Cannon, 1929). This emergency response involves diffuse sympathetic activation and, among other things, releases epinephrine from the adrenal medulla, increases respiration, and thereby stimulates the conversion of lactic acid to glucose, and redistributes blood flow away from the viscera and skin and to the active skeletal muscles. Circulating epinephrine, in turn, releases glycogen from the liver, which is then available for consumption as glucose by the active muscles. These ad-

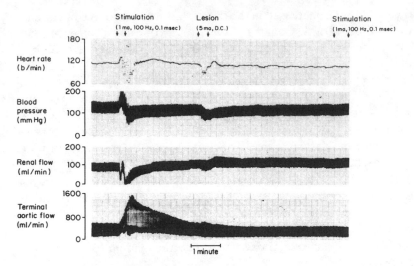

Figure 15.4. Hypothalamic defense response. Electrical stimulation of certain areas of the hypothalamus elicits a classic defense response involving widespread activation of the sympathetic nervous system. Thus, heart rate and blood pressure increase. Blood flow in striated muscle (terminal aortic flow) rises, while visceral blood flow (renal flow) decreases. Placement of a lesion in this brain region eliminates the subsequent elicitation of a defense response by hypothalamic stimulation. (From O.A. Smith et al., 1980, reproduced with permission.)

justments thus mobilize the organism's energy stores, redistribute blood away from the body surface (where cuts and scratches produce hemorrhaging in proportion to the amount of blood available) and toward the tissues most in need of substrate.

Recent work has challenged Cannon's view of the effects of epinephrine release (see Smith, 1973). Nevertheless, the idea that emotional stress is characterized by diffuse sympathetic activation is supported by experimental findings involving a number of species studied using a variety of different techniques. Only a few examples will be given.

The defense response. Electrical stimulation of the hypothalamus in the awake cat elicits hissing, snarling, arching of the back, exposure of the claws, piloerection, and attack, a behavioral reaction believed to be related to natural defense and accordingly is called the defense response (Hess, 1936). Accompanying this fit of rage is a pattern of autonomic activation that includes sympathetically mediated increases in blood pressure, heart rate, and cardiac output, together with visceral vasoconstriction and muscular vasodilation (Hilton, 1966). Similar autonomic changes can be elicited without behavioral signs in the anesthetized animal (Unvas, 1960; Folkow and Rubenstein, 1966). These autonomic changes coincide precisely with those proposed in the emergency hypothesis (Figure 15.4).

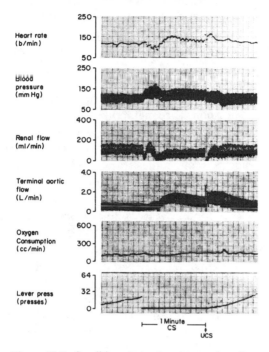

Figure 15.5. Conditioned emotional responses. Presentation of a 1-min conditioned stimulus (CS) to a baboon who is lever pressing for food reinforcement results in the suppression of this behavior. In addition, heart rate, blood pressure, blood flow to striated muscles (terminal aortic flow), and oxygen consumption rise, while visceral blood flow (renal flow) decreases. This classic defense response reflects the conditioned association that the animal has learned after having repeatedly experienced the CS as a reliable predictor of an unpleasant unconditioned stimulus (UCS). (From O.A. Smith et al., 1980, reproduced with permission.)

Conditioned emotional responses. Further support for the emergency hypothesis comes from studies of baboons subjected to classical fear conditioning while lever pressing for food reinforcement (A.O. Smith et al., 1980). In this standard conditioned anxiety or conditioned emotional response paradigm (Estes and Skinner, 1941), which is a staple in studies of brain and emotion, the onset of the conditioned stimulus (CS) in the classical conditioning phase of the experiment suppresses the operantly maintained lever pressing response. During the CS, the animal remains still while blood pressure and heart rate rise and while blood flow to the viscera (kidney in this case) falls and blood flow in the limbs, presumably reflecting skeletal vasodilation, increases (Figure 15.5).

The changes in blood flow have been further analyzed and appear to occur in two phases, a rapid initial response followed by a slower, longer lasting component. The rapid response is mediated by the sympathetic nerves innervating the vessels in the viscera and muscles. (Recall, sympathetic excitation typically results in a constriction of visceral vessels and a dilation of vessels in skeletal muscles.) The slower component, in contrast, is due to the effects of circulating epinephrine released from the adrenal medulla by the neural innervation of that gland during the first phase. Because adrenal epinephrine has largely the same synaptic effects as norepinephrine released by the sympathetic nerve terminals, the second component is similar to the first. However, because circulating hormones are metabolized slower than the same substance released into a synaptic cleft, the second response is prolonged. These changes, again, are consistent with the emergency model of emotion.

Acute mental stress. Another example comes from Brod's work using complex mental arithmetic as a stressful emotional situation (Brod et al., 1959). As in the other studies described, there are signs of diffuse sympathetic activation involving increases in blood flow to skeletal muscle and decreases in visceral blood flow in human subjects so stressed. Brod viewed these changes as reflecting an emotional reaction of purely psychogenic origin.

Motion or emotion. Diffuse sympathetic activation thus appears to characterize many acute and prolonged states of aversive emotional arousal. These changes are assumed to reflect the emotional state of the organism. However, it has been pointed out that similar if not identical autonomic changes occur during vigorous exercise (D. H. Cohen and Obrist, 1975). Might the autonomic changes during stress then simply reflect motor demands?

It goes without saying that autonomic changes will accompany movements produced during emotional reactions. The question is, however, whether there are autonomic changes that are directly related to the emotional state, independent of its motor expression. In the case of emotional reactions involving motor activity, such as aggressive behavior, fighting and fleeing, it is difficult to determine whether the autonomic changes represent primary responses or are instead secondary to movement. However, in the conditioned emotional response studies there is an active suppression of overt movement rather than an increase in movement during the emotional stimulus. Similarly, studies involving stressful mental arithmetic, where no detectable motor activity was observed, argue against the motor interpretation. Although it is possible that subtle covert muscular activity might in part account for ANS changes during emotion, it seems safe to conclude that diffuse sympathetic activation during aversive emotional arousal is, at least in part, directly related to psychogenic emotional factors.

Exceptions to the emergency hypothesis. In spite of the widespread generality of the emergency hypothesis, certain exceptions exist. For example, some species show the typical pattern of sympathetic excitation as a result of aversive classical conditioning, while others exhibit signs of parasympathetic dominance (e.g., slowing of the heart) (D. H. Cohen, 1980; Smith et al., 1980; Kapp, Pascoe, and Bixler, 1984; LeDoux, Sakaguchi, and Reis, 1984). This has led to the idea that there are individual and species typical coping strategies that influence the pattern observed. However, the experimental conditions have varied considerably in these studies, making comparisons difficult. Examples of uncontrolled variables include the intensity or aversiveness of the unconditioned stimulus, the organism's condition (restrained versus unrestrained body movements), the availability to avoid the aversive stimulus, and so forth. Moreover, within a species exhibiting sympathetic dominance, some individuals will show parasympathetic signs (D. H. Cohen and Obrist, 1975). These factors will have to be considered more carefully before accepting them as true exceptions to the emergency model of aversive emotional arousal. However, the evidence does at least favor the view that individual, species, and contextual factors can affect the autonomic response that will occur during emotional arousal.

Chronic stress. Thus far we have only considered the immediate or acute autonomic consequences of aversive emotional arousal. However, as Selye (1950) pointed out, the acute emergency reaction is but the initial response to stress. When the stress-evoking stimulus continues, as it often does in real life, the alarm phase is followed by a stage of resistance, characterized physiologically by activity of the pituitary–adrenal cortex axis, in which the organism attempts to adapt to the stress. Thus, ACTH released from the pituitary into the circulation stimulates the release of steroid hormones from the adrenal cortex. Finally, if the stress continues, exhaustion occurs, where adaptation is discontinued. Selye referred to these events, involving alarm, resistance, and exhaustion, as the general adaptation syndrome.

Selye's characterization of the response to stress is supported by considerable evidence (see Axelrod and Reisine, 1984). Animals subjected to forced immobilization, prolonged footshock, or crowded living conditions show elevations of plasma catecholamine levels. Moreover, adrenocortical steroid levels also rise, but, unlike catecholamines in blood, take several hours or even days to appear. The adaptive function of adrenocortical steroids is illustrated by the fact that conditioned stress reactions are enhanced by removal of the adrenal gland, an effect that is reversed by intravenous replacement of adrenal steroids (Weiss, 1970; Sakaguchi et al., 1984). Studies have shown how chronic stress, when allowed to continue, can lead to the long-term elevation of blood pressure in experimental animals (Henry et al., 1972) to gastric ulceration (Weiss, 1972), and to a breakdown in immune responses (Shavit et al., 1983). These animal models have their counterparts in human problems, such as hypertension, gastrointestinal disease, and cancer, each of which is believed to have a psychosomatic component.

e. Diversity of autonomic patterning: differentiation of emotions. Cannon's notion that the ANS responds with diffuse sympathetic activation during aversive emotional arousal is a useful and experimentally verified concept that seems to apply to many, though perhaps not all, situations involving the acute arousal of aversive emotions. However, two additional qualifications must be added. First, it is now known that the ANS is capable of patterning output to target organs. As a result, although most sympathetically innervated organs are activated simultaneously, the extent of activation can be differentially regulated. Second, studies measuring autonomic changes during different emotions have shown that the pattern of activation observed varies with the psychological nature of the emotion aroused.

Autonomic response patterning. In the early 1950s work by Lacey and his colleagues began to challenge the acceptance of the view that a single pattern of autonomic activation characterizes emotion (Lacey, 1950; Lacey et al., 1953; Lacey and Lacey, 1970):

The autonomic nervous system does indeed respond to experimentally imposed stress as a whole in the sense that all autonomically innervated structures seem to be activated, usually in the direction of sympathetic predominance. But it does not respond as a whole in the sense that all autonomically innervated structures exhibit equal increments or decrements of function. (Lacey et al., 1953)

In Lacey's studies, a number of physiological measures (systolic and diastolic blood pressure, heart rate, respiration, etc.) were obtained from human subjects performing different tasks, such as mental arithmetic. By examining the increases and decreases in each autonomic variable during the different tasks, they demonstrated widespread differences in sympathetic activation among individual subjects for a given task and highly reliable consistencies within individuals across the various tasks. These observations suggested that the autonomic nervous system has the capacity to respond in diverse ways that depend upon individual predispositions. Lacey's findings, however, did not allow the differentiation of emotions on the basis of sympathetic patterning.

Differentiation of emotions. One of the first successful attempts to experimentally differentiate emotions on the basis of their autonomic correlates was made by Ax (1953). He recorded seven physiological parameters simultaneously while subjects were placed in a "fear" or "anger" situation. Fear was induced by gradually increasing intermittent electric shocks to the finger and anger was induced by rude behavior on the part of one of the experimenter's colleagues. These emotional experiences were later verified in interviews with the subjects. The physiological measurements during the two conditions indicated that fear and anger do indeed have different autonomic manifestations.

Ax suggested that the anger pattern resembled the effects of intravenous injection of EPI and the fear pattern intravenous NE. A similar suggestion was made the next year by Funkenstein (1955), who differentiated "anger in" (fear) and "anger out" (anger) on the basis of multiple physiological measurements. The Ax–Funkenstein catecholaminergic hypothesis of fear and anger, if true, would have some important implications for the biology of emotion since EPI and NE have different (though overlapping) origins.

Many studies have examined the urinary and plasma levels of NE and EPI under a variety of emotional and nonemotional conditions in an effort to assess the Ax–Funkenstein hypothesis. Excellent reviews of this work have been provided by Frankenhauser (1971) and G. P. Smith (1974). The conclusion of these authors is that the two catecholamines are not good differentiators of fear and anger and, in general, are relatively insensitive indicators of emotional arousal, especially when the intensity of the emotional reaction is relatively weak.

Although Ax's and Funkenstein's suggestion that fear and anger might be differentially related to peripheral catecholaminergic mechanisms has not been borne out, this does not reflect negatively on their observations indicating that the two emotions can be differentiated by other physiological measurements. However, their findings have also suffered from methodological flaws in the experiments.

In a recent study designed to overcome the problems of earlier studies, Ekman et al. (1983) reexamined whether physiological measures might differentiate emotions (Figure 15.6). Not only were they able to show clear differences between negative emotions (fear and anger) but also between negative and positive emotions (anger and happiness).

f. Summary. Diffuse sympathetic activation accompanies many states of acute, intense aversive emotional arousal, as suggested by Cannon's emergency hypothesis. Such changes prepare the organism to face adversity over the short run. Although most organs innervated by sympathetic nerves will be activated, the particular pattern of activation will depend on the emotion aroused, the motor activity involved, and possibly the context and factors specific to the individual and species.

2. Visceral feedback and emotional determination. The preceding discussion has examined the role of the peripheral nervous system, particularly its autonomic division, in the bodily expression of emotion. In this section, we examine the question of whether these autonomic changes, by way of feedback to the brain, actually contribute to emotional determination.

a. Emotional determination. Emotional determination refers to the mechanisms that specify the nature of an emotional reaction. What, in other words, determines whether fear, anger, or joy will be experienced in a given situation?

As we have seen different autonomic patterns accompany qualitatively different emotions. These autonomic patterns are regulated by the brain on

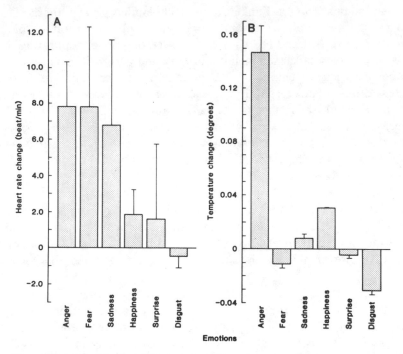

Figure 15.6. Physiological differentiation of emotion: Human subjects were asked to perform emotional facial expressions while heart rate (A) and finger temperature (B) were recorded. For heart rate, the changes associated with anger, fear, and sadness were significantly greater than those for happiness, surprise, and disgust. The finger temperature change during anger was significantly different from all other conditions. Thus, measurements of heart rate and skin temperature revealed that different emotions are accompanied by different physiological changes. (From Ekman, Levinson, and Friesen, 1983, reproduced with permission from the American Association for the Advancement of Science. Copyright 1983 by AAAS.)

the basis of its assessment of the nature of the emotional situation. The autonomic adjustments are thus consequences not causes of the value that is assigned to an emotional stimulus. Only by first identifying the emotional meaning of the stimulus situation can the brain set up the different patterns of autonomic activation. Visceral feedback, in other words, cannot determine the initial coding of stimulus significance.

On the other hand, once stimulus significance is assessed and on the basis of such assessment autonomic responses are expressed, visceral feedback could play a role in the modulation of the intensity and duration of the emotional reaction. Further, visceral feedback could play a role in conscious emotional experience, either as a nonspecific sign of arousal (as suggested by cognitive models of emotion) or as the determining factor of the feeling experienced (as suggested by the James–Lange model).

b. The nature of visceral feedback. There are two sources of visceral feedback from the periphery to the brain. The first is the visceral afferent system, the network of sensory nerves originating in receptors located in the internal organs and transmitting input concerning organ function to the CNS, much the same as the optic nerve transmits retinal input to the CNS. The visceral afferent system has received much less experimental attention than exteroceptive afferent systems and as a result is poorly understood. However, considerable work has been done on isolated visceral afferents, such as the afferent fibers of the vagus nerve (the primary input–output nerve of the parasympathetic system).

The second major source of visceral feedback is the endocrine system, which under the control of the ANS releases humoral substances into the systemic circulation. Some of these blood-borne hormones influence the CNS directly, crossing the blood–brain barrier and interacting with receptor sites in the brain. For example, steroids released from the ovaries, testes, and adrenal cortex have specific CNS sites of action (McEwen, 1981), many of these being in areas known to play a role in emotion. Other substances, such as EPI released from the adrenal medulla, are not believed to cross the blood–brain barrier under normal physiological conditions. However, there is now evidence that the brain may have select control over the blood–brain barrier so that its permeability to circulating substances can be regulated to meet the needs of specific situations (C. Iadecola, personal communication). Nevertheless, many peripheral substances do not have their CNS effects so directly, but instead activate peripheral organs innervated by visceral afferents and thereby stimulate neural feedback to the brain.

The interaction of neural and humoral feedback to the brain is indeed complex, particularly with respect to the possible roles of these factors in emotion. Studies attempting to evaluate the influences of neural and/or humoral feedback have taken one of two general approaches: either disrupting feedback or stimulating feedback.

c. Effects of disrupting visceral feedback

Methods of disrupting feedback. One of the most commonly used approaches in studies attempting to correlate neural structure and function involves the inference of function from deficits produced by interfering with the normal operation of the structure, either through lesions or drug manipulations. With regards to the peripheral nervous system and emotion, various procedures have been used to separate the periphery from the brain and thereby block either the autonomic responses that generate feedback, the sensory feedback mechanism, or both.

Sympathectomy, separation of the sympathetic nervous system from the brain, can be produced surgically by sectioning the sympathetic chain or by sectioning the spinal cord above the level where the sympathetic nerves depart from the CNS. Sympathectomy can also be produced functionally, by pharmacological interruption of transmission at either ganglionic or postganglionic sites: ganglionic blockers such as hexamethonium interrupts cholinergic transmission in the sympathetic (and parasympathetic) ganglia; postganglionic blockers, such as guanethidine, affect noradrenergic transmission; 6-hydroxydopamine is a specific toxin to postganglionic noradrenergic nerve terminals. Manipulations involving postganglionic mechanisms spare the adrenal medulla, which does not have postganglionic nerves. Nevertheless, postganglionic blockage of α- and β-adrenergic receptors antagonizes the effects of adrenal catecholamines at target organs. Moreover, the adrenal medulla can be surgically removed. Adrenal demedullation has, in fact, been used frequently in studies of emotion. All of the above procedures leave the parasympathetic nervous system unaltered. Its function can be disrupted by sectioning of the vagus nerve, which, as mentioned above, carries the parasympathetic efferents as well as vagal afferents, or by blockade of parasympathetic neurotransmission with antagonists to cholinergic muscarinic receptors. Finally, other organ systems, such as the pituitary–adrenal cortex axis, have been subjected to lesion and pharmacological manipulation in studies of emotion.

Consequences of disruption. Sherrington, in 1900, examined the contributions of visceral feedback to emotion by studies of dogs subjected to spinal cord transection. Concerning one of his female subjects, Sherrington remarked, "Her anger, her joy, her disgust, and when provocation arose, her fear, remained as evident as ever." Cannon et al. (1927) performed similar experiments and drew similar conclusions. However, others rightly argued that evidence of this type simply shows that visceral feedback is unnecessary for the behavioral expression of emotion (Angell, 1916; Perry, 1927). Whether emotional experience is altered is not easily evaluated in subhuman animals.

More recently, Hohmann (1966) examined this problem in humans with spinal cord injury. His patients fell into five groups depending on the location of the lesion. He found that the higher the lesion, and thus the greater the separation of the sympathetic nervous system from the brain, the greater the reduction in emotional feeling. The patients still behaved angrily or fearfully or happily when appropriate, and by their own reports still felt angry, fearful, and happy. Nevertheless, they said that the intensity of their feelings were greatly reduced.

These findings, though fascinating, interesting, and important, must be cautiously interpreted. Unlike animal studies, the lesions in this experiment were left purely to chance, and there is no histological verification of the extent of the damage. Moreover, even in the cases with the highest lesions, the vagus nerve was intact. Thus, the experiment can only be used to evaluate the effects of feedback reduction rather than elimination.

Schachter (1975) interpreted Hohmann's findings to mean that, in the absence of normal autonomic feedback to the brain, behavior that appears emotional will not be experienced as emotional. In fact, however, the patients readily recognized their emotional states. They did not confuse anger and fear and showed no reduction of their feelings of sentimentality. The main change was in the intensity of some experiences.

That peripheral feedback intensifies emotional reactions is also suggested by studies which have assessed the effects of sympathectomy on aversive conditioning in animals. Wynne and Solomon (1955), for example, found that sympathectomized dogs could achieve a normal level of avoidance conditioning, but acquisition was prolonged and extinction was facilitated. DiGiusto and King (1972) obtained similar results with sympathectomized rats. These findings are consistent with the view that sympathectomy reduced the level of "fear" aroused and as a result interfered with normal contribution of fear to aversive conditioning (see section entitled "Emotional Experience" such as concerning the problems involved in attributing conscious emotional feeling states such as fear to animals). However, the fact that the animals eventually learned indicates that sympathectic activation is not necessary for aversive learning to take place.

In contrast, the effects of removing the adrenal medulla are less clear-cut (see G. P. Smith, 1974). Several studies have failed to find changes in aversive conditioning in demedullated animals (Moyer, 1958; Moyer and Bunnell, 1958; Leshner, Brookshire, and Stewart, 1971). Others, however, have found that demedullation prolonged aversive conditioning (Levine and Soliday, 1962; Pare and Cullen, 1971). That these latter effects are due to removal of the adrenal medulla and not to incidental damage to the adrenal cortex or to other factors is indicated by the fact that replacement of EPI by intravenous infusion restores normal acquisition rates in demedullated rats (Corner and Levine, 1969). Martinez (1982), in more recent studies, has found adrenal effects in some tasks and not in others and has suggested that the adrenal medulla is mainly involved in aversive learning when the task involves high stress.

The effects of vagal blockade, produced by muscarinic cholinergic blockers or by vagotomy, on aversive conditioning have also been examined. Conflicting results have been obtained in different studies (DiVietti and Porter, 1969, Albiniak and Powell, 1980).

Interestingly, surgical vagotomy has been shown to attenuate the rewarding effects of electrical brain stimulation (Ball, 1974). This finding raises the possibility that peripheral systems are activated by electrical brain stimulation and that feedback to the brain concerning the peripheral activity is necessary for brain stimulation reward to be fully effective (Halperin and Pfaff, 1982).

In summary, it appears that attenuation of interactions between the peripheral nervous system and the brain reduces the intensity of emotional reactions but does not alter the appropriateness of emotional reactions and felt experiences to the emotional situation. However, the difficulty in completely separating the viscera from the brain complicates drawing firm conclusions. As our knowledge of ANS organization grows, indicating that peripheral transmission and humoral release may involve many more pathways and substances than originally thought, we must be especially suspicious of any approach that claims, on the basis of having blocked feedback from the viscera to the brain, that visceral feedback does not affect emotion.

d. Effects of stimulating visceral feedback. Just as function can be inferred by observing the consequences of disrupting organ activity, so can it be inferred from the effects of activating the organ, either by electrical stimulation of the motor nerves innervating the organ or by pharmacological excitation of the organ. Moreover, the effects of organ activation can be mimicked by infusion of the hormones that are normally released when the organ is naturally activated.

In an early attempt to initiate emotions by manipulating visceral activity, Maranon (1924) questioned humans about their experiences after intravenous infusion of EPI. While the vast majority of subjects reported feeling the physical effects of the treatment, which mimics natural sympathetic activation, relatively few described their experiences as emotional. Those who did experience emotional feelings described them as "as if" feelings. They felt as if angry or sad but did not actually feel angry or sad. Later studies reported similar effects with either EPI or NE infusions (Landis and Hunt, 1932; Frankenhauser, 1975). As with studies involving sympathetic disruption, studies involving sympathetic stimulation suggest that peripheral factors mainly contribute to the intensity of emotional reactions.

Results from animal studies further suggest that peripheral stimulation contributes to the intensity of emotional reactions. Epinephrine infusions seem to enhance performance and reduce aversive thresholds in standard tests of fear conditioning, such as conditioned emotional response and avoidance learning tasks (see G. P. Smith, 1974). However, such effects are not limited to stimulation of peripheral catecholaminergic systems, as infusion of ACTH, vasopressin, and various opioid compounds also affect aversive learning (see Martinez, 1982).

There is also evidence that parasympathetic activation affects emotionality. Corwin and Slaughter (1979), for example, found that electrical stimulation of the vagus nerve disrupted conditioned emotional responses. When the CS was presented during bar pressing, the rate of bar pressing was reduced in control rats, indexing the emotional consequences of CS presentation. However, vagal stimulation reduced bar press suppression, suggesting that the conditioned state of fear was reduced. They interpreted this finding in terms of competition between the sympathetic and parasympathetic systems. While the CS normally elicits sympathetic activation, vagal stimulation produces a state of parasympathetic predominance. By preventing the sympathetic response, fear was reduced, thus suggesting that the intensity of fear reactions normally depends on the relative activity in the peripheral sympathetic and parasympathetic systems.

In summary, it is clear that peripheral manipulations affect emotion, especially the intensity of emotional reactions. The peripheral systems, though, are extremely complex, involving mind-boggling interactions among circulating hormones and neural systems. Much more work is needed in this area if we are to ever hope to begin to understand the actual mechanisms involved in any one emotional situation, much less emotion in general.

e. Conclusions: visceral changes and emotion. On the basis of what we know today, Cannon's attack on the James–Lange theory was not wholly justified. The viscera do have the capacity to respond in select ways. Moreover, different visceral changes occur during qualitatively different emotional states. Nevertheless, there is little positive evidence to support the James–Lange notion that visceral and other bodily changes actually determine emotional feelings. It is clear that visceral feedback affects the intensity of emotional expression and experience, but probably in a nonspecific fashion. On the other hand, the periphery is proving to be extremely complex, and given its capacity for diverse, patterned responding, together with the expanding ways in which peripheral information can reach the brain, it seems, after 100 years, that it is still too early to put James and Lange to rest. At the same time, we are hardly ready to retire Cannon's CNS alternative to peripheral determination of emotion, for as we shall now see, much of the informational coding underlying the various processes we refer to as "emotional" takes place in select and specialized systems in the brain.

The modern search for the emotional brain

1. The concept of an emotional brain. William James (1884), it will be recalled, rejected the idea that there are special emotional centers in the brain. Emotion, in his view, was just another function of sensory and motor areas of the cortex. Today, we accept as a matter of fact that there are emotional centers. Since most of these centers are found in what is now called the limbic system, the limbic system is often described as the emotional system of the brain. We will have much to say about the role of various limbic regions in emotion, but first it is worthwhile to consider how the limbic system came to be viewed as the emotional brain.

a. Early studies. Shortly after publication of James' view that emotion is a normal function of cortex, a number of investigators reported that clear signs of emotion were present in experimental animals surgically deprived of cortex (Bechterev, 1887; Woodworth and Sherrington, 1904). These animals exhibited emotional reactions appropriate to various forms of sensory stimulation, indicating that emotional expression, and perhaps other aspects of emotion, are not dependent upon cortex.

In the years that followed, considerable work was devoted to the identification of subcortical emotional mechanisms. The basic strategy was to ablate the brain at various levels from the cortex down until all signs of emotion were eliminated. Cannon and Britton (1925), for example, found that in the absence of cortex, cats spontaneously exhibited fits of rage and could be excited into rage by innocuous stimulation, unless lower areas such as the thalamus were included in the ablation. These and other findings led Cannon to propose that subcortical areas (especially thalamic areas) are the seat of the emotions and that spontaneous and inappropriate discharge of the thalamus was normally inhibited by the cerebral cortex (Cannon, 1927). Subsequent work by Bard (1928) and Bard and Rioch, (1937), however, pointed toward the hypothalamus rather than the thalamus as being crucial for the expression of emotion. This conclusion was strengthened by the work of Hess (1936), who systematically explored the effects of electrical stimulation of discrete brain regions and consistently observed that hypothalamic stimulation elicited well-organized emotional behaviors (attack, defense, and flight reactions) and autonomic responses characteristic of natural emotional reactions (signs of diffuse sympathetic activation). The hypothalamus was thus coming to be regarded as an area that organized the discharge of impulses underlying the somatic and autonomic manifestations of emotion.

b. The Papez circuit. It was in this context that Papez in 1937 proposed a brain mechanism of emotion centered around the hypothalamus. Papez's objective was to answer the question of how afferent sensibilities, which pass through the thalamus en route to the cerebral cortex, acquire emotional coloration. To do this he drew upon clinical observation and scant anatomical facts and suggested that at the thalamus afferent pathways split into three routes: to the cortex, to the basal ganglia, and to the hypothalamus. While the cortical route represented the "stream of thought" and the projection to the basal ganglia the "stream of movement," the input to the hypothalamus constituted the "stream of feeling" (Figure 15.7).

Impulses reaching the hypothalamus were transmitted in one of two directions: downstream (toward the peripheral nervous system) or upstream (toward the cortex). Projections down toward the brainstem and spinal cord allowed sensory events to elicit autonomic and behavioral reactions without

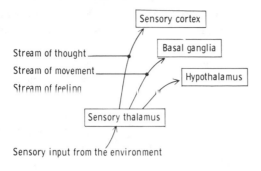

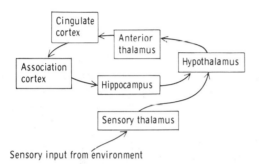

Figure 15.7. The Papez model of emotion: Papez (1937) postulated that sensory input transmitted from the environment to the thalamus is subsequently relayed along different pathways mediating separate psychological functions: to the cortex (the stream of thought), the basal ganglia (the stream of movement), and the hypothalamus (the stream of feeling) (top diagram). From the hypothalamus input to the "stream of feeling" enters a processing circuit, now known as "the Papez loop," and emotional coloration is added to raw sensory input (bottom diagram). Many of the anatomical connections postulated by Papez have now been confirmed. Although the functional implications of the model remain incompletely verified, the Papez hypothesis has been extremely influential.

the opportunity or necessity for the intervention of the higher regions of the forebrain, especially the cortex. This was in essence an extension of the models proposed by Cannon (1927) and Bard (1928).

More important for Papez were the upward projections from the hypothalamus to the cortex. This projection involved connections between the mammillary bodies of the hypothalamus to the anterior thalamic nucleus, which in turn sends projections to cingulate cortex. Papez viewed the cingulate cortex as the cortical receiving area for emotion, just as striate cortex is the receiving area for vision. By way of connections between cingulate cortex and other cortical areas, "psychic" processes occurring in cortex are given emotional flavor. Thus. visual impulses reaching striate cortex are charged with emotion when the outflow of striate cortex and cingulate cortex converges. Finally, by way of connections of various cortical areas to the hippocampus and thereby back to the hypothalamus, cortical processes are able to cause the hypothalamus to discharge downward in the expression of emotion. The hypothalamus could, therefore, be activated in two ways: directly by raw sensory input diverging from the sensory projection pathway at the thalamic level or by higher psychological processes of cortical origin. These various connections form a neural loop that has come to be known as the Papez circuit (Figure 15.7).

Papez's model was brilliant. He managed to salvage both cortical and subcortical theories of emotion, attributing emotional expression to the hypothalamus and emotional experience to the cortex, and though based on relatively little data, the anatomical connections proposed, with some modifications, have withstood the test of time.

c. The visceral brain and limbic system. The brain structures included in the Papez circuit are closely related to what Broca in 1878 called the great limbic lobe. This lobe surrounds the brainstem and is present in all mammals, although in varying degrees. Because of its involvement in olfaction, the limbic lobe was sometimes referred to as the rhinencephalon or smell brain. However, after Papez, it was clear that smell was not the only function of these areas. Nevertheless, it was Paul MacLean's (1949) early model that really provided us with the notion that the limbic lobe and related subcortical structures constitute a functional system, which he referred to as the "limbic system," involved in emotion. Though he borrowed heavily from Papez, MacLean also added some new twists to the theory.

MacLean referred to the limbic system as the "visceral brain" because of the relation of these regions to autonomic functions. He viewed the visceral brain as the system that integrated sensory input from the internal organs and autonomic nervous system with input from the external environment.

Especially important in MacLean's scheme was the hippocampus, which he saw as the place of visceral and environmental sensory integration and thus the seat of emotional experience. Much of MacLean's work (see MacLean, 1970) has thus been devoted to examining the inputs to the hippocampus from exteroceptive sensory systems such as vision. The hippocampus was for MacLean what the cingulate gyrus was for Papez: the cortical receiving area for emotion. Its role was to correlate internal and external cues and to discharge its evaluations through the hypothalamus.

d. The limbic system today. Our knowledge of the anatomy and physiology of the limbic system has expanded tremendously over the past several decades as a result of the development of new techniques for studying the structure and function of the nervous system. This material is presented in a number of excellent reviews (Nauta, 1972; Swanson, 1981; Issacson, 1982). Here, we will only identify the areas that constitute the limbic system and briefly summarize some of the limbic interconnections and the input and output pathways relevant to the studies of brain and emotion that will follow.

The limbic system consists of both cortical and subcortical forebrain structures, as initially described in the papers of Broca, Papez, and MacLean. The decision to include a structure in the limbic system is somewhat arbitrary and has to do with the proximity to, fiber connections with, and overlap of functions with other areas that are accepted as belonging to the limbic system. Consequently, what is a limbic area for one may not be for others.

The cortical limbic areas are, in general, not part of the neocortex but instead belong to more primitive zones with less laminar definition than neocortex. The cingulate, insular, entorhinal, pyriform, hippocampal, and retrosplenial cortical areas are generally viewed as limbic regions. However, the orbitofrontal region of the neocortex is also a limbic region. Subcortical limbic areas include the septum, amygdala, hypothalamus, some portions of the basal forebrain, and the anterior and mediodorsal thalamic nuclei. Each of these areas can be further divided into subnuclei with unique fiber relations and functions. Moreover, rich interconnections exist within the various subdivisions of a given area and between the subdivisions of different areas.

Exteroceptive sensory inputs to the limbic system allow information from the environment to influence limbic activity. Evoked sensory responses can, in fact, be recorded in many limbic areas in response to sensory stimulation (Kaada, 1951; O'Keefe and Bouma, 1969; MacLean, 1970). Such responses reflect the transmission of sensory input to limbic areas.

Exteroceptive sensory information reaches limbic areas by at least two routes (Figure 15.8). The first involves corticofugal (descending) connections from sensory processing areas of the neocortex. Each of the neocortical sensory systems sends projections to the hippocampus and amygdala by way of intermediary cortical association zones (van Hoesen, Pandya, and Butters, 1972; Turner, Mishkin and Knapp, 1980). By way of connections of these areas to other limbic areas, sensory input is distributed throughout the limbic system.

The second source of exteroceptive sensory input is by way of connections that diverge from ascending sensory projection systems before reaching neocortical areas (Figure 15.8). Although Papez speculated about these, only recently have they been discovered and implicated in emotional processing. Thus, visual and auditory areas in the thalamus, in addition to projecting to neocortical receiving areas, also send fibers to the hypothalamus and amygdala (Ebner, 1969; Swanson, Cowan, and Jones, 1974; Norita and Kawamura, 1980; LeDoux, Sakaguchi, and Reis, 1984). Moreover, direct retinal projections to the hypothalamus (Pickard and Silverman, 1981) and possibly the

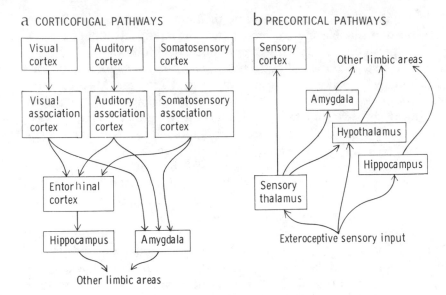

Figure 15.8. Exteroceptive sensory input to the limbic system. Sensory input from the external environment reaches the limbic system by several routes. (a) Sensory nuclei in the thalamus relay incoming signals to sensory receptive areas of the cortex. By way of corticocortical connections, the input is transmitted to modality specific association areas and then by way of transitional cortical areas to the hippocampus and amygdala. Each of these areas has extensive connections with other limbic structures. (b) In addition to relaying incoming signals to cortical receiving areas, some thalamic areas send projections directly to the amygdala and hypothalamus. There is also evidence for direct projections from peripheral receptor organs (retinal ganglion cells) to the hypothalamus and possibly the hippocampus (see text).

hippocampus (MacLean and Creswell, 1970) have been observed. These pathways allow crude sensory inputs to reach limbic areas rapidly and without neocortical intervention, a point to which we will return below.

Interoceptive or viscerosensory inputs also reach limbic areas. The hypothalamus and amygdala, for example, receive inputs from the nucleus tractus solitarii (NTS) (Richardo and Koh, 1978; Smith and DeVito, 1984), the brainstem site of termination of afferents arising in the abdominal viscera and traveling in the vagus nerve (Sumal et al., 1982). Other visceral afferent systems may influence limbic activity as well, but the anatomy and physiology of these systems is not well understood.

Outputs of limbic areas to autonomic control regions are numerous. Most of the more relevant projections for our concern here include pathways that project to the hypothalamus, which in turn has connections to brainstem areas controlling the ANS (Figure 15.9). The hypothalamus also has direct con-

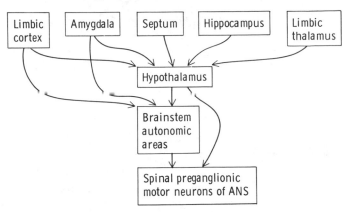

Figure 15.9. Circuits linking limbic areas to autonomic control regions: Information processing occurring in limbic areas can influence the autonomic nervous system (ANS) through several pathways. Most limbic areas are connected to the hypothalamus, either directly or through other limbic areas. The hypothalamus, in turn, is richly connected with brainstem autonomic control regions, as well as with the sympathetic preganglionic neurons in the intermediolateral column of the spinal cord. In addition, some limbic areas, such as the amygdala and insular cortex, have direct connections with brainstem regions involved in autonomic regulation.

ncctions with spinal preganglionic motor neurons (Saper et al., 1976). Inputs to the hypothalamus from limbic regions allow activity in these areas to influence autonomic activity. However, the amygdala (Schwaber et al., 1978) and insular (Saper, 1982) areas also have direct connections to brainstem autonomic regions, such as the nucleus of the solitary tract. These pathways will be considered further when we discuss the central neural control of emotional expression.

Limbic regions are richly interconnected, thus supporting the notion that limbic areas function as a system. These connections can only be briefly mentioned in the most general way. For more details see Brodal (1982). Papcz's classic notion that the hypothalamus (especially the mammillary bodies) projects to the anterior thalamus is now well established. Moreover, connections between the anterior thalamus and the cingulate cortex have been demonstrated. Cingulate cortex, in turn, has connections with neocortical areas involved in higher-order perceptual and cognitive processing, again, as postulated by Papez. The septum and hippocampus are closely related, as are the amygdala and the medial dorsal thalamus. Outputs of most limbic subsystems (such as the septohippocampal system, the amygdala–mediodorsal thalamus system) are distributed to other limbic areas as well as to hypothalamic exit points.

e. The limbic system as the seat of emotion: useful or misguided concept?
Today the limbic system in neurological code-talk means the emotional brain.
An important issue concerns whether this widespread view is useful.

There is little question that areas of the limbic system are involved in
emotion. However, some limbic areas subserve functions that seem to have
little obvious relation to emotion [such as the role of the hippocampus in
spatial functions, see O'Keefe and Nadel (1978)] and some nonlimbic areas
contribute to emotional behavior. Anatomists have questioned the value of
the limbic system notion (Swanson, 1981; Brodal, 1982). In this vein, it is
probably more useful to build up a concept of an emotional system empirically,
from the bottom up, rather than from an a priori conception of limbic circuits
or emotional circuits; that is, understanding the circuits underlying specific
emotions and seeking out the overlap of those circuits for different emotions
is likely to be more fruitful than trying to force limbic circuitry on specific
emotions. Nevertheless, as will be obvious, the notion of the limbic system
as a general emotional system has proved to be a powerful idea because
studies of emotion always seem to lead to the limbic system. The danger,
though, is one of false security. To say that emotion is a limbic function only
transfers our ignorance from a field called brain and emotion to a field called
limbic system function.

2. Emotional functions of the brain. As we have seen, the various functions
related to emotion fall into three broad categories: evaluation, expression,
and experience. Each of these is uniquely related to brain mechanisms and
must be accounted for by any model of brain and emotion.

Failure to recognize the distinctions between the evaluative, expressive,
and experiential aspects of emotion has been the source of much confusion
throughout the history of studies aimed at understanding the neural foundations of emotion. Today, many studies involving brain manipulations claim
to alter such emotions as fear in animals, but are unclear as to whether fear
refers to a conscious state experienced by the animal or instead is used as a
metaphor to describe the behavioral consequences of the manipulation
performed.

In the previous consideration of the peripheral nervous system and emotion,
we concentrated on the expressive and experiential facets of emotion. These
have, in fact, traditionally been the main topics discussed with respect to
emotion, for the long-standing debate over whether peripheral emotional
expression is the basis of emotional experience, or vice versa, has tended to
overshadow the importance of the evaluative component. However, as Descartes and James recognized, for any emotional reaction to an environmental
stimulus to take place, the significance of the stimulus must first be assessed.

Only on the basis of such an evaluation can the brain regulate the peripheral nervous system in a manner that is appropriate to the meaning of the stimulus. Stimulus evaluation precedes and constitutes the basis of emotional expression.

Some might say that evaluation is really just a part of emotional experience, that the conscious experience of stimulus meaning *is* the evaluation of stimulus meaning. However, this view is unacceptable, at least as the whole story, for two reasons. First, the significance of a stimulus is often evaluated at subconscious levels (Zajonc, 1980; Marcel, 1983a; LeDoux, 1984), with only the consequences of evaluation reaching consciousness. Second, we are barely sure, scientifically speaking, that humans are consciously aware, and at best we have circumstantial evidence for consciousness in subhuman creatures. Yet, there is little doubt that even the lowest of the vertebrates (and invertebrates as well) are guided in their behavioral interactions with their environment by the value or significance of stimulus events. The mechanisms underlying the evaluation of stimulus significance must be viewed independently from the mechanisms that give rise to conscious emotional experiences.

In the following, we will examine the neural substrates of emotional expression, evaluation, and experience. We begin on the output side, considering how the brain links up with the autonomic nervous system in the expression of emotion. We then turn to a consideration of how incoming sensory input is evaluated and thereby guides emotional expression. Finally, we will examine the troubling problem of feelings, or emotional experience.

3. Emotional expression. Any given emotion is expressed in a great number of ways. Studies, therefore, seldom attempt to characterize the expression of emotion completely, but instead use a limited set of reliable, easily measured, and quantifiable indicators. In this regard, cardiovascular changes have become particularly popular as emotional markers. This popularity is in part due to the detailed understanding of the anatomy and physiology of peripheral cardiovascular control, which allows insights into the synaptic mechanisms that mediate cardiovascular changes at peripheral end organs, as well as to the recent surge of studies that are illuminating the basic reflex circuits underlying the central control of the cardiovascular system (O. A. Smith, 1974; Chalmers, 1975; Reis, 1980). In this section, we will, therefore, examine how the brain normally regulates the peripheral cardiovascular system and will then turn to what is known about the CNS control of the cardiovascular system during emotional arousal.

a. Central cardiovascular control. The cardiovascular system, like other peripheral autonomic systems, is under both tonic and phasic influence from the CNS. A continuous flow of descending impulses from the brain prevents blood pressure and heart rate from falling to dangerously low levels. However, above and beyond the tonic maintenance of cardiovascular functions, the system must be phasically adjusted to meet the specific tissue needs at a given time.

Tonic regulation of blood pressure and heart rate – the vasomotor center. That the peripheral cardiovascular system is under tonic control by the brain is readily demonstrated: Sectioning of the spinal cord at high cervical levels results in a dramatic fall in blood pressure and heart rate. However, if the brainstem is sectioned above the medulla, blood pressure and heart rate are not affected. This fact, which has been known since the last century, has given rise to the idea that there exists, somewhere in the medulla, a brain region involved in the tonic maintenance of blood pressure and heart rate. Attempts to localize the critical region have until recently been unsuccessful. Work over the past several years, however, has indicated that the tonic vasomotor center is located in the rostral ventrolateral medulla (Kumada, Dampney, and Reis, 1979; Dampney, Kumada, and Reis, 1979; Dampney and Moon, 1980; Ross et al., 1984). This area contains the C1 adrenaline-synthesizing neurons (Hökfelt et al., 1974) and projects to spinal preganglionic neurons (Amendt et al., 1979; Ross et al., 1984). Tonic vasomotor control thus appears to involve adrenergic neurons in the rostral ventrolateral medulla.

Reflex modulation of blood pressure – the baroreceptor reflex. Tonic vasomotor control is often overridden by momentary demands. Thus, blood pressure may rise or fall relative to the natural resting level in the process of blood flow redistribution. However, control mechanisms exist that function to return the system to the resting state. One of the best understood of these is the baroreceptor reflex (Figure 15.10).

Located in the large arteries of the thoracic and neck regions of the body, but especially in the carotid artery and in the ascending aorta, are stretch receptors called baroreceptors (Guyton, 1981). When blood pressure rises, resulting in changes in the shape of the vessel walls, these receptors are mechanically disturbed and thus activated. The sensory nerves attached to baroreceptors are visceral afferents of the vagus. These terminate in the brainstem in the nucleus of the solitary tract [nucleus tractus solitarii, (NTS)]. Recent studies have implicated glutamate as the neurotransmitter in vagal afferents terminating in NTS (Talman, Granata, and Reis, 1984). However, other possible transmitters have been localized in NTS and may also contribute to baroreflex function.

Impulses reaching NTS by way of vagal afferents activate vagal parasympathetic efferents, which then slow the heart rate. When blood pressure rises, baroreceptors thus send messages that result in a reflexive reduction in heart rate. Since the mean arterial pressure is in part determined by the output of

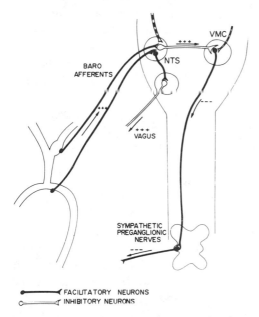

Figure 15.10. The baroreceptor reflex circuit. Receptors located on certain blood vessels detect changes in constriction and thus changes in blood pressure. Afferent nerves from the baroreceptors terminate in the brain in the nucleus of the solitary tract (NTS). The NTS projects to a nearby region that gives rise to efferent fibers of the vagus nerve. When the vagal nerves are excited by baroreceptor activation, the heart rate is slowed. Projections from NTS to the vasomotor center (VMC) can inhibit the activity of sympathetic nerves innervating the blood vessels. Together the slowing of the heart and the inhibition of the nerves innervating the blood vessels result in a lowering of systemic blood pressure. (From Chalmers, 1975, reproduced with permission from the American Heart Association, Inc.)

the heart, the net effect of baroreceptor reflex activation is to control rises in blood pressure.

Forebrain modulation of the cardiovascular system. Electrical stimulation of many areas throughout the forebrain produces changes in blood pressure and heart rate (see Korner, 1979; Smith and DeVito, 1984). Most, although not all, of these are limbic regions. Stimulation of some areas results in increases in blood pressure or heart rate and decreases result from stimulation of other areas. Natural activation of these areas is believed to play a role in the phasic adjustment of blood pressure and heart rate to meet specific situational demands from the internal or external environment.

Although the manner in which higher centers control the cardiovascular system are complex and poorly understood, certain facts have emerged. First, the hypothalamus, by way of connections with brainstem areas, can modulate cardiovascular activity. Second, the hypothalamus also has connections with spinal preganglionic neurons directly controlling the sympathetic nerves innervating the blood vessels and the adrenal medulla. Third, by way of connections with the pituitary gland, the hypothalamus can control the release of vasoactive substances, particularly vasopressin, into the circulation. Fourth, the hypothalamus receives projections from many limbic regions and is a way station through which these regions have their cardiovascular effects. Finally, some limbic regions bypass the hypothalamus in controlling the cardiovascular system. Examples include the insular cortex and central nucleus of the amygdala, which have direct projections to brainstem cardiovascular centers.

b. Central pathways controlling the cardiovascular system during emotional arousal. The brain mechanisms involved in controlling the cardiovascular expression of emotion have mainly been studied in two ways. First, many early experiments employed electrical brain stimulation techniques to elicit emotional behavior, particularly aggressive behavior, and examined cardiovascular concomitants. Second, studies have used conditioning techniques to couple cardiovascular changes to the emotional significance of sensory stimuli.

Stimulation of defense and attack pathways. Earlier we discussed the defense response, a complex behavioral reaction involving the assumption of threatening and defensive postures and other visible signs of aggression. The animal will randomly attack any object in site. Accompanying this behavioral display are a variety of cardiovascular changes (see Figure 15.3) mediated by diffuse activation of the sympathoadrenomedullary system. The defense response can be elicited by brain lesions, natural stimuli, or pain, as well as electrical brain stimulation (see Reis, 1974). The neural pathways underlying brain stimulation-induced defense are, however, the best understood and will be focused on here.

The defense response is classically elicited by stimulating the medial aspects of the hypothalamus (Hess, 1936). From there, the pathway descends to the central gray and subcollicular areas of the midbrain and then to a region around the floor of the fourth ventricle in the medulla (Abrahams, Hilton, and Zborzyna, 1960). Presumably, these pathways either link up with medullary cardiovascular regions and/or descend directly to spinal preganglionic neurons in the mediation of the peripheral autonomic changes. That hypothalamic projections to the brainstem are involved in the cardiovascular defense response is indicated by the fact that stimulation of the hypothalamic defense area inhibits neural activity in the baroreceptor reflex arc (Hilton, 1966; Mancia and Zanchetti, 1981). The manner in which the somatomotor manifestations of defense are mediated through descending pathways linking up with efferent motor pathways of the pyramidal or extrapyramidal systems is poorly understood.

The defense response can also be elicited by stimulation of the amygdala

(Ursin and Kaada, 1960). Studies by Hilton and Zbroyzna (1963) demonstrated that the amygdala-elicited defense response is mediated by projections to the hypothalamus from the amygdala.

Stimulation of lateral regions of the hypothalamus elicits a different form of aggressive behavior called quite biting attack (Wassman and Flynn, 1962). In contrast to the defense response, quite biting is directed at a specific goal object, such as suitable prey, and is believed to be predatory or food oriented. Further distinguishing quite biting attack is the lack of consistent autonomic correlates. The pathway underlying quite biting descends, probably through the medial forebrain bundle, to the midbrain (Siegel and Edinger, 1981). As with defensive behavior, linkage with somatomotor efferent systems is not understood.

Electrical stimulation of various limbic regions in conjunction with hypothalamic stimulation will inhibit or enhance quite biting (Egger and Flynn, 1962; Siegel and Flynn, 1968; Siegel and Skog, 1970). These findings suggest that limbic structures modulate attack behavior by influencing hypothalamic neural activity. Siegel and Edinger (1981) have presented evidence indicating that the inhibition of attack by the limbic forebrain involves projections from individual limbic areas to the medial dorsal thalamus. This structure then sends short projections to the midline thalamic regions, which in turn innervate lateral hypothalamic regions controlling quite biting.

Studies by Bandler (1982) have questioned the role of the hypothalamus in aggressive behavior. Using glutamate, an excitatory amino acid that activates cells but not passing fibers when injected in the brain, he failed to elicit rage by hypothalamic injection, suggesting that studies eliciting the rage response by electrical stimulation were activating fibers passing through the hypothalamus and terminating elsewhere. Bandler further demonstrated that injection of glutamate into the midbrain elicited rage and that the same midbrain region is the site of termination of fibers passing through the hypothalamus. These interesting findings raise important questions concerning the forebrain organization of rage that await answers.

Neural pathways and cardiovascular conditioned emotional responses. A number of investigators have used aversive classical conditioning procedures to examine the central efferent control of the cardiovascular system during emotional arousal. The ongoing approach taken by several laboratories will be briefly described.

Smith and his colleagues have been systematically examining the cardiovascular concomitants of emotional conditioning for a number of years (see Figure 15.4). Recently, they have identified an area in the hypothalamus that when lesioned blocks the expression of blood pressure, heart rate, and blood flow conditioned responses, but does not affect the conditioned suppression of bar pressing behavior (O. A. Smith et al., 1980). Moreover, other cardiovascular changes, such as those during exercise, are not affected, indicating that this region is specifically involved in the expression of conditioned emotional cardiovascular responses. It is of particular interest that anatomical tracing studies have shown that this region has direct connections with spinal preganglionic motor neurons of the autonomic nervous system and with brainstem cardiovascular areas as well (DeVito and Smith, 1982).

In studies of heart rate conditioning in the pigeon, Cohen has gone a long way toward mapping the underlying pathways linking the retina to the heart (D. H. Cohen, 1974, 1980). On the output side, it has been shown that conditioned tachycardia is mediated by excitation of sympathetic neurons in the last cervical and upper thoracic segments of the spinal cord and by inhibition of vagal efferents (Gold and Cohen, 1981). On this foundation, studies by Cabot, Goff, and Cohen (1981) sought to identify the final pathway connecting these peripheral systems to the brain. They implicated descending projections through the midline medulla, possibly involving projections from the raphe nucleus to the spinal cord.

Work by Kapp and his associates has begun to uncover the central pathways involved in the conditioning of bradycardia in restrained rabbits (see Kapp et al., 1984). They demonstrated that lesions of the central nucleus of the amygdala blocked the development of conditioned bradycardia, whereas stimulation of the same region elicits bradycardia. Moreover, manipulations of β-adrenergic or opiate receptors in the central amygdala also disrupted conditioned bradycardia (Gallagher et al., 1981). Combined with work showing projections from central amygdala to NTS (Schwaber et al., 1978), these latter findings suggest that opiate and catecholaminergic interactions in central amygdala are involved in conditioned bradycardia through descending projections that modulate vagal output to the heart.

In summary, much is now known about the neural pathways that link the brain and the autonomic periphery. This information has provided the foundation for studies examining the role of central efferent systems in emotional expression.

4. Evaluation of sensory input. Much of life consists of interactions with the environment. This is not to say that all behavior is a reactive response to the environment. However, even spontaneous behavior, once started, is usually guided by sensory input.

In mediating environmental interchanges, the brain is constantly called upon to evaluate the significance of information received. The neural pathways and physiological mechanisms involved in the reception of sensory stimuli and transmission through the brain in the buildup of perceptual images have been heavily investigated. Although less is known about how sensory information is evaluated, the pieces of the puzzle are beginning to fall into place.

a. The nature of emotional evaluation. Stimulus evaluation can be described as a process by which sensory input is compared with stored information or knowledge (James, 1884; Bowlby, 1969). Such knowledge can be bestowed by either experience or inheritance.

Inherited knowledge dominates environmental interactions in the lower vertebrates and invertebrates. Specific elements in the environment trigger preprogrammed adaptive behavior patterns by activating prewired (innate) neural circuits. An essential component of these circuits is a mechanism for recognizing as significant key stimulus features, such as colors that signal a predator or smells that signal a suitable mate, and organizing complex responses to these and not other stimuli.

Mammalian behavior is less influenced by inherited knowledge and instead is more under the control of information acquired through experience. Environmental events that should be approached or avoided are thus recognized by the recollection of past associations of the event with either other stimulus events or with behavioral consequences. Nevertheless, even in the higher mammals and humans, inborn programs of action triggered by sensory events can be demonstrated. This is especially true of unconditioned reflexes, such as withdrawal responses to painful stimuli, but may also apply to stimuli that are innocuous in their physical consequences but represent (symbolically) something of significance.

b. Neural pathways and emotional evaluation
Corticofugal pathways and stimulus significance: In 1937, Kluver and Bucy observed that monkeys sustaining large lesions involving the temporal lobe area of the neocortex and underlying subcortical structures, especially the amygdala, manifested a complex behavioral syndrome involving changes in sexual, feeding, and emotional behavior. Later studies using smaller, systematically placed lesions have pointed to amygdala damage as the origin of these behavioral changes (Weiskrantz, 1956; Jones and Mishkin, 1972). Moreover, these studies have indicated that the underlying basis for the changes involves a dissociation between the sensory–perceptual and affective qualities of stimuli. Thus, following amygdala lesions, monkeys are no longer threatened by the presence of human observers and other previously threatening objects, will eat feces, raw meat, and junk objects, and attempt to copulate with animals of the same sex or even of a different species (Jones and Mishkin, 1972). Objects are still perceived and emotional reactions are still expressed, but these functions are no longer coupled in the brain.

The neuroanatomical basis for sensory–affective dissociation produced by amygdala lesions thus appears to involve an interruption of the chain of events through which emotional behavior is guided by sensory input. In the visual system, sensory input reaching striate cortex is relayed along modality-specific neocortical pathways in the sequential build up of visual object perception. Mishkin and colleagues have demonstrated the importance of the sequential transfer of information from striate cortex, to prestriate cortex, to inferotemporal cortex in visual discrimination learning. While lesions of any one of these areas blocks discrimination learning, the inferotemporal lesion does so without producing blindness or other sensory deficits. The animals can see objects, but new meanings cannot be assigned to objects through stimulus–reward association. Such findings suggested to Jones and Mishkin (1972) that inferotemporal cortex might link the visual system to areas of the brain involved in assigning significance to sensory stimuli. This hypothesis was confirmed in their studies by showing that interruption of pathways from the inferotemporal cortex to the amygdala blocked the formation of stimulus–reward association. Interruption of pathways to the hippocampus or frontal cortex from inferotemporal cortex did not have such effects. The amygdala, in the view of Mishkin and Aggleton (1981), thus assigns meaning to cortically processed visual information. By way of projections from the amygdala to the hypothalamus and basal forebrain, behavior is guided by stimulus significance.

Much less is known about the role of corticofugal projections from other sensory modalities in forming sensory–affective associations. However, Turner et al. (1980) have shown that the somatosensory and auditory association areas of the neocortex, just as the visual areas, have sequential anatomical links with the amygdala. The amygdala, in this view, is a common, modality-independent way station that assigns significance to cortical sensory inputs and discharges through the hypothalamus in the guidance of behavior on the basis of the emotional significance of environmental events.

Modality-specific association areas of the neocortex also project to the hippocampus by way of multisynaptic projections involving the entorhinal cortex (van Hoesen, Pandya, and Butters, 1972). The hippocampus has been implicated in a variety of cognitive functions, including complex memory formation and spatial perception (see Issacson, 1982). It is thus of interest that recent studies (Poletti, Kliot, and Boytin, 1984) have shown that the output of the hippocampus can functionally activate the hypothalamus by way of connections through the amygdala, in addition to the more commonly described pathway involving the fornix. The amygdala may thus play a role in the emotional evaluation of higher cognitive information as well as sensory–perceptual events.

Subcortical sensory pathways and emotional evaluation. It is generally assumed that sensory pathways project in an obligatory fashion through the thalamus to the cortex and that emotional processing occurs when, as Mishkin's work suggests, sensory information is transmitted to limbic areas, particularly to the amygdala, by way of descending corticofugal pathways (MacLean, 1949; Penfield, 1958; Geschwind, 1965; Gray, 1982). However, a number of studies have now demonstrated that the hypothalamus, amygdala, and caudate nucleus, at least, receive sensory input from thalamic sensory stations (see earlier discussion of limbic organization). Might these projections play some role in emotional processing?

In order to answer this question, we began a series of investigations using auditory fear conditioning as a model (LeDoux et al., 1984). Rats so conditioned exhibit increases in mean arterial pressure (MAP) and decreases in somatic activity when exposed to the CS (Figure 15.11). In confirmation of the results of earlier studies, we found that large cortical ablations that included modality-specific auditory cortex failed to affect the establishment of conditioned fear responses. In contrast, lesions of the medial geniculate nucleus (the primary acoustic thalamocortical relay nucleus) or the inferior colliculus (the major auditory area in the midbrain) abolished auditory fear conditioning (Figure 15.12). These findings suggest that for an aversive meaning to be assigned to an acoustic stimulus, the stimulus must be transmitted to the thalamic level and must then be relayed to a subcortical rather than a cortical target.

Using axonal transport tracing techniques, we then identified possible subcortical targets of medial geniculate (MG) neurons and demonstrated that the medial division of MG (MGm) projects, in addition to neocortical areas, to a striatal field (STR) involving the dorsal archistriatum (amygdala) and the overlying neostriatum (posterior caudate nucleus). Subsequent studies examined the effects of interruption of these projections and found that in order for fear conditioning to take place, intact connections from MG to STR must exist (Figure 15.13). It is not yet possible to distinguish the contributions of the archistriatum and neostriatum to these effects. Nevertheless, the data indicate that auditory fear conditioning involves the relay of acoustic input from MG to either the posterior caudate nucleus, the posterior dorsal amygdala, or both.

Several interesting points concerning the anatomy of the MG projection to STR should be noted. First of all, not all aspects of MG project to STR. The projection is mainly from the medial division. This division represents the cochlea only weakly, at least when compared with the principal division, and is best viewed as an adjunct projection system (Graybiel, 1973). Second, striatal areas also receive inputs from subcortical areas of the visual system (Graybiel, 1973). However, the overlap of the visual and auditory projections into the striatum are not known. Third, STR receives inputs from thalamic areas involved in pain transmission. These observations suggest the hypothesis

AUTONOMIC RESPONSES   BEHAVIORAL RESPONSES

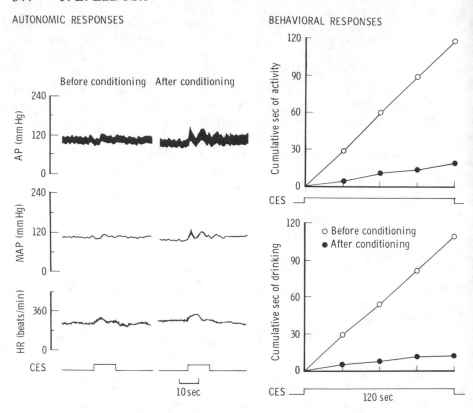

Figure 15.11. Emotional conditioning in freely moving rats. Rats chronically instrumented for cardiovascular recording were subjected to classical fear conditioning trials where a pure auditory tone lasting 10 sec was repeatedly paired with a brief delivery of mild footshock. The next day, while the rat was resting in its home cage the acoustic conditioned emotional stimulus (CES) was presented alone. As shown in the left panel, after conditioning the CES elicited large increases in arterial pressure (AP), mean arterial pressure (MAP), and heart rate (HR). In separate tests, the effects of CES presentation on behavior were assessed (right panel). The CES suppressed both exploratory activity (top) and drinking (bottom), after but not before conditioning. Thus, following classical conditioning, an otherwise insignificant stimulus acquires the capacity to evoke a powerful fear reaction consisting of sympathetically mediated increases in cardiovascular activity and the suppression of ongoing behavior.

that STR is involved in the neural integration of aversive UCSs with auditory and visual CSs and may thus constitute a subcortical mechanism of stimulus–reinforcement association. However, much more work is needed on this problem.

Parallel processing of stimulus significance. Neural pathways have thus been identified that seem to be involved in the evaluation of stimulus significance. The neocortical pathway is clearly necessary when emotional significance must be assigned to complex, highly discriminated perceptual information. The thalamic sensory neurons that allow for such discriminations, in fact, mainly

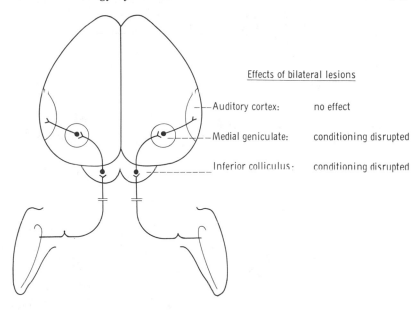

Figure 15.12. Effects of lesions of the auditory pathway on emotional conditioning. The conditioned emotional response complex illustrated in Figure 15.11 is expressed normally in rats with lesions of the auditory cortex, but is abolished by lesions of areas afferent to the auditory cortex, including the medial geniculate body and the inferior colliculus (see LeDoux et al., 1984). These data indicate that the emotional learning pathway involves the transmission of auditory signals from the periphery through the inferior colliculus to the medial geniculate. The medial geniculate then relays the signal to a target other than auditory cortex (see Figure 15.13).

project to neocortex. The pathways that diverge from ascending projection systems in the thalamus, in contrast, are more limited in their capacity to analyze stimulus information. Neurons along these pathways are only capable of mediating the emotional evaluation of simple, rather crude stimulus features. Although such features are seldom encountered alone in daily life, they are necessary components of more complex events. When sensory input reaches the thalamus, it is thus projected to cortex by way of primary projection systems for further analysis, but it is simultaneously relayed by adjunct systems to subcortical areas.

Two implications of these parallel emotional processing channels should be considered. First, the subcortical areas that receive thalamic inputs also receive neocortical inputs. The two pathways thus converge. The thalamic pathway, though, is several synapses shorter. Input reaching target areas such as the amygdala may therefore prime the area to receive the better analyzed neocortical inputs, providing a crude picture of what is to come, narrowing

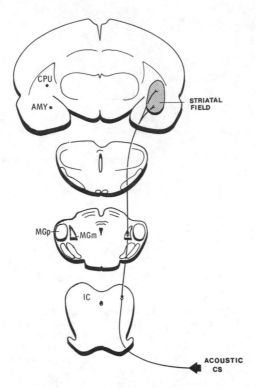

Figure 15.13. Emotional learning pathway. Through anatomical and behavioral studies, several synaptic links in an emotional learning pathway have been identified. Although the auditory cortex is the main known target of fibers leaving the medial geniculate (MG), our behavioral studies suggested that conditioning must be mediated by subcortical rather than cortical projections of the MG (see Figure 15.12 and LeDoux et al., 1984). Anatomical studies then demonstrated that neurons within the MG, particularly within the medial division (MGm), project to a striatal field involving portions of the neostriatum (caudate-putamen, CPU) and archistriatum (amygdala, AMY). Subsequent behavioral studies found that interruption of the connections linking the MG to the striatal field disrupted conditioning. These various findings indicate that auditory fear conditioning in rats is mediated by acoustic signals transmitted through the inferior colliculus (IC) to the medial MG and then relayed to the striatal field. Additional links in the pathway are expected to involve brain areas receiving projections from the CPU or AMY.

the affective possibilities, and perhaps even organizing possible and actual responses. Once the neocortical information arrives, further selections and if necessary corrections could be made.

Second, thalamic input to subcortical areas may allow environmental events to control the autonomic nervous system and behavior independent of the higher cognitive and conscious control mechanisms of the neocortex. Such projections might, in other words, constitute a pathway to unconscious aspects of the brain and could conceivably be a basis of unconscious behavior patterns and psychosomatic disturbances. These, however, are ideas that are to date untested.

c. Reward systems and stimulus evaluation. In 1954 Olds and Milner reported that delivery of electric stimuli to certain areas of the rat brain had effects comparable to the delivery of natural reinforcers. A monumental outflow of studies followed and has served to elevate brain stimulation reward to the level of being one of the best characterized phenomena in behavioral neuroscience. The literature in this area is massive. Fortunately, excellent summaries are available (Wanquier and Rolls, 1976; Olds, 1977). Our interest here is in the relationship between reward systems and stimulus evaluation. Other aspects of brain stimulation reward and emotion will not be discussed (see Pankseep, 1981).

Rolls (1976) has classified reward sites in the brain into three groups: sites related to the hypothalamus and medial forebrain bundle and closely associated with the catecholamine-containing cell groups of the brainstem and their ascending projection pathways, limbic sites, and other sites (such as olfactory bulb and central gray). Animals will work to obtain stimulation in any of these regions, but the parameters controlling the responses are different in each case.

Rolls has speculated that one of the functions of limbic reward sites, particularly amygdala sites, is related to learning to associate environmental events with rewards and punishments. In this view, connections from the medial forebrain bundle system (especially from the hypothalamus) provide the amygdala with input concerning whether reward has occurred, and connections from amygdala back to the hypothalamus informs the medial forebrain bundle system about which sensory events have been associated with a reward. This hypothesis overlaps considerably with and extends the concepts developed above concerning the role of the amygdala and its neural relations in evaluating sensory input.

d. Evaluation of viscerosensory input. Just as brain systems evaluate inputs from the external world, so are signals from within the body assessed. Much less is known about the neural pathways through which visceral afferent information is processed, but recent work has begun to identify, at least, vagal afferent pathways and their second-order forebrain projections.

The vagus nerve contains afferents from many of the visceral tissues of the abdominal cavity and terminates in NTS. As noted earlier, interruption of vagal transmission disrupts performance in emotional learning tasks and interferes with rewarding brain stimulation. These effects are presumably mediated by connections between NTS and forebrain areas. It is thus of interest that the portion of NTS that receives vagal afferents overlaps with the A2 catecholamine-containing cell group (Sumal et al., 1982). This region, in turn, sends projections to the central nucleus of the amygdala, as well as to other areas (Richardo and Koh, 1978). Moreover, stimulation of the vagus nerve affects activity in central amygdala (Radna and MacLean, 1981). On the basis of these and other observations, Kapp et al., (1984) have suggested that central amygdala is a key area for viscerosensory processing. Given the evidence already presented implicating the amygdala in evaluating exteroceptive inputs, the amygdala would seem to be a likely candidate for evaluating and integrating inputs from the external and internal environments.

Normally, viscerosensory information does not reach conscious levels. Adam (1978) has, for example, examined the effects of direct stimulation of visceral organs in human and animal subjects and has shown that although viscero-sensory inputs can have marked influences on behavior and are subject to control by conditioning processes, only in special situations are humans consciously aware of visceral functioning.

The fact that viscerosensory inputs are so resistant to conscious processing would seem to be a point against peripheral models of emotional experience. Nevertheless, it is clear that peripheral systems, probably through influences on limbic areas, can affect emotional behavior. The evidence to date mainly suggests that peripheral systems have nonspecific activating or arousing effects. However, the nonspecific nature of these systems may reflect our ignorance more than it does the nature of the mechanisms involved.

e. Summary. The evaluation of the affective significance of afferent input is a crucial functional link in the chain of events through which sensory information from the external world and from within the body come to guide behavior. With respect to emotional behavior, areas in the forebrain, particularly within the limbic system, appear to play a crucial role in emotional evaluation. Certain limbic areas, such as the amygdala, receive exteroceptive and interoceptive inputs and send outputs to the brain stem areas controlling the ANS. Much more work is needed, however, to understand the integrated role of these various networks. For example, we know little about the central integration of interoceptive and exteroceptive inputs. Moreover, we have little information concerning the specificity of evaluative systems with respect to the sensory modality of the incoming signal. We do not even know if the assignment of different meanings to a given stimulus involves different neural pathways or instead involves variation in the activity of a common pathway. These are topics for future research.

5. Emotional experience. The problem of experience is the enduring and troubling problem of consciousness. How phenomenal experiences, such as feelings, relate to brain mechanisms is poorly understood. There is no systematic body of evidence relating neural function to feelings. In this section, we will, therefore, not review experimental findings but instead will present a conceptual scheme for viewing conscious feelings. The objective is to relate speculations concerning possible neuroanatomical substrates of conscious feelings to our more solid understanding of how the brain mediates emotional expression and evaluation.

Most of what follows will mainly apply to emotional experience in humans. This restriction is necessary because we have no way of truly knowing whether nonhuman organisms have conscious experiences.

a. The problem of consciousness. Ever since Descartes formulated our modern concept of consciousness as a nonphysical, internal representational system, philosophers and scientists have struggled to resolve the chasm between what is mental and what is physical. The mind–body problem (the problem of relating "the mental" to "the physical") will not be examined here. Numerous theoretical and speculative discussions concerning the mind–body problem, as seen by philosophers, psychologists, and neuroscientists are available (see Ryle, 1949; Hook, 1960; Eccles, 1965; Sperry, 1968; Popper and Eccles, 1977; Searle, 1978; Rorty, 1979). For the sake of discussion, we assume that it is at least possible that "the mental" can be understood in terms of "the physical." Otherwise, the pursuit of a cognitive neuroscience would be empty.

Consciousness, phenomenal experience, feelings, and the like are particularly resistant to experimental study. Though the supreme and founding subject of psychology, consciousness, and all of its mental ramifications, was banished from the lips and laboratories of psychologists for many years, now the inevitable has happened. Consciousness has returned as a legitimate topic for discussion and experimentation. Significantly contributing to this comeback were experimental observations on brain function that seemed to demand explanation in mentalistic terms: Moruzzi and Magoun's (1949) discovery of the arousal and activating functions of the reticular core of the brainstem; Olds and Milner's (1954) finding "pleasure centers" in the rat brain; Heath's (1954) elicitation of "pleasurable feelings" from human brains during electrical stimulation; Penfield's (1958) finding that memories were "reexperienced" during stimulation of the human temporal lobe; Myers and Sperry's (1953) demonstration that sectioning of the corpus callosum in cats produced functionally independent mental systems on the two sides of the brain; Gazzaniga, Bogen, and Sperry's (1962) demonstration that in split-brain humans the talking left hemisphere was not aware of (could not describe) activities taking place in the speechless right hemisphere.

Today, for better or worse, consciousness is with us. The difficulties are no easier to overcome this time around. Clearly, neuroscience has not matured to the point of being able to account for consciousness in cellular or molecular terms. However, the effort must be continued, no matter how dismal the immediate prospects for final answers, for the problem of consciousness is a key aspect of the problem of life.

b. The problem of feelings. Feelings constitute a class of conscious experiences. For lack of a better definition, feelings might be described as those conscious experiences having to do with the individual's welfare. While on the surface such a view might (or might not) seem reasonable, it suffers from using one vague term (consciousness) to account for another (feelings). Unless we can specify more clearly what consciousness is, explaining feelings by reference to conscious experiences will not take us very far.

One widely held (though not universally accepted) view is that consciousness (self-awareness) is closely tied to natural language systems (see Griffin, 1976; LeDoux, Wilson, and Gazzaniga, 1979; Luria, 1982). Some of the evidence supporting this view includes: (a) organisms with well-developed language systems (i.e., humans) exhibit self-awareness, a capacity that is difficult to demonstrate in other organisms; (b) in split-brain humans the conscious properties of the language dominant hemisphere are typically robust; the minor hemisphere, lacking language, seems cognitively deficient and far less, if at all, self-aware; (c) in the few rare instances where language is present in both hemispheres of split-brain patients, each half-brain seems to possess an independent and well-developed sense of self (see LeDoux et al., 1979).

Though admittedly correlational, the close coupling of language and conscious awareness provides a significant clue to the mechanisms of consciousness, that is, understanding the brain substrates of language may help us to understand the neural foundations of consciousness. Though the relation between brain and language is hardly a simple nut to crack, considerable information is available (Zurif and Blumstein, 1978). At present, our understanding of the organization of language systems in the brain may be our best approach to the systems of consciousness.

How does using language as a first approximation of consciousness help us relate feelings to brain mechanisms? This requires another conceptual leap. We defined feelings earlier as conscious experiences having to do with the individual's welfare. We must now translate this statement to read, "linguistic coding having to do with the individual's welfare." But linguistic coding of what? Here is the leap. Feelings result when language systems receive input from limbic areas. The implication of this view is that by understanding the neural relations between the emotional and language systems of the brain we can gain an understanding of how brain circuits might mediate feelings.

Before turning to the neural substrates of language, it should be noted that there are many ways to argue against the idea that consciousness is related

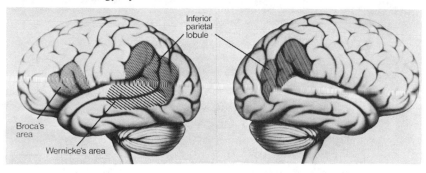

Figure 15.14. Language areas of the brain. In most right-handed adults, language processing is mediated by areas in the left cerebral hemisphere. An interior region, Broca's area, is involved in speech production and syntactic comprehension. The semantic content of spoken and written language is processed by posterior regions known as Wernicke's area and the inferior parietal lobule (IPL). In the right hemisphere the IPL appears to mainly be involved in mediating spatial interactions with the environment. (From LeDoux, 1982, reproduced by permission of S. Karger AG, Basel.)

to language. Many of these can be short-circuited by the further stipulation that consciousness is not so much a function of language as of the cognitive capacities that make language possible. Thus, though one is without the capacity to speak or understand speech, he may nevertheless possess humanlike self-awareness. The same argument applies to human infants who have not yet mastered language, but exhibit awareness of self. The cognitive underpinnings that make language and consciousness possible do not necessarily correspond with the capacity to express and understand language.

With regards to animal consciousness, it seems clear that whatever conscious capacities are present in nonhumans, these capacities are either qualitatively different or represent large quantitative differences from human conscious capacities. The intent is thus not to deny any form of consciousness to animals but to point out that their capacities are very different from ours and that the difference has to do with the neurocognitive mechanisms that make language possible.

c. Neural connections between emotional and language systems. The language systems of the brain consist of an anterior cortical region (Broca's area) involved in syntactic comprehension and speech production and a group of posterior regions concerned with understanding the semantic content of spoken or written language (see Zurif and Blumstein, 1978). Our interest here is mainly in that part of the posterior language area occupying the inferior parietal lobule (IPL) (Figure 15.14). Although this region is present in rudimentary form in other primates, it is more developed in man. One of the most important characteristics of IPL is that it receives highly processed inputs

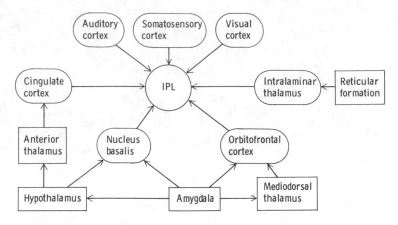

Figure 15.15. Inputs to the inferior parietal lobule (IPL). The IPL receives inputs from sensory receptive areas of the cerebral cortex, from cortical (cingulate, orbitofrontal) and subcortical (nucleus basalis) limbic regions, and from the reticular formation (by way of the intralaminar complex in the thalamus). This convergence of sensory, limbic, and reticular inputs renders IPL an interesting candidate for the mediation of conscious emotional experiences. Areas directly projecting to IPL are shown in ellipses, whereas indirect projections are indicated by rectangles. Drawing is based on the results of Mesulam et al. (1977).

from the cortical association areas of the various sensory modalities (Jones and Powell, 1970; Mesulam et al., 1977). Some have speculated that the convergence of such inputs in IPL was a neural precursor of the evolutionary development of language (Geschwind, 1965; LeDoux, 1982).

Recent studies have demonstrated that IPL, in addition to receiving highly processed sensory input, also receives direct projections from limbic areas (Mesulam et al., 1977). These include the cingulate cortex and the nucleus basalis (Figure 15.15). The cingulate cortex is part of the classic limbic circuit of Papez and receives inputs from other limbic regions. The nucleus basalis receives inputs from the amygdala and orbitofrontal cortex, as well as the hypothalamus. Other inputs to IPL of interest include projections from the reticular activating system by way of the intralaminar thalamus. Together, these various inputs allow IPL to serve as an integrative center, synthesizing information from exteroceptive sensory systems, limbic and viscerosensory areas, and from arousal systems. Such a pattern of inputs seems ideal for the mediation of conscious feelings.

d. Limbic-language circuits and emotion. Consider a hypothetical example of how conscious feelings might be generated by the circuits just described. Light reflected from some object in the environment excites the retina, which in turn generates impulses transmitted along the optic nerve to subcortical relay stations. At the thalamic level, the input is primarily projected to neocortical receiving areas, but at the same time is relayed directly to limbic areas. At the cortical level, the input then undergoes several levels of modality specific processing and then is transmitted over corticocortical pathways to modality independent cortical areas (IPL) for linguistic and conscious coding and over corticofugal pathways to limbic areas. Once the input enters limbic circuits, it is evaluated for its significance to the individual's welfare by comparison with stored information. The comparison process results in the assignment of some value to the stimulus input and either activates or suppresses the outflow of efferent activity in descending autonomic and somatic motor pathways. The results of the evaluative process are also relayed through limbic circuits toward cortical regions. Depending on the net excitatory and inhibitory activity impinging on these circuits, the results may or may not reach cortical areas, such as the cingulate region, and thereby be relayed to IPL. If the information does reach IPL, it is matched up with direct input to IPL from modality-specific cortical areas, coded linguistically, and represented in consciousness. The representation in consciousness by linguistic coding of stimulus evaluation gives rise to a feeling about the stimulus event, a conscious awareness of stimulus significance.

Not all evaluations of incoming events reach IPL. Those that do not (and most do not) constitute an important aspect of the domain of the unconscious mind.

Emotional expression can thus proceed without the intervention of conscious feelings. In the case where unconscious stimulus evaluation leads to emotional expression (autonomic and behavioral), but the limbic–IPL relay is not completed, a second chance for conscious awareness is available. Once peripheral responses are expressed, visceral and somatic feedback to the brain activates limbic and reticular circuits, facilitating nonspecific arousal. Cortical areas such as IPL are alerted that something significant is taking place, but do not have direct access to (awareness of) the critical information. Evaluation of the environmental context is then initiated, and the emotional significance of the event is supplied by cognitive rather than emotional processing systems. Because the cognitive evaluation is based on inference rather than on direct limbic input, it will often be imprecise or wrong. Nevertheless, such a system allows for the continuous construction of a unified sense of self, even though such unity defies the anatomy and physiology of the brain.

Conclusion

This chapter has focused on the neuroanatomical pathways underlying emotional processes. While a great deal is now known about these pathways, much more work is needed before we will be able to truly characterize the emotional circuitry of the brain. Some research strategies that, if followed, would help to fill the voids and build further on the existing foundation are

1. Broadening the range of emotional states available for neurobiological analysis by developing new animal models of various emotional processes
2. Development of animal models suitable for examining emotional phenomena which have emerged out of human studies, such as models for examining the biological viability of the attribution or cognitive/arousal model
3. Development of animal models for studying emotional experience, or, conversely, development of neurobiological techniques capable of relating emotional experience in humans to the microcircuitry and biochemistry of the brain
4. Identification of possible neurotransmitter candidates mediating information flow along known emotional pathways and determination of the role of the transmitters in emotional processing
5. Pursuit of the molecular basis of neurotransmission in emotional circuits
6. Extensions of basic findings concerning the anatomy, physiology, biochemistry, and molecular biology of emotional circuits to analyses of emotional disorders in man

16 A neurobiological view of the psychology of emotion

Joseph E. LeDoux

Everyone knows that physiological responses accompany emotional experiences. Our heart pounds when we are afraid. Tears fall in sadness. Blushing comes with embarrassment. It is thus not surprising that emotion has traditionally been an area in which psychological and biological approaches blend together. In fact, some of the most influential theories of emotion have drawn as heavily from anatomy and physiology as from psychology. This is true of contemporary theories, such as the cognitive/arousal theory, as well as of the classic view espoused by James and Lange, who were both physiologists by training. Unlike many areas of psychology, the psychology of emotion has not proceeded independently of biology.

In spite of this tradition of interplay between psychology and biology, we are nowhere near a unified approach to emotion. Neuroscientists have as yet not come up with a reasonable strategy for relating the microcircuitry and biochemistry of the brain to human conscious emotional experience. Although basic concepts and findings from biology, in general, as well as from neurobiology, are incorporated into psychological research and theories, these have often been used in a selective and uncritical way. History has paved the way for natural interactions between psychologists and neuroscientists interested in emotion, but the current interplay is sadly superficial. A unified approach to emotion is thus an unrealized potential rather than an ongoing endeavor.

In the following, I will attempt to point out some of the areas of overlap and divergence between the psychology and neurobiology of emotion and indicate the lessons each field has to offer the other. I will assume that the two preceding chapters are representative of the biology and psychology of emotion and will draw examples from these chapters to make my points.

The central questions about emotion concern the nature and origins of our conscious emotional experiences. It is, thus, reasonable that psychologists have focused on the phenomenal, experiential aspects of emotion, especially in humans. The aim of contemporary theories, such as the cognitive/arousal theory and the facial feedback theory, is primarily to account for the quality of emotional experiences.

355

Most work on emotion in neuroscience, in contrast to mainstream psychology, involves studies of experimental animals, organisms that have undocumented conscious capacities. Contemporary neurobiological research on emotion does not really address the problem of emotional experience. Instead, neuroscientists have been primarily concerned with the central and peripheral neural mechanisms that control the expression of emotion. While this emphasis might seem overly restrictive, it has had an important consequence. Neuroscientists studying emotional systems in the brain have been led to ask how the brain mediates the linkage of emotional responses to the stimuli that elicit them. Because not all sensory stimuli elicit emotional responses and because a given stimulus is not emotionally potent in all situations, neuroscientists have naturally assumed that the linkage must involve processes that evaluate the significance of incoming stimuli. This logic is apparent in the classic papers of Papez (1937) and MacLean (1949), as well as in contemporary work (Mishkin and Aggleton, 1981; LeDoux, Sakaguchi, and Reis, 1984).

The fact that psychologists and neuroscientists have emphasized different aspects of emotion in their studies does not reflect a fundamental disagreement about the nature of emotion. No neuroscientist would deny the centrality of emotional experience to a theory of emotion. Nevertheless, in the absence of techniques for objectively studying emotional experience in animals, or techniques for relating human emotional experience to the detailed organization and chemistry of the brain, neuroscience is not likely to make much progress in this area.

In spite of a dominant emphasis on the problem of emotional experience, the psychology of emotion has generally recognized that there is much about emotion that cannot be explained by models that focus exclusively on experience. For example, many models postulate primitive emotional systems that do not depend upon conscious control. Buck's readout model states that stimuli impinge upon motivational and emotional systems directly without cognitive and conscious mediation. Such processing channels allow stimuli to influence memory and behavior without conscious participation. Similarly, Lazarus' appraisal model invokes the concept of "primitive evaluative perceptions," which also occur outside of conscious awareness. Interestingly, recent studies by Zajonc (1980) have provided experimental evidence for processing systems that evaluate the affective significance of stimuli prior to and independent of the cognitive analysis of the perceptual content of the stimulus. This work obviously fits in well with studies of the neurobiological foundations of stimulus evaluation and creates a multitude of opportunities for interactions between psychologists and neuroscientists. If psychologists can precisely specify the functional subprocesses involved in stimulus evaluation, neuroscientists will have a better idea of what to look for and where to look in the brain.

One particularly interesting aspect of the psychological studies of affective processing in humans is the possibility that the underlying mechanisms are, at least in part, common to man and animal alike. If so, the phylogenetic continuity so often assumed to exist with respect to primitive emotional mechanisms might be demonstrated. Comparative studies of affective processing, using comparable test paradigms to examine different species, would be especially interesting.

An area of striking overlap in the two chapters concerns the emphasis on language mechanisms. It would be wrong, however, to draw the conclusion that this point of convergence reflects a true meeting of minds between psychologists and neuroscientists. It is not at all clear that most neuroscientists would agree with my speculative views on language, emotion, and consciousness. Similarly, Buck's emphasis on language would probably be somewhat controversial among psychologists.

It is encouraging, from the point of view of efforts to bring psychologists and neuroscientists together, that emotional psychologists have been willing to borrow concepts and findings from biology and use these inputs to buttress their arguments and constrain their theories. Not all areas of psychology have been so open minded. Unfortunately, however, in some instances, the items borrowed from biology have not been the items lent. For example, the cognitive theory of emotion assumes that autonomic changes during emotional arousal are uniform in all situations. This is clearly false. The sympathetic and parasympathetic nerves, together with autonomic hormones and numerous peripheral peptides, can set up a vast array of different patterns of peripheral activity that can then be transmitted to the brain, either directly or indirectly. While we cannot at present say with confidence that such feedback actually determines the quality of emotional experience, we certainly cannot reject this possibility. The cognitive theory needs reformulation. Whether the cognitive theory can survive the kind of restructuring that is required, given the facts about the periphery and its relation to the brain, remains to be seen.

Another area of concern involves the recent infiltration of ideas about cerebral lateralization into the psychology of emotion. It is now well established that certain processes are unequally represented in the two hemispheres of the human brain. This is especially true of language and spatial abilities, and to a lesser extent of some perceptual and perhaps emotional processes. Some have, on the basis of these facts, argued for different cognitive styles and specialized modes of information processing in the two hemispheres. Through evolution, the human left hemisphere has acquired an "analytic" mode of processing and the right a "synthetic" mode. Emotion, being a wholistic, synthetic function, is, therefore, lateralized to the right hemisphere.

Contrary to this somewhat popularized view, the evidence actually suggests that the similarities between the hemispheres outweigh the differences (see

LeDoux, 1982). The hemispheres in man, as in other animals, work together, not separately.

The idea of independent hemispheres, each specialized for different psychological processes, is fortunately on the wane among students of brain function. Functional localization is studied at the level of individual isolated neurons, and even at the subcellular level. Localization at the level of the half-brain is barely an improvement over the conclusion that a given process is a function of the brain.

The fact that functional differences can be produced using the experimental techniques of the cerebral lateralization field may nevertheless be useful at the psychological level of investigation. It provides a way of fragmenting the psychological function and thereby may reveal something about the process in question. However, I do not think that studies of cerebral lateralization are going to tell us much about the brain mechanisms of emotion.

As psychologists have been too eager to borrow, neuroscientists have been delinquent in their surveys of what psychology has to offer. It is probably safe to say that psychologists have broadened their work on emotion over the years in an attempt to account for the full range of human emotion. However, neuroscientists still focus almost exclusively on aversive emotions, such as fear and aggression. Moreover, the paradigms used by neuroscientists to study emotion have changed little over the years. Fresh input from psychologists is sorely needed.

In conclusion, psychologists and neuroscientists have much to offer each other in the quest to understand emotion. The tradition of interplay between the two fields should make the task of bringing the scientists together somewhat easier than in other areas where the two groups have had little or no interaction. However, achievements will not be attained by exchanging textbook generalities. A true collaboration must be sought. The facts exchanged must be specific, current, and subject to periodic critical evaluation. Theories must not be based on selected pieces of borrowed information simply because they fit. If psychological theories of emotion are to be truly constrained by biology, the inconsistencies must be accepted along with the consistencies. And neuroscientists must be open to developments in psychology and be willing to modify paradigms to test for new functional systems in the brain. Although interchanges of this type may prove to be initially painful, they will hopefully lead toward a unified psychobiological approach to emotion.

A psychologist's reply

Ross Buck

LeDoux and I clearly agree that psychologists studying emotion must be aware of the work of neuroscientists to provide a framework for their ideas, and that psychological theory and research may provide leads for neuroscientists in their progressive development of that framework. Even though these efforts may not quickly lead to an explanation of the nature and origin of conscious emotional experience, they will clarify the roles of emotion in adaptation and the maintenance of homeostasis and in social coordination and communication. In the end, I suspect that an understanding of these last functions will be necessary for the eventual understanding of emotional experience. I also agree with LeDoux that the concern of the neuroscientists for the phylogenetic continuity of emotion is important for the psychologist, and add that the concern of many psychologists for the ontology of emotion – its development within the individual – is an important subject that appears to be often neglected by neuroscience.

It is indeed interesting that both of our chapters have emphasized language. Our views may well be controversial, but I suspect that most would agree with the general point that the uniquely human aspects of emotion will not be understood without a consideration of language and its relationship with archaic structures and functions in the brain. This suggests in turn that there are grounds for new dialogues between psychology, neuroscience, and the communication sciences.

LeDoux states that cognitive theories of emotion have oversimplified the nature of peripheral physiological responding. This is the case for some but not all cognitive theories of emotion. Some allow for more complex peripheral responding, while others do not consider physiological processes at all. Clearly, this is an area of exciting and rapid development within neuroscience (particularly in studies of peptidergic systems and immune system functioning), and cognitive theorists should take note of these advances.

Regarding cerebral lateralization, I did not mean to suggest that the two hemispheres are independent or specialized for different psychological processes in any simple way. I suspect that most would agree with LeDoux that the similarities between the hemispheres outweigh the differences and that the hemispheres normally work together. Overall I am not as worried as LeDoux that psychologists will misrepresent or oversimplify the results of neuroscience: If this occurs, it should be self-correcting. Communication between the groups will, of course, both make oversimplification less likely and enhance self-correction.

On the other hand, I think it is extremely suggestive that the patterning of

359

lateralized processes can be interpreted with an analytic/synthetic distinction, because such a distinction has a long history in epistemological theory (Buck, 1984). LeDoux may be correct in saying that cerebral lateralization will not tell us much about brain mechanisms of emotion per se – that remains to be seen – but I think that there is clear evidence that cerebral lateralization must be considered in the higher-order processing of information from these mechanisms.

LeDoux states that there is a need in neuroscience for new paradigms to study emotion. I suggest that the recent research in emotional communication indicates that the precise description of emotional behavior can tell us much about the brain mechanisms involved. This is consistent with Teitelbaum's use of a dance notation system to relate the gait of rats lesioned in dopaminergic fiber tracts to the characteristic gait of humans with Parkinson's disease (Schallert et al., 1978), and with Vanderwolf's emphasis upon the detailed description of the behavioral effects of brain lesions (Vanderwolf and Robinson, 1981). It is important that such a paradigm can involve otherwise intact animals (or humans), freely moving in their natural environments.

17   A psychological view of the neurobiology of emotion

Ross Buck

Cognitive neuroscience and emotion

The discussion of our counterparts' chapter will cover three points, the first being whether we see a basis for interaction between cognitive psychologists and neuroscientists. The answer is clearly yes. LeDoux has contributed an excellent account of the development of theory relating to the biology of emotion and how this development has been, and continues to be, guided and influenced by advances in the understanding of biological mechanisms. Clearly, this process of understanding is now undergoing a revolution as new techniques for studying nervous system functioning become available. As we learn more about how information relevant to emotion is communicated between different parts of the nervous system, our theories can become more explicit and useful. The discussion of the role of the amygdala in what LeDoux terms "emotional evaluation" is a case in point, as is the discovery of direct projections from sensory to limbic areas and from limbic areas to posterior language areas. Because of this, there is a clear need for more interaction and communication between cognitive psychologists and neuroscientists. Those who prefer to approach emotion from a purely cognitive perspective ignore these biological mechanisms at their peril, for an understanding of these mechanisms can provide conceptual anchors in an area where language is notoriously slippery and imprecise.

Points of interaction

The Zajonc–Lazarus controversy

The second point concerns the key points of interaction between the approaches of cognitive psychologists and neuroscientists. It strikes me that one of LeDoux's major notions relates directly to the controversy between Zajonc and Lazarus noted in Chapter 14. This is the notion that the "evaluation of the significance of sensory input" from the external environment and body

361

is one of the major aspects of emotion. This seems directly analogous to Lazarus' notion of appraisal, and, as noted in Chapter 14, it is an area where the distinctions between attention, perception, emotion, and cognition are fuzzy indeed, and where important conceptual work remains to be done.

LeDoux's account of the role of the amygdala in effect provides us with a neurological model of appraisal. If the system is disrupted as in bilateral amygdalectomy, the sensory/perceptual qualities of stimuli appear to be divorced from their emotional qualities, even with such stimuli as feces, a burning match, a perhaps innately feared snake. LeDoux's account also suggests that there are different levels of the appraisal at the neurological level, that is, that the processing may take place at the limbic system level without cortical involvement, or that the cortex may be involved in interaction with the limbic system mechanism. Perhaps the former is appraisal as a "primitive evaluative perception" in Lazarus' (1984, p. 124) terms, while the latter is appraisal as a conscious, rational, symbolic process. If so, the controversy between Zajonc and Lazarus can be seen to involve the mainly semantic question of whether one defines the former process as cognitive or not.

Evaluation, appraisal, and filtering

Everyone agrees that something happens early in the emotion process that determines the impact of a particular stimulus for a particular individual, but the terms used – stimulus evaluation, appraisal, filtering – have different connotations. LeDoux defines stimulus evaluation as "a process by which sensory input is compared with stored information or knowledge," where the information or knowledge "can be bestowed by either experience or inheritance" (LeDoux, Chapter 15). I agree with this definition of the process, but suggest that the term filter is more appropriate here than the term evaluation.

Evaluation to me connotes an active cognitive process, and examination of the dictionary definition suggests that Lazarus' term appraisal has certain advantages. Evaluation is ascertaining the value or amount of, by trial, examination, or experiment. This appears to rule out inherited knowledge. Appraisal, in contrast, is setting a value on, or judging about quality, status, and so forth. This would appear to leave open the possibility of inherited knowledge as the basis of judgment. However, even appraisal has a decidedly cognitive ring.

In the model presented in Chapter 14, the handling of the initial stimulus was accomplished by a more passive concept: the notion of a stimulus filter. More active cognitive processes of appraisal and/or evaluation were seen to occur at a later stage. This filter fits LeDoux's definition and is consistent with the evidence of direct sensory input to limbic areas. The data presented by LeDoux suggest that the amygdala mechanism plays a central role in the affective filtering process. The hippocampus may also be involved, possibly mediating the effects of conditioned emotional responses. In this regard, a model of the interactions between the cortex, limbic system, hypothalamus, and reticular formation in the control of attention has been proposed by Pribram and McGuiness (1975; McGuiness and Pribram, 1980). The analyses of attention by Schneider and Shiffrin (1977) and Tucker and Williamson (1984) are also relevant here.

The notion of a filter avoids certain logical problems that are implied by terms such as evaluation and appraisal. If one assumes that the organism is in any way active in its initial processing of incoming stimuli, one quickly runs into the problem of explaining how the organism knows which incoming stimuli to process one way and which to process another. An infinite regress of processing mechanisms is avoided only if one assumes direct registration of initial stimuli. Gibsonian perceptual theorists have considered this issue, and their reasoning and conclusions should be useful to neuroscientists (cf. Gibson, 1966, 1979; Neisser, 1976; Shaw and Pittenger, 1977).

The notion of a filter might be clarified by an example. There is evidence that sensory information is processed at the cortex only if it is accompanied by nonspecific influences from subcortical systems (i.e., the reticular formation). Thus, we are not aware of the feeling of comfortable shoes because even though the sensory information about the feeling of those shoes is available to the brain (you are probably feeling your shoes as you read this), it is usually ignored. The reticular formation presumably does not *know* what the sensory stimuli are like, rather it uses certain decision rules to pass the information on or not. Thus, if the stimulation is weak or repetitive, it is ignored. The reticular formation thus is not an evaluator or an appraiser; it is a filter. Similar basic decision rules could be involved in attaching emotional relevance to incoming stimuli. In this way, it can be seen that we feel *more* than we know (Wilson, 1979); we feel *before* we know, and, in an important sense, feeling *determines what we know*.

The extent to which LeDoux's amygdala mechanism is most profitably regarded as an evaluator, appraiser, or filter remains to be seen. However, it seems clear that it must, at least, operate in conjunction with a stimulus filter. In this regard, it might be argued that just as the somatic sensory system is constantly informing the brain about such things as the feeling of one's shoes, one's clothes, or the pen in one's hand, motivational/emotional systems are constantly informing the brain about one's state of hunger, thirst, sexual arousal, happiness, sadness, anger, fear, and so forth. These states are noticed only if they are strong, or if one's attention is drawn to them (presumably cortical processes can influence the appropriate subcortical systems so that attention can be directed). The notion that such emotion and drive states are subjectively available only when they are strong is, I think, mistaken. They are simply ignored if they are weak or repetitive.

Knowledge by experience versus inheritance

The distinction between knowledge bestowed by experience and knowledge bestowed by inheritance is noted but not extensively discussed by LeDoux. One would presume that markedly different sorts of neural systems, with different evolutionary histories, would underlie knowledge by inheritance and experience. In the terminology used in Chapter 14, the behavior control systems built into the organism – reflexes, instincts, drives, primary affects, and cognitive motives – all involve knowledge by inheritance to some extent, but they are increasingly open to knowledge by experience as one goes through the list. One would also presume that different sorts of knowledge by experience would involve different neural systems (i.e., classical conditioning versus instrumental learning versus linguistic manipulation). All of this implies that there is a complex hierarchy of stimulus filtering, evaluation, and appraisal mechanisms that rely on inherited knowledge and knowledge by experience to different extents. Our increasing ability to tie our psychological concepts – attention, perception, emotion, cognition, knowledge – to specific neurochemical mechanisms will greatly increase our understanding of these processes.

Points of disagreement

The third point of discussion concerns the differences between the approaches of cognitive psychology and neuroscience. LeDoux and I clearly agree on most points, and there seem to be no fundamental or unbridgeable differences.

The nature of emotional experience

The major point of disagreement perhaps concerns our interpretations of emotional experience. LeDoux argues that consciousness is closely tied to natural language systems, and thus should be more or less unique to humans. I feel that the conscious experience of emotion and other subjective events has evolved hand-in-hand with the ability to have an internal cognitive representation of the environment, so that consciousness is widespread in the animals in direct proportion to the cognitive abilities of the species in question. LeDoux suggests that the neural basis of emotional experience involves input from the limbic system to the posterior language area, which in most persons is in the left cerebral hemisphere. I argue, in contrast, that most emotional experience is associated with right hemisphere mechanisms in most persons (Buck, 1984).

The evidence in this matter is, of course, difficult to interpret. LeDoux states that one important indicator of consciousness – self-awareness – is difficult to demonstrate in nonhumans, so that it may depend on a well-developed language system. However, self-awareness has been demonstrated in chimpanzees (Gallup, 1984). LeDoux also states that the (usually right) so-called minor hemisphere is cognitively deficient and far less if at all self-aware in comparison with the dominant hemisphere. However, the Sperry, Zaidel, and Zaidel (1979) study discussed in Chapter 14 suggests that the right hemisphere does in fact possess significant skills at self-awareness. LeDoux's position would seem to imply that emotional experience would be disrupted if the posterior language area were damaged. However, evidence suggests that it is right hemisphere damage that causes difficulties in emotional expression, while persons with left hemisphere damage may manifest, if anything, increased evidence of emotional responding (Buck and Duffy, 1980).

These are differences in the interpretation of the evidence, rather than fundamental differences in approach. The fact that we can meaningfully discuss the issue of consciousness at all is a measure of the potential usefulness of a cognitive neuroscience of emotion.

Communication and emotion

Another point of difference, if not disagreement, is that LeDoux takes a purely intraorganismic view of emotion, without directly considering its functions in the regulation of social behavior. It is traditional to regard emotion as an attribute of the individual. However, recent research suggests that the emotional system of the organism is not a closed system, even where the most basic biological phenomena are concerned. Thus, Hofer and his colleagues have demonstrated in studies of maternal deprivation how the presence of *specific* features of the mother have *specific* biological effects on the developing infant. For example, the mother's body warmth affects the levels of dopamine and norepinephrine in the infant's nervous system, and tactile stimuli normally provided by the mother affect the production of growth hormone (Hofer, 1984). Also, Harlow's studies suggest that the maturation of emotion systems within the brain is normally coordinated with the provision of social learning experiences necessary for the social control of those emotions. For example, fear and anger appear to come "on line" long after a young monkey has normally had positive loving experiences with his or her fellow monkey that provide a necessary emotional basis for social behavior (Harlow and Mears, 1983). This social regulation process appears to proceed via emotional communication between individuals, and its vital importance to us all is only beginning to be realized.

A neurobiologist's reply

Joseph E. LeDoux

Buck has provided a thoughtful and insightful discussion of my chapter on the neurobiology of emotion. It is encouraging, from the point of view of advancing serious interactions between psychologists and neuroscientists interested in emotion, that Buck and I have found much to agree about. In this final statement, I will focus on two of the points made in Buck's discussion.

Buck's preference for the term filtering over my use of stimulus evaluation reflects his desire to keep the organism passive in the early stages of emotional processing. I am more than willing to keep the conscious, thinking organism passive, for I believe that the process in question (filtering, stimulus evaluation, appraisal) often takes place in systems operating outside of conscious awareness. However, passivity stops there. Stimulus coding (comparison of incoming signals with acquired or inherited information stores) is, to me, an active process, at least on the neuronal level. The reticular formation example offered by Buck is not particularly helpful in this case. The reticular formation cannot be expected to "know" much about the stimulus because the stimulus information it receives represents the peripheral stimulus very poorly. In contrast, sensory input to limbic areas from sensory regions in the thalamus and cortex, conveys much more information about external stimuli. These systems organize and regulate complex somatic and visceral responses on the basis of stimulus meaning and must therefore "know" quite a lot about the stimulus, even when the conscious organism does not. The process, whatever we decide to call it, must, in my mind, be viewed as an active one.

With regards to emotional experience, two points need clarification. First, my feeling (and that is all anyone can have when this topic is considered) is that nonhuman organisms (especially but not exclusively the higher primates) do have conscious experiences. Hard scientific evidence is scant in this matter. However, whatever form of conscious awareness nonhumans do have appears to me to be very different from the conscious capacity made possible by the development of an advanced linguistic system and its cognitive baggage. The second and related point concerns the conscious capacities present in the right hemisphere of split-brain patients. When such right hemispheres lack extensive language representation, the level of conscious awareness seen is more like that which can be demonstrated in a chimpanzee than in a normal human.

In contrast, split-brain patients with bilateral language representation are very humanlike in self and social awareness in both hemispheres. These issues are discussed in detail elsewhere (LeDoux, Wilson, and Gazzaniga, 1979; Le-Doux, 1985).

While I thus accept nonhuman conscious awareness, I have focused on human consciousness in Chapter 15 in discussing emotional experience for two reasons. In the first place, I view nonhuman consciousness as very different from human consciousness. Second, if I am right in tying human consciousness to language, we may have a practical approach for studying emotional experience. If we can determine how affective signals reach language systems, we may be able to begin to analyze how language-related conscious systems transform those signals into experienced feelings.

18   Cognitive neuroscience: final considerations

William Hirst and Joseph E. LeDoux

Mental acts and processes critically depend on brain activities. If we are ever to understand how humans think, speak, perceive, attend, and emote, we must understand the relation between mind and brain.

We, the editors, began this book with the belief that the mysteries of mind and brain – the mind–body problem – would eventually be open to scientific investigation. We were uncertain, however, that the time was ripe. Much research has been conducted under the auspices of neuroscience, and the mind is beginning to open itself under the tutelage of cognitive scientists. Have these two sciences progressed to a stage where they can be woven together into a coherent account of the relation between mental processing and brain functioning? We did not expect that startling discoveries would come from this book; such things are rare indeed and are the stuff of serendipity. Our goal was more modest. We merely wanted to see whether given the opportunity and motivation to interact, neuroscientists and cognitive scientists could have meaningful discussions. This book was successful, then, if a substantive dialogue emerged.

Philosophers have been worried about the mind–body problem for about as long as there has been philosophy. They have carefully articulated the range of possible relations between the mind and brain and have offered various arguments in favor of one position over the other. Although most scientists take reductionism for granted, many philosophers are skeptical about the possibility of ever articulating the relation between mind and brain (see Churchland, 1984, for a readable review of the issues of the mind–body problem). These arguments are usually made "in principle," stressing that *logically* we are not *necessarily* compelled to believe that *every* mental action has a discernible brain correlate.

The working scientist is much more pragmatic than the worrying philosopher. He has a problem to solve and will use any method or any source of inspiration that he can get. If a neuroscientist can benefit from work in cognitive psychology, or vice versa, then it matters little whether such a symbiosis will work in the long run. It works now and will advance present research,

368

and that is enough. The philosophical arguments are beside the point. It may be that a connection between mental processing and neurological function can only be pushed so far. Or it may be that the scientist working inside his laboratory may prove the philosopher's skepticism unfounded, and show that whereas there may not necessarily be a mapping between mind and brain, in reality there is. The eventual outcome is not important. In this early stage of mutual exploration, pragmatism and mutual benefit rule.

What then is meant by "benefit?" Ultimately, the cognitive scientist would like to know whether the components of the mind have some physical realization. The relation desired here is similar to the preferred relation between linguistics and psychology. For many years, linguists have posited various grammatical rules and structures, and cognitive psychologists have tried to determine whether these rules and structures have "psychological reality." The worthiness of a linguistic theory did not depend on the results of experiments on "psychological reality," and the psychologists' theorizing about linguistic processing was not limited to what the linguist supplied. Nevertheless, it was considered foolhardy by many for the psychologist to ignore the work of the linguist and for the linguist to blind himself to the results of the psychologist. A meaningful dialogue was needed. What form this dialogue should take and how much success one could justly expect from it in the short-term is controversial (see Berwick and Weinberg, 1984). Nevertheless, despite the controversy, the dialogue continues (see Halle, Bresnan, and Miller, 1978). The results have furthered both linguistics and psychology.

In the dialogue between neuroscientists and cognitive scientists, cognitive scientists can be viewed in the place of the linguists. They are developing models of mental processing. The neuroscientists can offer tests of "neurological reality." Although there is no strong philosophical necessity for mental models to be neurologically real, it is clearly desirable. Neurological reality adds significance to the neurosciences and gives added credence to the cognitive sciences.

The discovery of neurological reality can take the form of dissociations or double dissociations of brain functions. The work of Milner on the amnesic H.M. is a classic example (see Chapter 10 by Schacter). H.M. showed an intact short-term memory and a long-term memory for events that occurred prior to onset of the amnesia. He could not remember, however, events that occurred after the onset of the amnesia. This finding was significant because it gave neurological reality to models of memory that were current in cognitive science, namely, those that posited discrete short-term and long-term memory mechanisms. Based on this model, it was posited that information entered H.M.'s intact short-term memory, but could not be transferred to his long-term memory.

Dissociations are not the only way to establish the neurological reality of a mental model. Neuroscience possesses a wide range of techniques, and each of these can contribute in their own way. Recordings of single neuronal units have added credence to feature models of perception (see Chapter 3 by Marrocco); studies of evoked related potentials have weighed on the issue of the location of the bottleneck in models of attention (see Chapter 7 by Robinson and Petersen); and recent neuroanatomical studies have revived the older psychological distinction between habits and memories (see Chapter 10 by Schacter).

Although in each of these examples, results from the neuroscientist's laboratory have provided support for models developed in the cognitive scientist's laboratory, the interaction between the two fields need not be so black and white. A dialogue between neuroscientist and cognitive scientist could be concerned with the immediate problems of advancing current research, providing meaningful theoretical frameworks for unexplained data, or outlining new areas of investigation. Such a dialogue does not have to be concerned with developing in-principled (philosophical) ways of mapping mind to brain or vice versa. Moreover, it need not be aimed at proving or disproving existent theorizing in the two fields. Neuroscientists seek a better understanding of the workings of the brain, and cognitive scientists may be able to help them. Similarly, cognitive scientists want to understand the workings of the mind, and neuroscientists might help them.

Sometimes the exchange centers around experimental technique. For instance, Robinson and Peterson (see Chapter 7) describe animal lesion work in their laboratory that adopted a paradigm developed by Posner to study attention in college students. Similarly, psychological studies of contrast sensitivity paved the way for using spatial frequency techniques to analyze visual system normal activity (see Chapter 2 by Hoffman and Chapter 3 by Marrocco).

At other times, the contributions of one field to another are more subtle and often appear only implicitly. Some involve the exchange of general concepts, if not explicit theories. Whereas this may be taken as obvious or trivial, it is nevertheless noteworthy that the neuroscientists whom we asked to contribute to this book did not have difficulty writing about "attention," "memory," "perception," and "emotion." Much of neuroscience is today organized around brain structure, with books coming out on the hippocampus, the thalamus, the basal ganglia, and so forth. Nevertheless, many neuroscientists are concerned with the biological underpinning of the mind. It is not surprising, then, that neuroscientists interested in the biology of the hippocampus are often compelled to think about the biology of memory. The facts achieve greater depth and broader significance when the latter is true.

In pursuing a functional organization of the brain, the neuroscientist often looks to psychology for clues as to what the brain's functions are and how they should be conceived. Traditional wisdom tells the neuroscientist that attention, perception, speech, memory, emotion, thought, and other common language terms refer to functions represented in the brain, but translating these terms into meaningful scientific concepts and discovering ways to subject them to experimental scrutiny is usually the business of psychology. The more precise the psychological definition of a particular process, the better the neuroscientist's chance of relating it to basic biological mechanisms. In searching for neurological reality, the biologist must know what to look for, and once he's found it, psychology only benefits when the news is shared.

In evaluating the success of this book, then, we had several concerns in mind:

1. Do the chapters indicate that psychology and neuroscience have helped each other shape research directions and ideas?
2. Do the chapters indicate that studies of the brain have given neurological reality to psychological processes?
3. Do the chapters raise questions, identify new areas of collaborative research, suggest research techniques or experiments, or clarify language that will help workers in both fields?

We now turn to the four areas explored in the book and examine them with the above concerns in mind.

Perception

It is clear from the contributions of Marrocco and Hoffman that although contemporary psychological and neurobiological approaches to perception differ in substantial ways, the two fields have interacted to advantage in the past and no doubt will continue to do so. Selected examples are described below.

Marrocco emphasized the widespread use of Fourier analysis in studies of single-unit activity in the visual system. Sine-wave gratings have, in fact, taken over as the dominant method for studying receptive field properties and have led to the development of new psychological theories of visual perception. The basic technique of spatial frequency stimulation, however, was not devised to study neural activity. As described by Hoffman, Campbell and Robinson's application of sine-wave gratings in psychophysical studies is responsible for the introduction of spatial frequency analysis to physiologists' studies.

Sine-wave gratings are not the only method of visual stimulation that psychophysics has contributed to neurophysiology. Physiological studies of color saturation, depth perception, visual masking, and other basic perceptual phenomena have utilized psychophysical approaches, making it possible to begin to localize, in the nervous system, the neurons responsible for different types of perceptual analyses. At the same time, the discovery that neural activity in some part of the brain corresponds with a particular perceptual process has added "neurological reality" to psychologically identified processes. In the case of color and motion perception, for example, it appears that there exist functionally specialized neurons located in different brain regions. The cells that determine whether an object is moving are different from those that analyze its color. The neurophysiological data, therefore, confirm the phenomenal impression and psychophysical finding that the basic mechanisms of color and movement perception are different.

Psychologists have also benefited from developments in neuroscience. For example, one of the major concepts to emerge from physiological studies of perception is that of parallel processing. The notion of X and Y systems originating in retinal cells and projecting in tandem to cortex has been particularly influential. As noted by Hoffman, the X and Y concepts have been used to explain several important perceptual phenomena, including visual masking and global precedence effects.

Another instance where psychology has borrowed from physiology involves the application of receptive field stimulation techniques in psychophysical studies of spatial perception. Such work has shown that the sensitivity profiles, as measured psychophysically, produce the same pattern of spatial interaction as receptive field analyses. Here, we see an instance of "psychological reality" being given to a neurophysiological observation.

Neurophysiological studies of perception have tended to focus on elemental forms of perception. Little work has been done on what Hoffman referred to as "top–down" perceptual processes. This is an important area for neurobiologists to consider. As psychological studies refine methods for studying the contribution of such top–down factors as context and memory in perception, physiologists will be in a better position to approach these problems. However, Marrocco is clearly correct in cautioning that such studies are going to be particularly difficult to approach from the single-unit point of view characteristic of most studies of visual physiology.

Emotion

Biological and psychological approaches to emotion are traditionally more integrated than is the case for the other functional topics discussed in this book. This is evident from Chapter 14 by Buck and Chapter 15 by LeDoux. The main question that arises from their contributions, however, is whether this apparent integration is real.

Buck's chapter on the psychology of emotion makes frequent references to physiological concepts. Many examples also indicate that psychologists have often used physiological manipulations as either dependent or independent variables. Such studies have played a key role in the development of contemporary models of emotion. Consider the cognitive arousal model.

The cognitive arousal model emphasizes physiological arousal in the initiation and regulation of emotion. The arousal has usually been assumed to be nonspecific, varying little from emotion to emotion. The quality of the emotion is believed to be set by the cognitive appraisal of the situation. This model was developed largely on the basis of studies where the arousal state of human subjects was manipulated by injection of epinephrine, which activates the sympathetic nervous system.

The interpretation of the data was based on the assumption, which was once widely accepted by physiologists, that the autonomic nervous system responds in a single, predetermined way to all emotion-producing situations. However, it is now well established that the autonomic nervous system is capable of producing patterned responses that are tailored to meet the metabolic demands imposed by the specific situation.

This illustration points out how important it is for psychologists and physiologists to communicate, especially when one field is borrowing data or concepts from another. While psychologists and neuroscientists cannot be expected to keep up with every change in each other's field, close contact in the design of experiments and in the interpretation of data is a viable and practical solution to the informational gap separating the two fields.

However, other barriers also exist. The emphasis of neuroscientists on animal studies and of psychological approaches to emotion on human models creates a conceptual distance that is difficult to bridge. Whether the underlying processes and mechanisms of emotion are different in animals and man or whether the differences are more semantic in nature is a crucial issue to resolve. The emphasis of psychological theorists such as Plutchick (1980) and Buck on evolutionary mechanisms is thus important. The development of animal models of the emotional processes being studied in humans and the development of physiological techniques to study the human brain mechanisms that have been identified in animal work would also be important steps in the right direction.

Perhaps one of the most promising signs for new interactions between psychologists and neuroscientists involves the emergence of information-processing approaches to emotion. In contrast to the traditional concern of psychologists with the mechanisms of human conscious emotional experiences, recent work by Zajonc (1980) and Bower (1981) has attempted to understand how the emotional significance of a stimulus is evaluated, applying the methods of cognitive psychology to the problem of affective processing. What is important here is that such analyses can be performed across species, are compatible with electrophysiological and anatomical methods used in neuroscience, and fit well within ongoing research traditions in neuroscience, as described by LeDoux. The opportunity for real multidisciplinary interactions thus exist in studies of affective processing.

Attention

The dialogue between cognitive scientists and neuroscientists is only beginning to develop when it comes to attention. Its infancy is not surprising. Cognitive scientists have not decided what attention is. As Hirst makes clear, there are a host of models, none of which can stand on its own, and all of which need more work before they even approach a fully explicated theory. Neuroscientists, on the other hand, are confronted with a plethora of different kinds of attention – visual–spatial, auditory, neglect – and with each kind of attention the implication of another brain structure.

In the few instances in which there has been a simple experimental design and a set of clear, unambiguous results, then there has been a good exchange between the fields. We have in mind the work of Posner and his colleagues along with the neuroscientific follow-up of Robinson and his colleagues. These instances have, however, been few.

The lack of agreement among cognitive theorists and the diversity of results of neuroscientists should not be viewed too glumly, however. A careful reading of the attention section in this book points to several areas of possible collaboration. First, various theorists of attention have posited an attentional filter or resource pool(s). The neurological reality of these constructs could be tested. As stated in the discussion sections on attention, Hillyard and his colleagues interpreted their early work with event-related potential (ERP) in the manner consistent with stimulus filtering, but other explanations are possible. Holtzman et al. suggested that their work provided support for a resource approach to the study of attention, but again, further research is needed. Nevertheless, these two examples show that if neuroscientists took the constructs of attentional theory more seriously, they might (1) establish their neurological reality and (2) in succeeding or failing, provide support for the various competing cognitive theories.

Hirst suggests another area of collaboration when he notes that neuroscientists have not studied divided attention. By and large, neuroscientists have emphasized situations in which subjects are led to expect a stimulus in some location and have examined facilitation effects. Kinsbourne has suggested a "neurological" model of divided attention, but, as far as we know, there has been no serious experimental work on the topic.

Another area of collaboration centers on the concepts of automaticity and effort, both of which are central to the study of attention. Despite their importance, no one has yet determined whether an automatic version of, for example, visual scanning differs qualitatively from effortful visual scanning. Clearly, if we better understood changes in the brain that occur with practice, perhaps using techniques such as PET scanning, blood flow, or ERP, we might be better able to address this issue.

Finally, Hirst raises the issue of the role consciousness plays in processing: Is all information impinging upon the sensorium fully processed? Or is full processing reserved for a selected few? Work on neglect, extinction, and blindsight might resolve this issue, but at present, this work is only beginning. Traditional work in cognitive psychology may inform the neuroscientist how to proceed.

Memory

Chapter 10 by Schacter and Chapter 11 by Gabriel, Sparenborg, and Stolar cover a wide range – from details about the neurotransmitters involved in classical conditioning to discussions of the feeling-of-knowing accompanying recall failures. This diversity was expected. Both cognitive scientists and neuroscientists have failed to answer fundamental conceptual questions. Cognitive scientists cannot always decide what memory is, or what different types of memory there are. In the past, they have suggested separate memories for linguistic and perceptual skills, motor skills, classical and instrumental conditioning, short-term retention, long-term retention, autobiographical events, facts, word meaning, and so on. On the other hand, neuroscientists, as Gabriel et al. indicated, cannot decide whether to adopt a functional or reductionistic approach to memory. As a consequence, even within the field of neuroscience, there are subgroups that fail to find common ground.

Although this diversity, and lack of agreement, may frustrate workers in both fields, it appears not to preclude discussions across fields. Neuroscientists have borrowed many of the experimental paradigms developed by psychologists. A large number of these are taken from the learning literature. Indeed, most neuroscience experiments on memory involve either classical or instrumental conditioning. This reliance on the vocabulary and techniques of conditioning and learning shows clearly how one field, in this instance, psychology, can shape the research directions and ideas of another field, neuroscience. What would neuroscience use as a model if psychologists in the 1940s and 1950s had not made the learning and conditioning vocabulary synonymous with memory?

Gabriel et al. indicate that neuroscientists are becoming increasingly sophisticated about the limitations of their studies of conditioning. Cognitive science many years ago showed decisively that there was more to memory than can be encompassed by even the most complex learning theory. Kandel may have modeled short-term sensitization in *Aplysia* (which in itself is quite impressive), but his rather detailed model may not have anything to say about animal *memory*, let alone human memory. Amnesics possess little or no memory, but can nevertheless be classically conditioned. Clearly, there is a long step from the study of sensitization or even classical conditioning to a reductionistic model of memory.

If neuroscientists are to make any progress in moving from their current work with conditioning and learning paradigms to more complex memory paradigms, they must appreciate the relation between an experiment on memory and an experiment on, for example, instrumental conditioning. This relation should presumably be plotted by cognitive scientists, but, unfortunately, they have not provided either a detailed or clear answer. When the field of psychology moved away from the study of learning toward a study of memory, there was no systematic evaluation of the relation between the "old" work and the "new." The arguments for this shift emphasized that theories of learning could not account for the complexity of human memory and, consequently, should not be pushed too far. Research was then directed toward modeling the more cognitive aspects of memory.

Cognitive science would do the field of neuroscience a great service if it articulated the limitation of conditioning as a model of memory and then stated the relation between conditioning and memory. Such a discussion between neuroscience and cognitive science would indicate how far models such as Kandel's, Lynch's, and Gabriel's can be extended, what their limitations as neurological substantiations of a cognitive phenomenon are, and what can be done to expand them so that they account for a wider range of phenomena. This dialogue should involve such fundamental questions as: To what extent can nonhuman species be said to memorize and remember? What would be an objective test of memory? Is there a "raw material of memory" that can be separated from associating, thinking, perceiving, and emoting? To what extent should we assume that traces or engrams are formed of past events? Do the engrams/traces of some kinds of memory (autobiographical, for instance) differ from those of other kinds of memory (skills, for instance)? Is the "storehouse" metaphor, in which memory is likened to a repository of engrams or traces, appropriate? It is heartening that a dialogue between neuroscientists and cognitive scientists could involve such fundamental questions.

Schacter rightly points out that there is one area in which the dialogue between neuroscience and cognitive science not only has potential, but is bearing fruit. As already indicated, cognitive scientists have suggested a host of distinctions between different kinds of memories. Neuroscientists could and have devoted significant energy to exploring the neurological reality of these distinctions. The work on what Cohen and Squire (1980) introduced as the distinction between procedural and propositional memory is a good example of how the interaction between neuroscience and cognitive science can work.

Cohen and Squire suggested that the distinction may be neurologically real because human anterograde amnesics are able to learn to read mirror images of words and to solve a puzzle called the Tower of Hanoi, even though they fail to encode specific instances of either the words or the puzzle solution. Gabriel reviewed the evidence under the terms limbic and nonlimbic memory; Schacter has adopted Cohen and Squire's terminology.

The distinction that Cohen and Squire drew is an old one: James suggested it when he contrasted habits with memories. Other psychologists from Bruner (1969) to Tulving (1972) have offered similar distinctions, and philosophers such as Ayers (1956), Russell (1921), and Ryle (1949) have argued persuasively for something along the same line. Each of these distinctions differs slightly, and, as yet, traditional experimental work on neurologically intact subjects has failed to provide any decisive evidence in support of one distinction over the others or even support for proposals that some amalgam represents a functional and psychologically real dichotomy.

Cohen and Squire's finding, then, is important because it indicates that amnesia may be a perfect model from which to refine a distinction that scholars have been struggling toward for many years. The recent findings that amnesics also show normal priming may force Cohen and Squire to revise their articulation (see Chapter 10 by Schacter for a discussion of the priming work). The articulations of other scholars may prove more successful, or a new conceptualization may be needed. Whatever the eventual outcome, a distinction offered initially by psychologists and philosophers may have a neurological basis. The work in neuroscience provides in turn a basis for refining the distinction, and the interchange between the two fields will presumably continue until a neurologically real and precise articulation emerges, or the initial and present enthusiasm proves misguided.

Conclusion

For us, this book nicely demonstrates the fruitfulness of a continuing dialogue between cognitive scientists and neuroscientists. Many of the instances in which a strong relation between the two fields is discussed have probably been appreciated for many years by scientists in both disciplines. Our enthusiasm springs not from the facts or models that the authors reviewed, but from the questions that they raised. It was quite possible that researchers in the two fields would discover little in common. More likely, neuroscientists and cognitive scientists may have proferred questions of mutual interest, but would then find them too broad and vague to help them in structuring their everyday needs. These two scenarios did not emerge. Rather, as our summaries indicate, there are many different grounds for discussion, and many quite specific questions to be answered. If neuroscientists and psychologists continue to talk, we might find both fields redirecting their interests to a common set of issues, results, and theories.

References

Abrahams, V. C., Hilton, S. M., and Zborzyna, A. W. (1960). Active muscle vasodilation by stimulation of the brainstem: its significance for the defense reaction. *Journal of Physiology (London), 154,* 491–513.

Adam, G. (1978). Visceroception, awareness, and behavior. In G. E. Schwartz and D. Shapiro (eds.). *Consciousness and Self Regulation.* New York: Plenum.

Adey, W. R., Walter, D. O., and Lindsley, D. F. (1962). Subthalamic lesions. *Archives of Neurology, 6,* 194–207.

Albiniak, B. A., and Powell, D. A. (1980). Peripheral autonomic mechanisms and Pavlovian conditioning in the rabbit (*Oryctolagus cuniculus*). *Journal of Comparative Physiology and Psychology, 94,* 1101–1113.

Albrecht, D. G., De Valois, R. L., and Thorell, L. G. (1980). Visual cortical neurons: are bars or gratings the optimal stimuli? *Science, 207,* 88–90.

Aleksidze, N., Potemska, A., Murphy, S., and Rose, S. P. R. (1981). Passive avoidance training in the young chick affects forebrain α-bungarotoxin and serotonin binding. *Neuroscience Letters, Suppl. 7,* 244.

Alkon, D. L. (1982). A biophysical basis for molluscan associative learning. In C. D. Woody (ed.). *Advances in behavioral biology,* Vol. 26. *Conditioning: representation of involved neural functions.* pp. 147–170, New York: Plenum.

(1984). Calcium-mediated reduction of ionic currents: a biophysical memory trace. *Science, 226,* 1037–1045.

Allman, J. M., and Kaas, J. (1974). The organization of the second visual area (V II) in the owl monkey: a second order transformation of the visual hemifield. *Brain Research, 76,* 247–265.

Allport, D. A. (1970). Parallel encoding within and between elementary stimulus dimensions. *Perception and Psychophysics, 10,* 104–108.

(1979). Conscious and unconscious cognition: A computational metaphor for the mechanism of attention and integration. In L.-G. Nilsson (ed.). *Perspectives on memory research.* Hillsdale, N.J.: Erlbaum.

(1980). Attention and performance. In G. L. Claxton (ed.). *New Directions in Cognitive Psychology.* London: Routledge & Kegan Paul.

Allport, D. A., Antonis, B., and Reynolds, P. (1972). On the division of attention: A disproof of the single channel hypothesis. *Quarterly Journal of Experimental Psychology, 24,* 225–235.

Altman, J., Brunner, R. L., and Bayer, S. A. (1973). The hippocampus and behavioral maturation. *Behavioral Biology, 8,* 557–596.

Amacher, M. P. (1964). Thomas Laycock, I. M. Sechenov, and the reflex arc concept. *Bulletin of the History of Medicine, 38,* 168–183.

Amendt, K., Czachurski, J., Dembowsky, K., and Seller, H. (1979). Bulbospenal projections to the interomedolateral cell column: a neuroanatomical study. *Journal of the Autonomic Nervous System, 1*, 103–117.

Andersen, R. A. and Mountcastle, V. B. (1983). The influence of the angle of gaze upon the excitability of the light-sensitive neurons of the posterior parietal cortex. *Journal of Neuroscience, 3*, 532–548.

Anderson, J. R. (1976). *Language, memory, and thought*. Hillsdale, N.J.: Erlbaum.

——— (1978). Arguments concerning representations for mental imagery. *Psychological Review, 85*, 249–277.

——— (1980). *Cognitive psychology and its implications*. New York: Freeman.

Anderson, J. R., and Bower, G. H. (1973). Human associative memory. Washington, D.C.: Winston.

——— (1983). *The architecture of cognition*. Cambridge, Mass.: Harvard University Press.

Anderson, R. C., Pichert, J. W., Goetz, E. T., Schallert, D. L., Stevens, K. V., and Trollip, S. R. (1976). Instantiation of general terms. *Journal of Verbal Learning and Verbal Behavior, 15*, 667–679.

Andrew, R. J. (1963). The origin and evolution of the calls and facial expressions of the primates. *Behavior, 20*, 1–109.

Andrew, R. J. (1965). The origins of facial expressions. *Scientific American, 213*, 88–94.

Angell, J. R. (1916). A reconsideration of James' Theory of emotion in light of recent criticisms. *Psychological Review, 23*, 259.

Arkes, H. R., and Freedman, M. R. (1984). A demonstration of the costs and benefits of expertise in recognition memory. *Memory & Cognition, 12*, 84–89.

Averill, J. R. (1980). On the paucity of positive emotions. In K. R. Blankenstein, P. Pliner, and J. Polivy (eds.), *Assessment and modification of emotional behavior*. New York: Plenum.

Averill, J. R., Opton, E. M., Jr., and Lazarus, R. S. (1969). Cross cultural studies of psychophysiological responses during stress and emotion. *International Journal of Psychiatry, 4*, 33–102.

Ax, A. (1953). The physiological differentiation between fear and anger in humans. *Psychosomatic Medicine, 15*, 433–442.

Axelrod, J., and Reisine, T. D. (1984). Stress hormones: their interaction and regulation, *Science, 224*, 452–459.

Ayers, A. J. (1956). *The problem of knowledge*. New York: Penguin.

Baddeley, A. D. (1976). *The psychology of memory*. New York: Basic Books.

——— (1978). The trouble with "levels": a re-examination of Craik and Lockhart's framework for memory research. *Psychological Review, 85*, 139–152.

——— (1982). Amnesia: A minimal model and an interpretation. In L. S. Cermak (ed.), *Human memory and amnesia*. Hillsdale, N.J.: Erlbaum.

Bahrick, H. P., and Karis, D. (1982). Long-term ecological memory. In C. R. Puff (ed.), *Handbook of research methods in human memory and cognition*. New York: Academic Press.

Bahrick, H. P., Bahrick, P. O., and Wittlinger, R. P. (1975). Fifty years of memories for names and faces: a cross-sectional approach. *Journal of Experimental Psychology: General, 104*, 54–75.

Baizer, J. S., Robinson, D. L., and Dow, B. M. (1977). Visual responses of area 18 neurons in awake, behaving monkey. *Journal of Neurophysiology, 40*, 1024–1037.

Baker, F. H., and Malpeli, J. G. (1977). Effects of cryogenic blockade of visual cortex on responses of lateral geniculate neurons in the monkey. *Experimental Brain Research, 29*, 433–444.

Balaz, M. A., Capra, S., Kasprow, W. J., and Miller, R. (1982). Latent inhibition of the conditioning context: further evidence of contextual potentiation of retrieval in the absence of appreciable context-US associations. *Animal Learning & Behavior, 10*, 242–248.

Ball, G. G. (1974). Vagotomy: effect on electrically elicited eating and self-stimulation in the lateral hypothalamus, *Science, 184*, 484–485.

Bandler, R. J. (1970). Cholinergic synapses in the lateral hypothalamus for the control of predatory aggression in the rat. *Brain Research, 20*, 409–424.

(1982). Induction of "rage" following microinjection of glutamate into midbrain but not hypothalamus of cats, *Neuroscience Letters, 30*, 183–188.

Barclay, J. R., Bransford, J. D., Franks, J. J., McCarrell, N. S., and Nitsch, K. (1974). Comprehension and semantic flexibility. *Journal of Verbal Learning and Verbal Behavior, 13*, 471–481.

Bard, P. (1928). A diencephalic mechanism for the expression of rage with special reference to the sympathetic nervous system. *American Journal of Physiology, 84*, 490–515.

Bard, P., and Rioch, D. McK. (1937). A study of four cats deprived of neocortex and additional portions of the forebrain. *Bulletin of Johns Hopkins Hospital, 60*, 73–147.

Barlow, H. B. (1972). Single units and sensation: a neuron doctrine for perceptual psychology. *Perception, 1*, 371–394.

Bartlett, F. C. (1932). *Remembering*. Cambridge: Cambridge University Press.

Baughman, R. W., and Gilbert, C. D. (1979). Aspartate and glutamate as possible neurotransmitters of cells in layer 6 of the visual cortex. *Nature (London), 287*, 848–850.

Beaton, R., and Miller, J. M. (1975). Single cell activity in the auditory cortex of the unanesthetized, behaving monkey: correlation with stimulus controlled behavior. *Brain Research, 100*, 543–562.

Bechterev, W. (1887). Die Bedeutung der Sehhügel auf Grund von experimentellen und pathologischen Paten. *Virchows Archiv, 110*, 102–159.

Beck, A. T. (1976). *Cognitive therapy and the emotional disorders*. New York: International Universities Press.

Beck, J. (1981). Textural segmentation. In J. Beck (ed.), *Organization and representation in perception*. Hillsdale, N.J.: Erlbaum.

Bender, D. B. (1982). Receptive field properties of neurons in the macaque inferior pulvinar. *Journal of Neurophysiology, 48*, 1–17.

Bender, M. B., and Diamond, S. P. (1965). An analysis of auditory perceptual defects with observations on the localization of dysfunction. *Brain, 87*, 675–686.

Benevento, L. A., and Rezak, M. (1976). The cortical projections of the inferior pulvinar and adjacent lateral pulvinar in rhesus monkey (*Macaca mulatta*): an autoradiographic study. *Brain Research, 108*, 1–24.

Benjamin, R. M., Jackson, J. C., and Golden, G. T. (1978). Cortical projections of the thalamic mediodorsal nucleus in the rabbit. *Brain Research, 141*, 251–265.

Bennett, E. L., Diamond, M. C., Krech, D., and Rosenzweig, M. R. (1964). Chemical and anatomical plasticity of the brain. *Science, 146*, 610–619.

Bennett, T. L. (1975). The electrical activity of the hippocampus and processes of attention. In R. L. Isaacson and K. H. Pribram (eds.), *The Hippocampus*, Vol. II, *Neurophysiology and Behavior*, pp. 71–97, New York: Plenum.

Benowitz, L. (1974). Conditions for the bilateral transfer of monocular learning in chicks. *Brain Research, 65*, 203–213.

Benowitz, L. I., Bear, D. M., Rosenthal, R., and Mesulam, M. (1983). Sensitivity to nonverbal communication after unilateral brain damage. *Cortex, 19*, 5–12.

Benson, D. A., and Hienz, R. D. (1978). Single-unit activity in the auditory cortex of monkeys selectively attending left vs. right ear stimuli. *Brain Research, 159*, 307–320.

Benson, H., Herd, J. A., Morse, W. H., and Kelleher, R. T. (1969). Behavioral induction of arterial hypotension and its reversal. *American Journal of Physiology, 217*, 30–34.

Benson, H., Shapiro, D., Turskey, B., and Schwartz, G. E. (1971). Decreased systolic blood pressure through operant conditioning techniques in patients with essential hypertension. *Science, 173*, 740–742.

Beresford, W. A. (1962). A Nauta and gallocyanin study of the corticolateral geniculate projection in the cat and monkey. *Zeitschrift für Hirnforschung, 5*, 210–228.

Berger, P. A., and Barchas, J. D. (1981). In G. J. Siegel, R. W. Abers, B. W. Agranoff, and R. Katzman (eds.). *Basic neurochemistry*, pp. 759–774. Boston: Little, Brown.

Berger, T. W. (1984). Long-term potentiation of hippocampal synaptic transmission accelerates behavioral learning. *Science, 224*, 627–630.

Berger, T. W., and Orr, W. B. (1982). Role of the hippocampus in reversal learning of the rabbit nictitating membrane response. In C. D. Woody (ed.), *Advances in Behavioral Biology*, Vol. 26. *Conditioning: Representation of Involved Neural Functions*, pp. 1–12, New York: Plenum.

Berger, T. W., and Thompson, R. F. (1978). Neuronal plasticity in the limbic system during classical conditioning of the rabbit nictitating membrane response. I. The hippocampus. *Brain Research, 145*, 323–346.

Berger, T. W., Milner, T. A., Swanson, G. W., Lynch, G. S., and Thompson, R. F. (1980). Reciprocal anatomical connections between the anterior thalamus and cingulate-retrospenial cortex in the rabbit. *Brain Research, 201*, 411–417.

Berson, D. M., and Graybiel, A. M. (1983). Organization of the striate-recipient zone of the cat's lateralis posterior–pulvinar complex and its relations with the geniculostriate system. *Neuroscience, 9*, 337–372.

Berwick, R. C., and Weinberg, A. S. (1984). *The grammatical basis of linguistic performance*. Cambridge, Mass.: MIT Press.

Best, M. R., and Best, P. J. (1976). The effects of state of consciousness and latent inhibition on hippocampal unit activity in the rat during conditioning. *Experimental Neurology, 51*, 564–573.

Biederman, I. (1972). Perceiving real world scenes. *Science, 177*, 77–80.
  (1981). On the semantics of a glance at a scene. In M. Kubovy and J. R. Pomerantz (eds.), *Perceptual organization*. Hillsdale, N.J.: Erlbaum.

Biederman, I., Mezzanotte, R. J., and Rabinowitz, J. E. (1982). Scene perception: Detecting and judging objects undergoing relational violations. *Cognitive Psychology, 14*, 143–177.

Biederman, I., Teitelbaum, R. C. and Mezzanotte, R. J. (1983). Scene perception: A failure to find a benefit from prior expectancy or familiarity. *Journal of Experimental Psychology: Learning, Memory, and Cognition, 9*, 411–429.

Birt, D., and Woody, C. D. (1983). Patterns of response to a behavioral US among neurons of the sensorimotor cortex of awake and anesthetized cats. *Society for Neuroscience Abstracts, 9*, 330.

Black, A. H., Nadel, L., and O'Keefe, J. (1977). Hippocampal function in avoidance learning and punishment. *Psychological Bulletin, 84*, 1107–1129.

Blake, M. (1973). Prediction of recognition when recall fails: Exploring the feeling of knowing phenomenon. *Journal of Verbal Learning and Verbal Behavior, 12*, 311–319.

Blakemore, C. B. (1970). The representation of three-dimensional visual space in the cat's striate cortex. *Journal of Physiology, London, 209*, 155–178.

Blakemore, C. B., and Campbell, F. W. (1969). On the existence of neurons in the human visual system selectively sensitive to the orientation and size of retinal images. *Journal of Physiology, London, 203*, 237–260.

Blakemore, C. B., and Tobin, E. A. (1972). Lateral inhibition between orientation detectors in the cat's visual cortex. *Experimental Brain Research, 15,* 439–445.

Blakemore, C. B., Garner, E. T., and Sweet, J. A. (1972). The site of size constancy. *Perception, 1,* 111–119.

Blanchard, R. J., Blanchard, C. D., and Fial, R. A. (1970). Hippocampal lesions in rats and their effect on activity, avoidance, and aggression. *Journal of Comparative and Physiological Psychology, 71,* 92–102.

Blum, J. S., Chow, K. L., and Pribram, K. H. (1950). A behavioral analysis of the organization of the parieto-temporo-preoccipital cortex. *Journal of Comparative Neurology, 93,* 53–100.

Borod, J., and Caron, H. (1980). Facedness and emotion related to lateral dominance, sex and expression type. *Neuropsychologia, 18,* 237–241.

Borod, J. C., Caron, H. S., and Koff, E. (1981). Assymetry in positive and negative facial expressions: sex differences. *Neuropsychologia, 19,* 819–824.

Bower, G. H. (1967). A multicomponent theory of the memory trace. In K. W. Spence and J. T. Spence (eds.), *The psychology of learning and motivation.* Vol. 1, New York: Academic Press.

(1981). Mood and memory. *American Psychologist, 36,* 129–148.

Bower, G. H., and Gilligan, S. G. (1979). Remembering information related to one's self. *Journal of Research in Personality, 13,* 420–432.

Bowlby, J. (1969). *Attachment and loss.* London: Chatto & Windus/Hogarth and Institute of Psychoanalysis.

Bowmaker, J. H., Dartnall, H. J. A., and Mollon, J. D. (1980). Microspectrophotometric demonstration of four classes of photoreceptor in an old world primate, *Macaca fascicularis. Journal of Physiology, London, 298,* 131–143.

Bowsher, D. A. (1965). The anatomicophysiological basis of somatosensory discrimination. *International Review of Neurobiology, 8,* 35–60.

Boynton, R. M., and Gordon, J. (1965). Bezold–Brücke hue shift measured by color-naming technique. *Journal of the Optical Society of America, 55,* 78–86.

Bradley, P., and Horn, G. (1981). Imprinting, a study of cholinergic receptor sites in parts of the chick brain. *Experimental Brain Research, 41,* 121–123.

Bransford, J. D., McCarrell, N. S., Franks, J. J., and Nitsch, K. E. (1977). Toward unexplaining memory. In R. Shaw and J. Bransford (eds.), *Perceiving, acting, and knowing.* Hillsdale, N.J.: Erlbaum.

Brecha, N. (1983). Retinal neurotransmitters: histochemical and biochemical studies. In P. C. Emson (ed.), *Chemical neuroanatomy,* New York: Raven.

Breitmeyer, B. G. (1975). Simple reaction time as a measure of the temporal response properties of transient and sustained channels. *Vision Research, 15,* 1411–1412.

(1980). Unmasking visual masking: A look at the "why" behind the veil of "how." *Psychological Review, 87,* 52–69.

Breitmeyer, B. G. and Ganz, L. (1976). Implications of sustained and transient channels for theories of visual pattern masking, saccadic suppression, and information processing. *Psychological Review, 83,* 1–36.

Broadbent, D. C. (1954). The role of auditory localization in attention and memory span. *Journal of Experimental Psychology, 47,* 191–196.

Broadbent, D. E. (1957). A mechanical model for human attention and immediate memory. *Psychological Review, 64,* 205–215.

(1958). *Perception and communication.* Oxford: Pergamon.

(1971). *Decision and stress.* New York: Academic Press.

Broca, P. (1878). Anatomie comparée des circonvolutions cérébrales. Le grand lobe limbique et la scissure limbique dans la série des mammifères. *Rev. Anthrop., 1,* 385–498.

Brod, J., Fencl, V., Hejl, Z., and Jirka, J. (1959). Circulatory changes underlying blood pressure elevation during acute emotional stress (mental arithmetic) in normotensive and hypertensive subjects. *Clinical Science, 18*, 269–279.

Brodal, A. (1982). *Neurological anatomy*. New York: Oxford University Press.

Brons, J. F., and Woody, C. D. (1980). Long-term changes in excitability of cortical neurons after pavlovian conditioning and extinction. *Journal of Neurophysiology, 44*, 605–615.

Brooks, B. A., and Fuchs, A. F. (1975). Influence of stimulus parameters on visual sensitivity during saccadic eye movement. *Vision Research, 15*, 1389–1398.

Brooks, D. N., and Baddeley, A. D. (1976). What can amnesic patients learn? *Neuropsychologia, 14*, 111–122.

Brooks, L. R. (1968). Spatial and verbal components of the act of recall. *Canadian Journal of Psychology, 22*, 349–368.

Brown, I. D. and Poulton, E. C. (1961). Measuring the spare mental capacity of car drivers by a subsidiary task. *Ergonomics, 4*, 35–40.

Brown, R., and Kulik, J. (1977). Flashbulb memories. *Cognition, 5*, 73–99.

Bruner, J. S. (1969). Modalities of memory. In G. A. Talland and N. C. Waugh (eds.), *The pathology of memory*. New York: Academic Press.

Buck, R. (1976). *Human motivation and emotion*. New York: Wiley.

(1979). Individual differences in nonverbal sending accuracy and electrodermal responding: The externalizing-internalizing dimension. In R. Rosenthal (ed.), *Skill in nonverbal communication: Individual differences*. Cambridge, Mass.: Oelgeschlager, Gunn & Hain.

(1980). Nonverbal behavior and the theory of emotion: The facial feedback hypothesis. *Journal of Personality and Social Psychology, 38*, 811–824.

(1982). A theory of spontaneous and symbolic expression: Implications for facial lateralization. Paper presented at the symposium, Asymmetries in Facial Expression: Method and Meaning, International Neuropsychology Society convention, Pittsburgh, February 4.

(1983). Emotional development and emotional education. In R. Plutchik and H. Kellerman (eds.), *Emotions in early development*. New York: Academic Press.

(1984). *The communication of emotion*. New York: Guilford Press.

(1985). Prime theory: An integrated view of motivation and emotion. *Psychological Review, 92*, 389–413.

Buck, R., and Duffy, R. (1980). Nonverbal communication of affect in brain-damaged patients. *Cortex, 16*, 351–362.

Buck, R., Miller, R. E., and Caul, W. F. (1974). Sex, personality, and physiological variables in the communication of emotion via facial expression. *Journal of Personality and Social Psychology, 30*, 587–596.

Buck, R., Savin, V., Miller, R. E., and Caul, W. F. (1972). Nonverbal communication of affect in humans. *Journal of Personality and Social Psychology, 23*, 362–371.

Buiserret, P., and Maffei, L. (1977). Extraocular proprioceptive projections to the visual cortex. *Experimental Brain Research, 28*, 421–425.

Buiserret, P., and Singer, W. (1983). Proprioceptive signals from extraocular muscles gate experience-dependent modifications of receptive fields in the kitten visual cortex. *Experimental Brain Research, 51*, 443–450.

Bullier, J., and Henry, G. H. (1979a). Ordinal position of neurons in cat striate cortex. *Journal of Neurophysiology, 42*, 1251–1263.

(1979b). Neural path taken by afferent streams in striate cortex of the cat. *Journal of Neurophysiology, 42*, 1264–1270.

(1979c). Laminar distribution of first-order neurons and afferent terminals in cat striate cortex. *Journal of Neurophysiology, 42*, 1271–1281.

Bunt, A. H., Hendrickson, A. E., Lund, J. S., Lund, R. D., and Fuchs, A. F. (1975).

Monkey retinal ganglion cells: morphometric analysis and tracing of axonal projections, with a consideration of the peroxidase technique. *Journal of Comparative Neurology, 164*, 265–286.

Burkhalter, A., and van Essen, D. C. (1982). Processing of color, form, and disparity in visual areas V2 and VP of ventral extrastriate cortex in the macaque. *Neuroscience Abstracts, 8*, 811.

Bushnell, M. C., Duncan, G. H., Dubner, R., and Fang, H. L. (1984). Activity of trigeminothalamic neurons in medullary dorsal horn of awake monkeys trained in a thermal discrimination task. *Journal of Neurophysiology, 52*, 170–187.

Bushnell, M. C., Goldberg, M. E., and Robinson, D. L. (1981). Behavioral enhancement of visual responses in monkey cerebral cortex. I. Modulation in posterior parietal cortex related to selective visual attention. *Journal of Neurophysiology, 46*, 755–772.

Byrne, J., Castellucci, V., and Kandel, E. R. (1974). Receptive fields and response properties of mechanoreceptor neurons innervating siphon skin and mantle shelf in *Aplysia. Journal of Neurophysiology, 37*, 1041–1064.

Cabot, J. B., Goff, D. M., and Cohen, D. H. (1981). Enhancement of heart rate responses during conditioning and sensitization following interruption of ruphe-spinal projections. *Journal of Neuroscience, 1*, 760–770.

Campbell, F. W., and Robson, J. G. (1968). Application of Fourier analysis to the visibility of gratings. *Journal of Physiology, 197*, 551–566.

Campbell, R. (1978). Asymmetries in interpreting and expressing a posed facial expression. *Cortex, 14*, 327–342.

Cannon, W. B. 1915. *Bodily changes in pain, hunger, fear, and rage.* New York: Appleton.
(1927). The James–Lange theory of emotion: A critical examination and an alternative theory. *American Journal of Psychology, 39*, 106–124. [Reprinted in M. Arnold (ed.) *The Nature of emotion.* Baltimore: Penguin, 1968.]
(1929). *Bodily changes in pain, hunger, fear, and rage*, 2nd ed. New York: Appleton.
(1931). Against the James–Lange and thalamic theories of emotion. *Psychological Review, 31*, 281–295.

Cannon, W. B., and Britton, S. W. (1925). Pseudoaffective meduliadrenal secretion. *American Journal of Physiology, 72*, 283–294.

Cannon, W. B., Lewis, J. T., and Britton, S. W. (1927). The dispensability of the sympathetic division of the autonomic system. *Boston Medical and Surgical Journal, 197*, 514.

Carmon, A., and Nachshon, I. (1972). Ear asymmetry in perception of emotional and nonverbal stimuli. *Acta Physiologica, 37*, 351–357.

Carpenter, M. (1982). *Human neuroanatomy.* Baltimore: Williams & Wilkins.

Carroll, M. and Kirsner, K. (1982). Context and repetition effects in lexical decision and recognition memory. *Journal of Verbal Learning and Verbal Behavior, 21*, 55–64.

Castellucci, V., Pinsker, H., Kupferman, I., and Kandel, E. R. (1970). Neuronal mechanisms of habituation and dishabituation of the gill-withdrawal reflex in *Aplysia. Science, 167*, 1745–1748.

Cermak, L. S. (ed.). (1982). *Human memory and amnesia.* Hillsdale, N.J.: Erlbaum.

Cermak, L. S., and Craik, F. I. M. (eds.). (1979). *Levels of processing in human memory.* Hillsdale, N.J.: Erlbaum.

Cermak, L. S., Talbot, N., Chandler, K., and Wolbarst, L. R. (1985). The perceptual priming phenomenon in amnesia. *Neuropsychologia, 23*, 615–622.

Chalmers, J. P. (1975). Brain amines and models of experimental hypertension. *Circulation Research, 36*, 469–480.

Chalupa, L. M., Coyle, R. S., and Lindsley, D. B. (1976). Effect of pulvinar lesions on visual pattern discrimination in monkeys. *Journal of Neurophysiology, 39*, 354–369.

Chase, W. G., and Ericsson, K. A. Skilled memory. (1981). In J. R. Anderson (ed.), *Cognitive skills and their acquisition*. Hillsdale, N.J.: Erlbaum.

Cheesman, J., and Merickle, P. M. (1984). Priming with and without awareness. *Perception and Psychophysics, 36,* 387–395.

Cherry, C. (1957). *On human communication: A review, a survey, and a criticism*. New York: Wiley.

Chiesi, H., Spillich, G., and Voss, J. F. (1979). Acquisition of domain-related information in relation to high and low domain knowledge. *Journal of Verbal Learning and Verbal Behavior, 18,* 257–273.

Churchland, P. M. (1984). *Matter and consciousness*. Cambridge, Mass.: MIT Press.

Cicone, M., Wapner, W., and Gardner, H. (1980). Sensitivity to emotional expressions and situations in organic patients. *Cortex, 16,* 145–147.

Clark, D. M., and Teasdale, J. D. (1982). Diurnal variation in clinical depression and accessibility of memories of positive and negative experiences. *Journal of Abnormal Psychology, 91,* 87–95.

Cleland, B. G., and Levick, W. R. (1974a). Brisk and sluggish concentrically organized ganglion cells in the cat's retina. *Journal of Physiology, London, 240,* 421–456.

(1974b). Properties of rarely encountered types of ganglion cells in the cat's retina and an overall classification. *Journal of Physiology, London, 240,* 457–492.

Cofer, C. C. (1967). Conditions for the use of verbal associations. *Psychological Bulletin, 68,* 1–12.

Cohen, D. H. (1974). The neural pathways and informational flow mediating a conditioned autonomic response. In L. V. DiCara (ed.). *Limbic and autonomic neurons system research*. New York: Plenum.

(1980). The functional neuroanatomy of a conditioned response. In Thompson, R. F., Hicks, L. H., and Shvyrkov, V. B. (eds.). *Neural mechanisms of goal-directed behavior and learning*. New York: Academic Press.

(1982). Central Processing time for a conditioned response in a vertebrate model. In C. D. Woody (eds.). *Advances in behavioral biology*, Vol. 26, *Conditioning: representation of involved neural functions*, pp. 517–534, New York: Plenum.

Cohen, D. H., and Obrist, P. A. (1975). Interactions between behavior and cardiovascular system. *Circulation Research, 37,* 693–706.

Cohen, N. J. (1984). Amnesia and the distinction between procedural and declarative knowledge. In N. Butters and L. R. Squire (eds.), *The neuropsychology of memory*. New York: Guilford Press.

Cohen, N. J., and Squire, L. R. (1980). Preserved learning and retention of pattern-analyzing skill in amnesia: Dissociation of "knowing how" and "knowing that." *Science, 210,* 207–209.

Coltheart, M., Masterson, J., Byng, S., Prior, M., and Riddoch, J. (1983). Surface dyslexia. *Quarterly Journal of Psychology, 35A,* 469–495.

Corner, R. L., and Levine, S. (1969). The effects of adrenal hormones on the acquisition of signaled avoidance behavior. *Hormones and Behavior, 1,* 73–83.

Cornsweet, T.N. (1970). *Visual perception*. New York: Academic Press.

Corteen, R. S. and Dunn, D. (1972). Shock associated words in a nonattended message: A test for momentary awareness. *Journal of Experimental Psychology, 94,* 308–313.

Corteen, R. W. and Wood, B. (1972). Autonomic responses to shock-associated words in an unattended channel. *Journal of Experimental Psychology, 94,* 308.

Corwin, J. V., and Slaughter, J. S. (1979). Effects of vagal stimulation of the learning of specific and diffuse conditioned suppression. *Behavior and Neural Biology, 25,* 364–370.

Cowey, A., and Porter, J. (1979). Brain damage and global stereopsis. *Proceedings of the Royal Society of London, B 204,* 399–407.

Craik, F. I. M. (1979). Human memory. *Annual Review of Psychology, 30*, 63–102.
(1983). On the transfer of information from temporary to permanent memory. In
D. E. Broadbent (ed.). *Functional aspects of memory*. London: Royal Society.
Craik, F. I. M., and Lockhart, R. S. (1972). Levels of processing: A framework for
memory research. *Journal of Verbal Learning and Verbal Behavior, 11*, 671–684.
Craik, F. I. M., and Tulving, E. (1975). Depth of processing and the retention of
words in episodic memory. *Journal of Experimental Psychology: General, 104*,
268–294.
Cramer, P. (1966). Mediated priming of associative responses: The effect of time lapse
and interpolated activity. *Journal of Verbal Learning and Verbal Behavior, 5*,
163–166.
Crovitz, H. F., and Schiffman, H. (1974). Frequency of episodic memories as a
function of their age. *Bulletin of the Psychonomic Society, 4*, 517–518.
Crow, T. and Offenbach, N. (1983). Modification of the initiation of locomotion in
Hermissenda crassicornia: behavioral analysis. *Brain Research, 271*, 301–310.
Crowder, R. G. (1976). *Principles of learning and memory*. Hillsdale, N.J.: Erlbaum.
Cynader, M., and Regan, D. (1978). Neurons in cat parastriate cortex sensitive to
the direction of motion in three-dimensional space. *Journal of Physiology, London, 274*, 549–569.
Dampney, R. A. L., and Moon, E. A. (1980). Role of ventrolateral medulla in
vasomotor response to cerebral ischemia. *American Journal of Physiology, 239*,
H349–358.
Dampney, R. A. L., Kumada, M., and Reis, D. J. (1979). Central neural mechanisms
of the cerebral ischemic response: characterization, effect of brainstem and cranial
nerve transection, and stimulation by electrical stimulation of restricted regions
of medulla oblongata in rabbit. *Circulation Research, 44*, 48–62.
Daniels, J. D., and Pettigrew, J. D. (1975). A study of inhibitory antagonism in cat
visual cortex. *Brain Research, 93*, 41–62.
Darian-Smith, I., Johnson, K. O., and Dykes, R. W. (1973). "Cold" fiber population
innervating palmar and digital skin of the monkey: responses to cooling pulses.
Journal of Neurophysiology, 36, 325–346.
Darwin, C. (1872). *Expressions of the emotions in man and animals*. London: Murray.
Daugman, J. G. (1980). Two-dimensional spectral analysis of cortical receptive field
profiles. *Vision Research, 20*, 847–856.
Davis, J. W. and Gillette, R. (1978). Neural correlate of behavioral plasticity in
command neurons of *Pleurobranchaea*. *Science, 199*, 801–804.
Day, M. D. (1979). *Autonomic pharmacology*. Edinburgh: Churchill Livingston.
Dean, P. (1979). Visual cortex ablation and thresholds for successively presented
stimuli in rhesus monkey: II. Hue. *Experimental Brain Research, 35*, 69–83.
De Groot, A. D. (1965). *Thought and choice in chess*. New York: Basic Books.
DeKosky, S. T., Heilman, K. M., Bowers, D., and Valenstein, E. (1980). Recognition
and discrimination of emotional faces and pictures. *Brain and Language, 9*, 206–
214.
Delgado, J. M. R. (1969). *Physical control of the mind*. New York: Harper & Row.
de Monasterio, F. M. (1978a). Properties of concentrically organized X and Y ganglion
cells of macaque retina. *Journal of Neurophysiology, 41*, 1394–1417.
(1978b). Center and surround mechanisms of opponent-color X and Y ganglion
cells of macaques. *Journal of Neurophysiology, 41*, 1418–1434.
(1978c). Properties of ganglion cells with atypical receptive-field properties in retina
of macaques. *Journal of Neurophysiology, 41*, 1435–1449.
de Monasterio, F. M., and Schein, S. J. (1982). Spectral bandwidths of color-opponent
cells in the geniculostriate pathway of macaque monkeys. *Journal of Neurophysiology, 47*, 214–224.

Desmond, J. E. and Moore, J. W. (1982). A brainstem region essential for classically conditioned but not unconditioned nictitating membrane response. *Physiology and Behavior, 28*, 1029–1033.

Denenberg, V. H. (1981). Hemispheric laterality in animals and the effects of early experience. *Behavioral and Brain Sciences, 4*, 1–49.

Denenberg, V. H., Garbanati, J., Sherman, G., Yutzey, D. A., and Kaplan, R. (1978). Infantile stimulation induces brain lateralization in rats. *Science, 201*, 1150–1152.

Denny-Brown, D., Meyer, J. S., and Horenstein, S. T. (1952). The significance of perceptual rivalry resulting from parietal lesion. *Brain, 75*, 433–471.

Derry, P. A., and Kuiper, N. A. (1981). Schematic processing and self-reference in clinical depression. *Journal of Abnormal Psychology, 90*, 286–297.

Desimone, R. (1982). Symposium on Prestriate Cortex. *Society for Neuroscience Abstracts, 8*, 220.

Desimone, R., and Schein, S.J. (1983). Receptive field properties of neurons in visual area V4 of the macaque. *Society for Neuroscience Abstracts, 9*, 153.

Desimone, R., Albright, T. D., Gross, C. G., and Bruce, C. (1984). Stimulus-selective properties of inferior temporal neurons in the macaque. *Journal of Neuroscience, 4*, 2051–2062.

Desmedt, J. E., and Robertson, D. (1977). Differential enhancement of early and late components of the cerebral somatosensory evoked potentials during forced-paced cognitive tasks in man. *Journal of Physiology (London), 271*, 761–782.

Desmond, J. E. and Moore, J. W. (1982). A brain stem region essential for the classically conditioned but not unconditioned nictitating membrane response. *Physiology and Behavior, 28*, 1029–1033.

Deutsch, J. A., and Deutsch, D. (1963). Attention: Some theoretical considerations. *Psychological Review, 70*, 80–90.

De Valois, K. K., and Marrocco, R. T. (1973). Single cell analysis of saturation discrimination in the macaque. *Vision Research, 13*, 701–711.

De Valois, K. K., and Tootell, R. E. (1983). Spatial-frequency-specific inhibition in cat striate cortex. *Journal of Physiology. London, 336*, 359–376.

De Valois, R. L., and De Valois, K. K. (1980). Spatial vision. *Annual Review of Psychology, 31*, 309–341.

De Valois, R. L., Abramov, A., and Jacobs, G. H. (1966). Analysis of response patterns of monkey LGN cells. *Journal of the Optical Society of America, 56*, 966–977.

De Valois, R. L., Abramov, I., and Mead, W. R. (1967). Single-cell analysis of wavelength discrimination at the lateral geniculate cells in the macaque, *Journal of Neurophysiology, 30*, 415–433.

De Valois, R. L., Albrecht, D. G., and Thorell, L. G. (1978). Cortical cells: bar detectors or spatial frequency filters? In S. J. Cool and E. L. Smith (eds.) *Frontiers of visual science*. Berlin: Springer.

(1982). Spatial frequency selectivity of cells in macaque visual cortex. *Vision Research, 22*, 545–559.

De Valois, K. K., De Valois, R. L., and Yund, E. W. (1979). Responses of striate cortex cells to grating and checkerboard patterns. *Journal of Physiology, London, 291*, 484–505.

De Valois, R. L., Yund, E. W., and Hepler, N. (1982). The orientation and direction selectivity of cells in macaque striate cortex. *Vision Research, 22*, 531–544.

DeVito, J. L., and Smith, O. A. (1982). Afferent projections to the hypothalamic area controlling emotional responses (HACER). *Brain Research, 252*, 231–226.

de Weert, C. M. M. (1979). Colour contours and stereopsis. *Vision Research, 19*, 555–564.

Diamond, M. C. (1967). Extensive cortical depth measurements and neuron size increases in the cortex of environmentally enriched rats. *Journal of Comparative Neurology, 131*, 357–364.

Diamond, S. P. and Bender, M. B. (1965). On auditory extinction and allocusis. *Transactions of the American Neurological Association, 90*, 154–157.

Diamond, R., and Rozin, P. (1984). Activation of existing memories in the amnesic syndrome. *Journal of Abnormal Psychology, 93*, 98–105.

Dichgans, J., and Jung, R. (1969). Attention, eye movements and motion detection: facilitation and selection in optokinetic nystagmus and railway nystagmus. In C. R. Evans and T. B. Mulholland (eds.) *Attention in neurophysiology*. London: Butterworth.

DiGiusto, E. L., and King, M. G. (1972). Chemical sympathectomy and avoidance learning in the rat. *Journal of Comparative Physiology and Psychology, 81*, 491–500.

DiVietti, T. L., and Porter, P. B. (1969). Modification of the autonomic component of the conditioned emotional response. *Psychological Reports, 24*, 951–958.

Dixon, N. F. (1971). *Subliminal perception: The nature of a controversy*. London: McGraw-Hill.

(1987). *Preconscious processing*. New York: Wiley.

Domesick, V. B. (1969). Projections from the cingulate cortex in the rat. *Brain Research, 12*, 296–320.

(1972). Thalamic relationships of the medial cortex in the rat, *Brain, Behavior and Evolution, 6*, 141–169.

Douglas, R. J. (1967). The hippocampus and behavior. *Psychological Bulletin, 67*, 416–422.

Dowling, J. E. (1979). Information processing in local circuits: the vertebrate retina as a model system. In F. O. Schmitt and F. G. Worden (eds.), *The neurosciences fourth study program*. Cambridge, Mass.: MIT Press.

Dowling, J. E. and Boycott, B. B. (1966). Organization of the primate retina: electron microscopy. *Proceedings of the Royal Society of London, B, 116*, 80–111.

Downey, J. E., and Anderson, J. E. (1915). Automatic writing. *American Journal of Psychology, 26*, 161–195.

Dreher, B., Fukada, Y., and Rodieck, R. W. (1976). Identification, classification, and anatomical segregation of cells with X-like and Y-like properties in the lateral geniculate nucleus of Old-World primates. *Journal of Physiology, London, 258*, 433–452.

Dreher, B., Leventhal, A. G., and Hale, P. T. (1980). Geniculate input to cat visual cortex: a comparison of area 19 with areas 17 and 18. *Journal of Neurophysiology, 44*, 804–826.

Duffy, R. J., Duffy, J. R. and Pearson, K. (1975). Pantomime recognition in aphasics. *Journal of Speech and Hearing Research, 18*, 115–132.

Dumais, S. T. (1979). Perceptual learning in automatic detection: processes and mechanisms. Unpublished doctoral dissertation, Indiana University.

Duncan, J. (1984). Selective attention and the organization of visual information. *Journal of Experimental Psychology: General, 113*, 501–517.

Eagle, M. (1959). The effects of subliminal stimuli of aggressive content upon conscious cognition. *Journal of Personality, 27*, 578–600.

Eason, R. G. (1981). Visual evoked potential correlates of early neural filtering during selective attention. *Bulletin of the Psychonomic Society, 18*, 203–206.

Eason, R. G., Harter, M. R., and White, C. T. (1969). Effects of attention and arousal on visually evoked cortical potentials and reaction times in man. *Physiology and Behavior, 4*, 283–289.

Ebbinghaus, H. (1885). *Uber das Gedächtnis.* Leipzig: Duncker and Humbolt.

Ebner, F. (1969). A comparison of primitive forebrain organization on metatherian and eutherian mammals. *Annals of the New York Academy of Science, 167,* 241–257.

Eccles, J. C. (1965). *The brain and the unity of conscious experience.* Cambridge University Press.

Efron, R., and Yund, E. W. (1974). Dichotic competition of simultaneous tone bursts of different frequency: I. Dissociation of pitch from lateralization and loudness. *Neuropsychologia, 12,* 249–256.

Egger, M. D., and Flynn, J. P. (1962). Amygdaloid suppression of hypothalamically elicited attack behavior. *Science, 136,* 43–44.

Eibl-Eibesfeldt, I. (1970). *Ethology: the biology of behavior.* New York: Holt, Rinehart & Winston.

Eich, J. E. (1980). The cue-dependent nature of state-dependent retrieval. *Memory & Cognition, 8,* 157–173.

(1984). Memory for unattended events: Remembering with and without awareness. *Memory & Cognition, 12,* 105–111.

Eich, J. M. (1982). A composite holographic associative recall model. *Psychological Review, 89,* 627–661.

Eidelberg, E. and Schwartz, A. S. (1971). Experimental analysis of the extinction phenomenon in monkeys. *Brain, 94,* 91–108.

Ekman, P. (1980). Assymetry in facial expression. *Science, 209,* 833–834.

Ekman, P. and Friesen, W. (1975). *Unmasking the face.* Englewood Cliffs, N.J.: Prentice-Hall.

Ekman, P., Friesen, W. V., and Ellsworth, P. (1972). *Emotion in the human face.* New York: Pergamon.

Ekman, P., Hagar, J., and Friesen, W. (1981). The symmetry of emotional and deliberate facial action. *Psychophysiology, 18,* 101–106.

Ekman, P., Levinson, R. W., and Friesen, W. V. (1983). Autonomic nervous system activity distinguishes among emotions. *Science, 221,* 1208–1210.

Ekman, P., Sorenson, E. R., and Friesen, W. V. (1969). Pan-cultural elements in the facial displays of emotion. *Science, 164,* 86–88.

Elkind, D. (1971). Cognitive growth cycles in mental development. In J. K. Cole (ed.) *Nebraska Symposium on Motivation.* Lincoln: University of Nebraska Press.

Elman, J. L., and McClelland, J. L. (1983). Speech perception as a cognitive process: the interactive activation model. Institute for Cognitive Science, technical report. No. 8302. University of California, San Diego.

Engle, F. L. (1971). Visual conspicuity, directed attention, and retinal locus. *Vision Research, 11,* 563–576.

Enroth-Cugell, C., and Robson, J. G. (1966). The contrast sensitivity of retinal ganglion cells of the cat. *Journal of Physiology, London, 187,* 517–552.

Eriksen, C. W., and Schultz, D. W. (1979). Information processing in visual search: a continuous flow conception of experimental results. *Perception and Psychophysics, 25,* 249–263.

Estes, W. K. and Skinner, B. F. (1941). Some quantitative properties of anxiety. *Journal of Experimental Psychology, 29,* 390–400.

Ewert, J.-P. (1980). *Neuroethology.* Berlin: Springer.

Faugier-Grimaud, S., Frenois, C., and Stein, D. G. (1978). Effects of posterior parietal lesions on visually guided behavior in monkeys. *Neuropsychologia, 16,* 151–168.

Feldman, J. A., and Ballard, D. H. (1983). Computing with connections. In J. Beck, B. Hope, and A. Rosenfeld (eds.), *Human and machine vision.* New York:

Academic Press.

Feustel, T. C., Shiffrin, R. M., and Salasoo, A. (1983). Episodic and lexical contributions to the repetition effect in word identification. *Journal of Experimental Psychology: General, 112,* 309–346.

Finch, D. M., Derian, E. L., and Babb, T. L. (1984). Excitatory projection of the rat subicular complex to the cingulate cortex and synaptic integration with thalamic afferents. *Brain Research, 301,* 25–37.

Fischer, B., and Boch, R. (1981a). Selection of visual targets activates prelunate cortical cells in trained rhesus monkey. *Experimental Brain Research, 41,* 431–433.

(1981b). Enhanced activation of neurons in prelunate cortex before visually guided saccades of trained rhesus monkeys. *Experimental Brain Research, 1981, 44,* 129–137.

Fisher, R. P., and Craik, F. I. M. (1977). The interaction between encoding and retrieval operations in cued recall. *Journal of Experimental Psychology: Human Learning and Memory, 3,* 701–711.

Fodor, J. A. (1975). *The language of thought.* Cambridge, Mass.: Harvard University Press.

1983. *The modularity of mind.* Cambridge, Mass.: MIT Press.

Folkow, B. and Rubinstein, E. H. (1966). Cardiovascular effects of acute and chronic stimulation of the hypothalamic defense area in the rat. *Acta Physiologica Scandanavica, 68,* 48–57.

Foster, K., Orona, E., Lambert, R.W., and Gabriel, M. (1980). Early and late acquisition of discriminative neuronal activity during differential conditioning in rabbits: specificity within the laminae of cingulate cortex and the antcroventral thalamus. *Journal of Comparative and Physiological Psychology, 94,* 1069–1086.

Fowler, C. A., Wolford, G., Slade, R., and Tassinary, L. (1981). Lexical access with and without awareness. *Journal of Experimental Psychology: General, 110,* 341–362.

Frankenhauser, M. (1971). Behavior and circulating catecholamines. *Brain Research, 31,* 241–262.

(1975). Experimental approaches to the study of catecholamine and emotion. In L. Levi (ed.). *Emotions: their parameters and measurement.* New York: Raven.

Franks, J. J., Bransford, J. D., and Auble, P. M. (1982). The activation and utilization of knowledge. In C. R. Puff (ed.), *Handbook of research methods in human memory and cognition.* New York: Academic Press.

Friedland, R. P., and Weinstein, E. A. (1977). Hemi-inattention and hemisphere specialization: Introduction and historical review. *Advances in Neurology, 18,* 1–31.

Friedman, A. (1979). Framing pictures: The role of knowledge in automatized encoding and memory for gist. *Journal of Experimental Psychology: General, 108,* 316–355.

Frisby, J. P. (1980). *Seeing: illusion, brain, and mind.* Oxford University Press.

Fu, K. (1974). *Syntactic methods in pattern recognition.* New York: Academic Press.

Fukuda, Y., and Stone, J. (1974). Retinal distribution and central projections of Y-, X-, and W-cells of the cat's retina. *Journal of Neurophysiology, 37,* 749–772.

(1976). Evidence of differential inhibitory influences of X- and Y-type relay cells in the cat's lateral geniculate nucleus. *Brain Research, 113,* 188–196.

Funkenstein, D. H. (1955). The physiology of fear and anger, *Scientific American, 192,* 74–80.

Fuster, J. M. (1980). *The prefrontal cortex: anatomy, physiology, and neuropsychology of the frontal lobe.* New York: Raven.

Gabriel, M. (1970). Intersession exposure of rabbits to conditioning apparatus, avoidance extinction, and intertrial behavior. *Journal of Comparative and Physiological Psychology, 72*, 244–249.

Gabriel, M., and Orona, E. (1982). Parallel and serial processes of the prefrontal and cingulate cortical systems during behavioral learning. *Brain Research Bulletin, 8*, 781–785.

Gabriel, M., and Saltwick, S. E. (1980). Rhythmic, theta-like unit activity of the hippocampal formation during acquisition and performance of avoidance behavior in rabbits. *Physiology and Behavior, 24*, 303–312.

Gabriel, M., Foster, K., and Orona, E. (1980a). Interaction of the laminae of cingulate cortex and the anteroventral thalamus during behavioral learning. *Science, 208*, 1050–1052.

Gabriel, M., Foster, K., Orona, E., Saltwick, S. E., and Stanton, M. (1980b) Neuronal activity of cingulate cortex, anteroventral thalamus and hippocampal formation in discriminative conditioning: Encoding and extraction of the significance of conditional stimuli. In J. Sprague and A. N. Epstein (eds.), *Progress in physiological psychology and psychobiology*, Volume 9, pp. 125–231, Academic Press: New York.

Gabriel, M., Lambert, R. W., Foster, K., Orona, E., Sparenborg, S., and Maiorca, R. R. (1983). Anterior thalamic lesions and neuronal activity in the cingulate and retrosplenial cortices during discriminative avoidance behavior in rabbits. *Behavioral Neuroscience, 97*, 675–696.

Gabriel, M., Sparenborg, S. P., Stolar, N., Ragsdale, D., and Klein, S. (1984a). Subicular lesions and discriminative neuronal activity in the cingulate cortex and anteroventral thalamus during learning in rabbits. *Society for Neuroscience Abstracts, 10*, 795.

Gabriel, M., Sparenborg, S. P., Stolar, N., Dreyzehner, J., and Colletier, P. (1984b). Opposite effects of cingulate cortical and subicular lesions on avoidance behavior in rabbits. *Society for Neuroscience Abstracts, 10*, 127.

Gabriel, M., Sparenborg, S. P., Colletier, P., and Tenner, J. (1985). Effects of separate and combined medial and anterior thalamic lesions on discriminative avoidance learning in rabbits. *Society for Neuroscience Abstracts, 11*.

Gainotti, G. (1969). Réactions "catastrophiques" et manifestations d'indifférence au cours des atteintes cérébrales. *Neuropsychologia, 7*, 195–204.

—— (1972). Emotional behavior and hemispheric side of the lesion. *Cortex, 8*, 41–55.

Galin, D. (1974). Implications for psychiatry of left and right cerebral specialization: a neurophysiological context for unconscious processes. *Archives of General Psychiatry, 31*, 572.

Gallagher, M., Kapp, B. S., Frysinger, R. C., and Rapp, P. (1980). Beta-adrenergic manipulation in amygdala central nucleus alters rabbit heart rate conditioning. *Pharmacology, Biochemistry and Behavior, 12*, 419–426.

Gallagher, M., Kapp, B. S., McNall, C., and Pascoe, J. P. (1981). Opiate effects in the amygdala central nucleus on heart rate conditioning in rabbits. *Pharmacology, Biochemistry and Behavior, 14*, 497–505.

Gallup, G. (1984). Do minds exist in species other than our own? Address delivered at the convention of the American Psychological Association, Toronto, August 27, 1984.

Garner, W. R. (1974). *The processing of information and structure.* Potomac, Md.: Erlbaum.

Garner, W. R., Hake, H. W., and Eriksen, C. W. (1956). Operationism and the concept of perception. *Psychological Review, 63*, 149–159.

Gattass, R., Oswaldo-Cruz, E., and Sousa, A. P. B. (1979). Visual receptive fields of units in the pulvinar of cebus monkey. *Brain Research, 160*, 413–430.

Gazzaniga, M. S. (1984). *Handbook of cognitive neuroscience.* New York: Plenum.

Gazzaniga, M. S., Bogen, J. E., and Sperry, R. W. (1962). Some functional effects of sectioning the cerebral commissures in man, *Proceedings of the National Academy of Science, 48*, 1765–1769.

Geisert, E. E., Langootmo, A., and Spear, P. D. (1981). Influence of the cortico-geniculate pathway on response properties of cat lateral geniculate neurons. *Brain Research, 208*, 409–415.

Geldard, F. (1972). *The human senses.* New York: Wiley.

Gershon, M. D., Dreyfus, D. F., and Rothman, T. P. (1979). The mammalian enteric nervous system: a third autonomic division. In S. Kalsner (ed.), *Trends in autonomic pharmacology*, Vol. 1, pp. 59–102.

Gervais, M. J., Harvey, L. O., Jr., and Roberts, J. O. (1984). Identification confusions among letters of the alphabet. *Journal of Experimental Psychology: Human Perception and Performance, 10*, 655–666.

Geschwind, N. (1965). The disconnexion syndromes in animals and man. *Brain, 88*, 237–294.

Geschwind, N. (1979). Specializations of the human brain. *Scientific American, 241*, 180–199.

Gibson, E. J., and Yonas, A. (1966). A developmental study of visual search behavior. *Perception and Psychophysics, 1*, 169–171.

Gibson, J. J. (1966). *The senses considered as perceptual systems.* Boston: Houghton Mifflin.

(1979). *The ecological approach to visual perception.* Boston: Houghton Mifflin.

Gilbert, C. D. (1977). Laminar differences in receptive field properties in cat primary visual cortex. *Journal of Physiology, London, 268*, 391–421.

Gilbert, C. D., and Wiesel, T. N. (1979). Morphology and intracortical projections of functionally characterised neurones in the cat visual cortex. *Nature (London), 280*, 120–125.

Gilchrist, A. L. (1979). The perception of surface blacks and whites. *Scientific American, 240*, 112–124.

(1980). When does perceived lightness depend on perceived spatial arrangement? *Perception and Psychophysics, 28*, 527–538.

Gilchrist, A. L., and Jacobsen, A. (1983). Lightness constancy through a veiling luminance. *Journal of Experimental Psychology; Human Perception and Performance, 9*, 936–944.

Gilchrist, A. L., Delman, S., and Jackobsen, A. (1983). The classification and integration of edges as critical to the perception of reflectance and illumination. *Perception and Psychophysics, 33*, 425–436.

Gildea, P. (1984). On resolving ambiguity: Can context constrain lexical access? Princeton University, unpublished doctoral dissertation.

Glucksberg, S., and Cowen, G. N. (1970). Memory for nonattended auditory material. *Cognitive Psychology, 1*, 149–156.

Gold, M. R., and Cohen, D. H. (1981). Modification of the discharge of vagal cardiac neurons during learned heart rate change. *Science, 214*, 345–347.

Goldberg, M. E. and Bushnell, M. C. (1982). Behavioral enhancement of visual responses in monkey cerebral cortex. II. Modulation in frontal eye fields specifically related to saccades. *Journal of Neurophysiology, 46*, 773–787.

Goldberg, M. E. and Wurtz, R. H. (1972). Activity of superior colliculus in behaving monkey. II. Effect of attention on neuronal responses. *Journal of Neurophysiology, 35*, 560–574.

394   *References*

Gormezano, I., Kehoe, E. J., and Marshall, B. S. (1983). Twenty years of classical conditioning research with the rabbit. In J. M. Sprague and A. N. Epstein (eds.), *Progress in psychobiology and physiological psychology*, Vol. 10, pp. 197–275, New York: Academic Press.

Graf, P., and Mandler, G. (1984). Activation makes words more accessible but not necessarily more retrievable. *Journal of Verbal Learning and Verbal Behavior, 23*, 553–568.

Graf, P., and Schacter, D. L. (1985). Implicit and explicit memory for new associations in normal and amnesic subjects. *Journal of Experimental Psychology, Learning, Memory, and Cognition, 11*, 501–518.

Graf, P., Mandler, G., and Haden, P. E. (1982). Simulating amnesic symptoms in normal subjects. *Science, 218*, 1243–1244.

Graf, P., Squire, L. R., and Mandler, G. (1984). The information that amnesic patients do not forget. *Journal of Experimental Psychology: Learning, Memory, & Cognition, 10*, 164–178.

Graham, N. (1980). Spatial frequency channels in human vision: Detecting edges without edge detectors. In C. S. Harris (ed.), *Visual coding and adaptability*. Hillsdale, N.J.: Erlbaum.

Grastyan, E., Lissak, K., Madarasz, I., and Donhoffer, H. (1959). Hippocampal electrical activity during the development of conditioned reflexes. *EEG and Clinical Neurophysiology, 11*, 533–557.

Gray, J. A. (1982). *The neuropsychology of anxiety*. New York: Oxford University Press.

Graybiel, A. M. (1973). The thalamo-cortical projection of the so-called posterior nuclear group: a study with anterograde degeneration methods in the cat. *Brain Research, 49*, 229–244.

Graybiel, A. M., and Ragsdale, C. W. (1982). Pseudocholinesterase staining in the primary visual pathway of the macaque monkey. *Nature (London), 299*, 439–442.

Greenblatt, S. H. (1965). The major influences on the early life and work of John Hughlings Jackson. *Bulletin of the History of Medicine, 39*, 346–376.

Greenough, W. T. (1984). Structural correlates of information storage in the mammalian brain: a review and hypothesis. *Trends in Neuroscience, 7*, 229–233.

Greenough, W. T. and Chang, F.-L. (1985). Synaptic structural correlates of information storage in mammalian nervous systems. In C. W. Cotman (ed.), *Neuronal plasticity*, pp. 335–372. New York: Guilford Press.

Greenough, W. T., Volkmar, F. R., and Fleihmann, T. B. (1976). Environmental effects on brain conductivity and behavior. In D. I. Mostofsky (ed.), *Behavior control and modification of physiological activity*, pp. 220–245. Englewood Cliffs, N.J.: Prentice-Hall.

Gregory, R. L. (1977). Vision with isoluminant colour contrast: a projection technique and observations. *Perception, 6*, 411–416.

Griffin, D. R. (1976). *The question of animal awareness*. New York: Rockefeller University Press.

Gross, C. G., and Mishkin, M. (1977). The neural basis of stimulus equivalence across retinal translation. In S. Harned, R. Doty, J. Jaynes, L. Goldberg, and G. Krauthamer (eds.). *Lateralization in the nervous system*. New York: Academic Press.

Gross, C. G., Bender, D. B., and Gerstein, G. L. (1979). Activity of inferotemporal neurons in behaving monkeys. *Neuropsychologia, 17*, 215–230.

Gross, C. G., Rocha-Miranda, C. E., and Bender, D. B. (1972). Visual properties of neurons in inferotemporal cortex of the macaque. *Journal of Neurophysiology, 35*, 96–111.

Guthrie, G., and Wiener, M. (1966). Subliminal perception or the perception of partial cues with pictorial stimuli. *Journal of Personality and Social Psychology, 3*, 619–628.

Guyton, A. C. (1981). *A textbook of physiology*. Philadelphia: Saunders.

Guzman, A. (1968). Decomposition of a visual scene into three-dimensional bodies. *AFIPS, Fall Joint Computer Conference, 33*, 291–304.

— (1969). Decomposition of a visual scene into three-dimensional bodies. In A. Graselli (ed.), *Automatic interpretation and classification of images*. New York: Academic Press.

Haenny, P. E., and Schiller, P. H. (1983). The behavioral significance of visual stimuli influences the responses of single cells in VI. *Investigative Ophthalmology and Visual Science, 24*, 106.

Haggard, M. P. and Parkinson, A. M. (1971). Stimulus and task factors as determinants of ear advantages. *Quarterly Journal of Experimental Psychology, 23*, 168–177.

Halle, M., Bresnan, J., and Miller, G. A. (1978). *Linguistic theory and psychological reality*. Cambridge, Mass.: MIT Press.

Halperin, R. and Pfaff, D. W. (1982). Brain-stimulation reward and control of autonomic function: are they related? In Pfaff, D. W. (ed.), *The physiological mechanisms of motivation*. New York: Springer-Verlag.

Hamori, J., Pasik, P., and Pasik, T. (1983). Differential frequency of P-cells and I-cells in magnocellular and parvocellular laminae of monkey lateral geniculate nucleus. An ultrastructural study. *Experimental Brain Research, 52*, 57–66.

Hansen, J. C., and Hillyard, S. A. (1983). Selective attention to multidimensional auditory stimuli. *Journal of Experimental Psychology: Human Perception and Performance, 9*, 1–19.

Harlow, H. F., and Mears, C. E. (1983). Emotional sequences and consequence. In R. Plutchik and H. Kellerman (eds.), *Emotion: theory, research, and experience*. Vol. 2, *Emotions in Early Development*. New York: Academic Press.

Hart, J. T. (1967). Memory and the memory-monitoring process. *Journal of Verbal Learning and Verbal Behavior, 6*, 685–691.

Harter, M. R., and Previc, F. H. (1978). Size-specific information channels and selective attention: Visual evoked potential and behavioral measures. *Electroencephalography and Clinical Neurophysiology, 45*, 628–640.

Harter, M. R., Aine, C., and Schroeder, C. (1982). Hemispheric differences in the neural processing of stimulus location and type: Effects of selective attention on visual evoked potentials. *Neuropsychologia, 20*, 421–438.

Hawkins, H. L., and Ketchum, R. D. (1980). The case against secondary task analyses of mental workload, Center for Cognitive and Perceptual Research, University of Oregon.

Hawkins, H. L., Rodriquez, E., and Reicher, G. M. (1982). Is time-sharing a general ability? Center for Cognitive and Perceptual Research, University of Oregon.

Hawkins, R. D., and Kandel, E. R. (1984). Is there a cell-biological alphabet for simple forms of learning? *Psychological Review, 91*, 375–391.

Heath, R. G. (1954). *Studies in schizophrenia*. Cambridge, Mass.: Harvard University Press.

— (1964). Pleasure responses of human subjects to direct stimulation of the brain: Physiologic and psychodynamic considerations. In R. G. Heath (ed.) *The role of pleasure in behavior*. New York: Hoeber Medical Division, Harper & Row.

Heggelund, P. (1981). Receptive field organization of simple cells in cat striate cortex. *Experimental Brain Research, 42*, 89–98.

Heilman, K. M., and Valenstein, E. (1972a). Auditory neglect in man. *Archives of Neurology, 26*, 32–35.

— (1972b). Frontal lobe neglect in man. *Neurology, 22*, 660–664.

Heilman, K. M., Pandya, D. N., and Geschwind, N. (1970). Trimodal inattention following parietal lobe ablations. *Transactions of the American Neurological Association, 95,* 259–261.

Heilman, K. M., Pandya, D. N., Karol, E. A., and Geschwind, N. (1971). Auditory inattention. *Archives of Neurology, 24,* 323–325.

Heilman, K. M., Scholes, R., and Watson, R. T. (1975). Auditory affective agnosia. *Journal of Neurology, Neurosurgery and Psychiatry, 38,* 69–72.

Heilman, K. M., Schwartz, H. D., and Watson, R. T. (1978). Hypoarousal in patients with neglect and emotional indifference. *Neurology, 28,* 229–232.

Heinemann, E. G. (1950). Simultaneous brightness induction as a function of inducing and test-field luminance. *Journal of Experimental Psychology, 50,* 89–96.

Hellige, J. B., Cox, B. J., and Litvac, L. (1979). Information processing in the cerebral hemispheres: Selective activation and capacity limitations. *Journal of Experimental Psychology: General, 108,* 251–279.

Henry, J. P., Ely, D. L., and Stevens, P. M. (1972). Changes in catecholamine-controlling enzymes in response to psychosocial activation of the defence and alarm reactions. *Physiology, Emotion, and Psychosomatic Illness.* Ciba Foundation Symposium No. 8. Amsterdam: Elsevier.

Hernández-Peón, R., Scherrer, H., and Jouvet, M. (1956). Modification of electric activity in cochlear nucleus during "attention" in unanesthetized cats. *Science, 123,* 331–332.

Hess, W. R. (1936). Hypothalamus und die Zantren des autonomen Nervensystems: Physiologie. *Archiv für Psychiatrie und Nervenkrankheiten, 104,* 548–557.

Hier, D. B., Davis, K. R., Richardson, E. P., and Mohr, J. P. (1977). Hypertensive putaminal hemorrhage. *Annals of Neurology, 1,* 152–159.

Hilgard, E. R., Atkinson, R. L., and Atkinson, R. C. (1979). *Introduction to psychology.* New York: Harcourt, Brace, Jovanovich.

Hillyard, S. A., and Munte, T. F. (1984). Selective attention to color and location: an analysis with event-related brain potentials. *Perception and Psychophysics, 36,* 185–198.

Hillyard, S. A., and Kutas, M. (1983). Electrophysiology of cognitive processing. *Annual Review of Psychology, 34,* 33–61.

Hillyard, S. A., Hink, R. F., Schwent, V. L., and Picton, T. W. (1973). Electrical signs of selective attention in the human brain. *Science, 182,* 177–179.

Hillyard, S. A., Picton, T. W., and Regan, D. M. (1978). Sensation, perception, and attention: analysis using ERPs. In E. Calloway, P. Tueting, and I. S. Koslow (eds.), *Event-related brain potentials in man.* New York: Academic Press.

Hillyard, S. A., Simpson, G. V., Woods, D. L., vanVoorhis, G., and Munte, T. (1984). Event-related brain potentials and selective attention to different modalities. In F. Reinoso-Suarez and C. Ajmone-Marson (eds.), *Cortical integration: basic, archicortical, and cortical association levels of neural integration,* Vol. II. IBRO Monograph Series, New York: Raven.

Hilton, S. M. (1966). Hypothalamic regulation of the cardiovascular system. *British Medical Bulletin, 22,* 243–248.

Hilton, S. M. and Zbrozyna, A. W. (1963). Amygdaloid region for defense reactions and its efferent pathway to the brain stem. *Journal of Physiology (London), 165,* 160–173.

Hinton, G. E. (1981). A parallel computation that assigns canonical object-based frames of reference. *Proceedings of the Seventh International Joint Conference on Artificial Intelligence.*

Hinton, G. E., and Anderson, J. A. (eds.) (1981). *Parallel models of associative memory*. Hillsdale, N.J.: Erlbaum.

Hirsh, R. (1974). The hippocampus and contextual retrieval of information from memory: a theory. *Behavioral Biology, 12*, 421–444.

Hirst, W. (1982). The amnesic syndrome: descriptions and explanations. *Psychological Bulletin, 91*, 435–460.

Hirst, W., and Kalmar, D. (1983). Selection in divided attention. Paper presented to the meeting of the American Psychological Association, Los Angeles.

(1984). Evaluating the multiple resource theory. Paper presented at the meeting of the American Psychological Association, Toronto, Canada.

Hirst, W., and Volpe, B. (1984). Automatic and effortful encoding with amnesia. In M. Gazzaniga (ed.), *Handbook of Cognitive Neuroscience*, pp. 369–386, New York: Plenum.

Hirst, W., Spelke, E., Reaves, C., Chaharack, G. and Neisser, U. (1980). Dividing attention without alternation or automaticity. *Journal of Experimental Psychology: General, 109*, 98–117.

Hocherman, S., Benson, D. A., Goldstein, M. H. Jr., Heffner, H. E., and Hienz, R. D. (1976). Evoked unit activity in auditory cortex of monkeys performing a selective attention task. *Brain Research, 117*, 51–68.

Hocherman, S., Itzhaki, A., and Gilat, E. (1981). The response of single units in the auditory cortex of rhesus monkeys to predicted and to unpredicted sound stimuli. *Brain Research, 230*, 65–86.

Hochstein, S., and Shapley, R. M. (1976). Quantitative analysis of retinal ganglion cell classifications. *Journal of Physiology, London, 262*, 237–264.

Hofer, M. A. (1984). Relationships as regulators: A psychobiologic perspective or bereavement. *Psychosomatic Medicine, 46*, 183–198.

Hoffman, J. E. (1980). Interaction between global and local levels of a form. *Journal of Experimental Psychology: Human Perception and Performance, 6*, 222–234.

Hohmann, G. (1966). Some effects of spinal cord lesions on experimental emotional feelings. *Psychophysiology, 3*, 143–156.

Hökfelt, T., Fuxe, K., Goldstein, M., and Johnansson, O. (1974). Immunohisto-chemical evidence for the existence of adrenaline neurons in the rat brain. *Brain Research, 66*, 235–251.

Holtzman, J. and Gazzaniga, M. S. (1982). Dual task interactions due exclusively to limits on processing resources. *Science, 218*, 1325–1327.

Holtzman, J. D., Sidtis, J. J., Volpe, B. T., Wilson, D. W. and Gazzaniga, M. S. (1981). Dissociation of spatial information for stimulus localization and the control of attention. *Brain, 104*, 861–872.

Holway, A. H., and Boring, E. G. (1941). Determinants of apparent visual size with distance variant. *American Journal of Psychology, 54*, 21–37.

Hook, S. (1960). Dimensions of mind. New York: Collier.

Horn, G. (1981). Neural mechanisms of learning: an analysis of imprinting in the domestic chick. *Proceedings of the Royal Society, Britain, 213*, 107–137.

Horn, G., McCabe, B. J., and Bateson, P. P. G. (1979). An autoradiographic study of the chick brain after imprinting. *Brain Research, 168*, 361–373.

Horowitz, L. M., and Manelis, L. (1972). Toward a theory of redintegrative memory: adjective-noun phrases. In G. H. Bower (ed.), *The psychology of learning and motivation*, Vol. 6, New York: Academic Press.

Houck, M. R., and Hoffman, J. E. (In press). Conjunction of color and form without attention. *Journal of Experimental Psychology: Human Perception and Performance*.

Hubel, D. H. and Wiesel, T. N. (1959). Receptive fields of single neurons in the cat's striate cortex. *Journal of Physiology, 148*, 574–591.

(1962). Receptive fields, binocular interaction, and functional architecture in the cat's visual cortex. *Journal of Physiology, London, 160*, 106–154.

(1965). Receptive fields and functional architecture of two nonstriate visual areas (18 and 19) of the cat. *Journal of Neurophysiology, 28*, 229–289.

(1968). Receptive fields and functional architecture of monkey striate cortex. *Journal of Physiology, London, 195*, 215–243.

(1970). Cells sensitive to binocular depth in area 18 of the macaque monkey cortex. *Nature (London), 225*, 41–42.

Hubel, D. H., Henson, C. O., Rupert, A., and Galambos, R. (1959). "Attention" units in the auditory cortex. *Science, 129*, 1279–1280.

Hubel, D. H., Wiesel, T. N., and LeVay, S. (1977). Plasticity of ocular dominance columns in monkey striate cortex. *Philosophical Transactions of the Royal Society of London, E, 278*, 377–409.

Hubel, D. H., Wiesel, T. N., and Stryker, M. P. (1977). Orientation columns in macaque monkey visual cortex demonstrated by the 2-deoxyglucose autoradiographic technique. *Nature (London), 269*, 328–330.

Hyvärinen, J. (1982). Posterior parietal lobe of the primate brain. *Physiological Reviews, 62*, 1060–1129.

Hyvärinen, J., and Poranen, A. (1974). Function of the parietal associative area 7 as revealed from cellular discharges in alert monkeys. *Brain, 97*, 673–692.

Hyvärinen, J., Poranen, A., and Jokinen, Y. (1980). Influence of attentive behavior on neuronal responses to vibration in primary somato-sensory cortex in the monkey, *Journal of Neurophysiology, 43*, 870–882.

Ikeda, H., and Sheardown, M. J. (1983). Transmitters mediating inhibition of ganglion cells in the cat retina: iontophoretic studies in vivo. *Neuroscience, 3*, 837–853.

Illing, R.-B. and Wässle, H. (1978). The retinal projection to the thalamus in the cat: a quantitative investigation and a comparison with the retinotectal pathway. *Journal of Comparative Neurology, 202*, 265–285.

Intraub, H. (1981). Rapid conceptual identification of sequentially presented pictures. *Journal of Experimental Psychology: Human Perception and Performance, 7*, 604–610.

(1985). Visual dissociation: an illusory conjunction of pictures and forms. *Journal of Experimental Psychology: Human Perception and Performance, 11*, 431–442.

Isaacson, R. L. (1982). *The limbic system.* New York: Plenum.

Isaacson, R. L., and Kimble, D. P. (1972). Lesions of the limbic system: their effects upon hypotheses and frustration. *Behavioral Biology, 7*, 767–793.

Izard, C. (1971). *The face of emotion.* New York: Appleton-Century-Crofts.

Jacobs, G.H. (1965). Effects of adaptation on the lateral geniculate response to light increment and decrement. *Journal of the Optical Society of America, 55*, 1535–1540.

Jacoby, L. L. (1983a). Perceptual enhancement: Persistent effects of an experience. *Journal of Experimental Psychology: Learning, Memory, and Cognition, 9*, 21–38.

(1983b). Remembering the data: Analyzing interactive processes in reading. *Journal of Verbal Learning and Verbal Behavior, 22*, 485–508.

(1984). Incidental vs. intentional retrieval: remembering and awareness as separate issues. In N. Butters and L. R. Squire (eds.), *The neuropsychology of memory.* New York: Guilford Press.

Jacoby, L. L., and Dallas, M. (1981). On the relationship between autobiographical memory and perceptual learning. *Journal of Experimental Psychology: General, 100*, 306–340.

Jacoby, L. L. and Witherspoon, D. (1982). Remembering without awareness. *Canadian Journal of Psychology, 36*, 300–324.

James, W. (1884/1968). What is an emotion? *Mind, 9*, 188–205. (Reprinted in M. Arnold, *The nature of emotion*. Baltimore: Penguin, 1968).
___ (1890). *The principles of psychology*. Vol. 1. New York: Holt.
Jarrard, L. E., Isaacson, R. L., and Wickelgren, W. O. (1964). Effects of hippocampal ablation and intertrial interval on acquisition and extinction of a runway response. *Journal of Comparative and Physiological Psychology, 57*, 442–445.
Johnson, K. O., and Lamb, G. D. (1981). Neural mechanisms of spatial tactile discrimination: neural patterns evoked by Braille-like dot patterns in the monkey. *Journal of Physiology, London, 310*, 117–144.
Johnson, K. O., Darian-Smith, I., and LaMotte, C. (1973). Peripheral neural determinants of temperature discrimination in man. A correlative study of responses to cooling pulses. *Journal of Neurophysiology, 36*, 347–370.
Johnson, M. (1983). A multiple-entry, modular memory system. In G. H. Bower (ed.), *The psychology of learning and motivation*, Vol. 17. New York: Academic Press.
Jones, B., and Mishkin, M. (1972). Limbic lesions and the problem of stimulus–reinforcement associations. *Experimental Neurology, 36*, 362–377.
Jones, E. G., and Powell, T. P. S. (1970). An anatomical study of converging sensory pathways within the cerebral cortex of the monkey, *Brain, 93*, 793–820.
Jones, G. V. (1976). A fragmentation hypothesis of memory: cued recall of pictures and sequential position. *Journal of Experimental Psychology: General, 105*, 277–293.
Judge, S. J., Richmond, B. J., and Wurtz, R. H. (1980). Vision during saccadic eye movements. I. Visual interactions in striate cortex. *Journal of Neurophysiology, 43*, 1133–1155.
Julesz, B. (1971). *Foundations of cyclopean perception*. University of Chicago Press.
Julesz, B., and Bergen, J. R. (1983). Textons: The fundamental elements in preattentive vision and perception of textures. *The Bell System Technical Journal, 62*, 1619–1645.
Jung, R. (1974). Neuropsychologie und Neurophysiologie des Kontur und Formsehens im Zeichnung und Malerei. In H. H. Wieck (ed.), *Psychopathologie musicher Gestaltungen*, Stuttgart: Shattauer Verlag.
Kaada, B. R. (1951). Somato-motor, autonomic and electrocoricographic responses to electrical stimulation of "rhinecephalic" and other structures in primates, cats and dog. *Acta Physiologica Scandinavica, 24*, Supp., 83.
Kaas, J. H., Guillery, R. W., and Allman, J. M. (1972). Some principles of organization in the dorsal lateral geniculate nucleus. *Brain, Behavior, and Evolution, 6*, 253–299.
Kaas, J. H., Nelson, R. J., Sur, M., and Merzenich, M. M. (1978). Organization of somatosensory cortex in primates. In F. O. Schmitt, F. G. Worden, G. Adelman, and S. G. Dennis (eds.), *The organization of the cerebral cortex*. Cambridge: MIT Press.
Kahneman, D. (1973). *Attention and effort*. Englewood Cliffs, N.J.: Prentice-Hall.
Kahneman, D., and Treisman, A. M. (1984). Changing views of attention and automaticity. In R. Parasuraman & D. R. Davies (eds.), *Varieties of attention*, pp. 29–62. Orlando, Fla.: Academic Press.
Kalil, R., and Chase, M. (1970). Corticofugal influence on activity of lateral geniculate neurons in the cat. *Journal of Neurophysiology, 33*, 459–474.
Kamin, L. J. (1969). Selective association and conditioning. In N. J. MacKintosh and W. K. Honig (eds.), *Fundamental issues in associative learning*, Halifax: Dalhousie University Press.
Kandel, E. R. (1976). *Cellular basis of behavior*, San Francisco: Freeman.

Kandel, E. R., and Schwartz, J. H. (1982). Molecular biology of learning: modulation of transmitter release. *Science, 218,* 433–443.

Kaneko, A. (1973). Receptive field organization of bipolar and amacrine cells in the goldfish retina. *Journal of Physiology, London, 235,* 133–153.

——— (1979). Physiology of the retina. *Annual Review of Neuroscience, 2,* 169–191.

Kantowitz, B. H. and Knight, J. L., Jr. (1976). On experimenter-limited processes. *Psychological Review, 83,* 502–507.

Kaplan, E., and Shapley, R. M. (1982). X and Y cells in the lateral geniculate nucleus of macaque monkeys. *Journal of Physiology, London, 330,* 125–142.

Kapp, B. S., Gallagher, M., Applegate, C. D., and Frysinger, R. C. (1982). The amygdala central nucleus: contributions to conditioned cardiovascular responding during aversive pavlovian conditioning in the rabbit. In C. D. Woody (ed.), *Advances in Behavioral Biology,* Vol. 26. *Conditioning: Representation of Involved Neural Functions,* pp. 581–600. New York: Plenum.

Kapp, B. S., Pascoe, J. P., and Bixler, M. A. (1984). The amygdala: a neuroanatomical systems approach to its contribution to aversive conditioning. In N. Butters and L. R. Squire (eds.), *The neuropsychology of memory.* New York: Guilford Press.

Karlsen, R. L., and Fonnum, F. Evidence for glutamate as a neurotransmitter in the corticofugal fibers to the dorsal lateral geniculate body and the superior colliculus in rats. *Brain Research, 151,* 457–467.

Kean, M.-L., and Nadel, L. (1982). On the emergence of cognitive neuroscience. In J. Mehler, E. C. T. Walker, and M. Garrett (eds.), *Perspectives on mental representation.* Hillsdale, N.J.: Erlbaum.

Keele, S. W., and Hawkins, H. L. (1982). Explorations of individual differences relevant to high level skill. *Journal of Motor Behavior, 14,* (No. 1), 3–23.

Keenan, J. M., and Baillet, S. D. (1980). Memory for personally and socially significant events. In R. Nickerson (ed.), *Attention and performance VIII.* Hillsdale, N.J.: Erlbaum.

Kelly, J. P., and van Essen, D. C. (1974). Cell structure and function in the visual cortex of cat. *Journal of Physiology, London, 238,* 515–547.

Kelso, J. A. S., Southard, D., and Goodman, D. (1979). On the nature of human interlimb coordination. *Science, 203,* 1029–1031.

Kennard, M. A. (1939). Alternations in response to visual stimuli following lesions of frontal lobe in monkeys. *Archives of Neurology and Psychiatry, 41,* 1153–1165.

Keys, W. R., and Robinson, D. L. (1979). Eye movement-dependent enhancement of visual responses in the pulvinar nucleus of the monkey. *Society for Neuroscience Abstracts, 5,* 791.

Kimble, D. P. (1968). The hippocampus and internal inhibition. *Psychological Bulletin, 70,* 285–295.

Kimble, D. P., and Dannen, E. (1977). Persistent spatial maze learning deficits in hippocampal-lesioned rats across a 7-week postoperative period. *Physiological Psychology, 5,* 409–413.

Kimble, D. P., and Gostnell, D. (1964). The role of cingulate cortex in shock avoidance behavior in the cat. *Journal of Comparative and Physiological Psychology, 58,* 38–46.

Kimble, D. P., and Kimble, R. J. (1970). The effect of hippocampal lesions on extinction and "hypothesis" behavior in rats. *Physiology and Behavior, 5,* 735–738.

Kimble, D. P., and Pribram, K. (1963). Hippocampectomy and behavior sequences. *Science, 139,* 401–407.

Kimura, D. (1958). Effects of selective hippocampal damage on avoidance behavior in the rat. *Canadian Journal of Psychology, 12,* 213–218.

Kinchla, R. A. (1980). The measurement of attention. In R. S. Wickerson (ed.), *Attention and performance*, Hillsdale, N.J.: Erlbaum.

Kinchla, R. A. and Wolfe, J. (1979). The order of visual processing: "top down," "bottom up," or "middle out." *Perception and Psychophysics, 25*, 225–231.

Kinchla, R. A., Solis-Macias, V., and Hoffman, J. E. (1983). Attending to different levels of structure in a visual image. *Perception and Psychophysics, 33*, 1–10.

Kinsbourne, M., and Hicks, R. (1978). Functional cerebral space. In J. Requin (ed.), *Attention and performance VII*, Hillsdale, N.J.: Erlbaum.

Kinsbourne, M. and Wood, F. (1975). Short-term memory processes and the amnesic syndrome. In D. Deutsch and J. A. Deutsch (eds.), *Short-term memory processes*, New York: Academic Press.

Kintsch, W. (1980). Semantic memory: A tutorial. In R. Nickerson (ed.), *Attention and performance VIII*. Hillsdale, N.J.: Erlbaum.

Kintsch, W., and Bates, E. (1977). Recognition memory for statements from a classroom lecture. *Journal of Experimental Psychology: Human Learning, and Memory, 3*, 150–159.

Kintsch, W., and van Dijk, T. A. (1978). Toward a model of text comprehension and production. *Psychological Review, 85*, 363–394.

Kirby, A. W. (1979). The effect of bicuculline and picrotoxin on X and Y cells in the cat retina. *Journal of General Physiology, 74*, 71–84.

Klapp, S. T. (1979). Doing two things at once: the role of temporal compatibility. *Memory and Cognition, 7*, 375–381.

Klatzky, R. L. (1980). *Human memory: structures and processes*. San Francisco: Freeman.

Klein, M. and Kandel, E. R. (1980). Mechanisms of calcium current modulation underlying presynaptic facilitation and behavioral sensitization in *Aplysia. Proceedings of the National Academy of Sciences, U.S.A., 77*, 6912–6916.

Klein, R. (1979). Does oculomotor readiness mediate cognitive control of visual attention? *Attention and Performance VIII*, Hillsdale, N.J.: Erlbaum.

Kluver, H., and Bucy, P. C. (1937). Psychic blindness and other symptoms following bilateral temporal lobectomy in rhesus monkeys. *American Journal of Physiology, 119*, 352–353.

Knibestol, M., and Vallbo, A. B. (1980). Intensity of sensation related to activity of slowly-adapting mechanoreceptive units in the human hand. *Journal of Physiology, London, 300*, 251–267.

Knight, R. T., Hillyard, S. A., Woods, D. L., and Neville, H. J. (1980). The effects of frontal and temporal–parietal lesions on the auditory evoked potential in man. *Electroencephalography and Clinical Neurophysiology, 50*, 112–124.

(1981). The effects of frontal cortex lesions on event-related potentials during auditory selective attention. *Electroencephalography and Clinical Neurophysiology, 52*, 571–582.

Kolers, P. A. (1957). Subliminal stimulation in problem solving. *American Journal of Psychology, 70*, 437–441.

(1975). Memorial consequences of automatized encoding. *Journal of Experimental Psychology: Human Learning and Memory, 1*, 689–701.

(1979). A pattern-analyzing basis of recognition. In L. S. Cermak & F.I.M. Craik (eds.), *Levels of processing in human memory*. Hillsdale, N.J.: Erlbaum.

Komatsu, H. (1982). Prefrontal unit activity during a color discrimination task with go and no-go responses in the monkey. *Brain Research, 244*, 269–277.

Konorski, J. (1967). *The integrative activity of the brain: an interdisciplinary approach.* University of Chicago Press.

Koriat, A., and Lieblich, I. (1974). What does a person in a "TOT" state know that a person in a "don't know" state doesn't know? *Memory & Cognition, 2*, 647–655.

Korner, P. L. (1979). Central nervous control of autonomic cardiovascular function. In *Handbook of physiology, the cardiovascular system*, Vol. 1. Bethesda: American Physiological Society.

Kruger, J., and Gouras, P. (1980). Spectral selectivity of cells and its dependence on slit length in monkey visual cortex. *Journal of Neurophysiology, 43*, 1055–1069.

Kubie, J. L., and Ranck, J. B., (1982). Tonic and phasic firing of rat hippocampal complex-spike cells in three different situations: Context and place. In C. D. Woody (ed.), *Advances in behavioral biology*, Vol. 26, *Conditioning: representation of involved neural functions*, pp. 89–98, New York: Plenum.

Kubie, J. L., Muller, R. U., and Ranck, J. B., Jr. (1983). Manipulations of the geometry of environmental enclosures control the spatial firing patterns of rat hippocampal neurons. *Society for Neuroscience Abstracts, 9*, 646.

Kuffler, S. W. (1953). Discharge patterns and functional organization of mammalian retina. *Journal of Neurophysiology, 16*, 37–68.

Kuffler, S. W., and Nichols, J. G. (1976). *From neuron to brain*. Sunderland: Sinauer Associates.

Kumada, M., Dampney, R. A. L., and Reis, D. J. (1979). Profound hypotension and abolition of the vasomotor component of the cerebral ischemic response produced by restricted lesions of the medulla oblongata: relationship to the so-called tonic vasomotor center. *Circulation Research, 45*, 63–70.

Kunst-Wilson, W. R., and Zajonc, R. B. (1980). Affective discrimination of stimuli that cannot be recognized. *Science, 207*, 557–558.

LaBerge, D. (1975). Acquisition of automatic processing in perceptual and associative learning. In P. M. A. Robbitt and S. Dornic (eds.) *Attention and performance V*. New York: Academic Press.

LaBerge, D., and Samuels, S. J. (1974). Toward a theory of automatic information processing in reading. *Cognitive Psychology, 6*, 293–323.

Lacey, J. I. (1950). Individual differences in somatic response patterns. *Journal of Comparative Physiology and Psychology, 43*, 338–350.

Lacey, J. I., and Lacey, B. C. (1970). Some autonomic-central nervous system interrelationships. In P. Black (ed.), *Physiological correlates of emotion*. New York: Academic Press.

Lacey, J. I., Bateman, D. E., and van Lehn, R. (1953). Autonomic response specificity: an experimental study. *Psychosomatic Medicine, 15*, 8–21.

Lachman, R., Lachman, J. L., and Butterfield, E. C. (1979). *Cognitive psychology and information processing: an introduction*. Hillsdale, N.J.: Erlbaum.

Lackner, J. R., and Garrett, M. F. (1972). Resolving ambiguity: Effect of biasing context in the unattended ear. *Cognition, 1*, 359–372.

Lambert, R. W., and Gabriel, M. (1982). Effects of midline limbic cortical lesions on discriminative avoidance behavior and neuronal activity in the neostriatum. *Society for Neuroscience Abstracts, 8*, 318.

Land, E. H., Hubel, D. H., Livingstone, M., Perry, S. H., and Burns, M. M. (1983). Color-generating interactions across the corpus callosum. *Nature (London), 303*, 616–618.

Landis, C., and Hunt, W. A. (1932). *The startle pattern*. New York: Farrar.

Lange, C.G. (1985). *Om Sindsbevaegelser et psyko. fysiolog. studie*. Copenhagen: Kronar.

Langely, J. N. (1921). *The autonomic nervous system*. Cambridge: Heffer.

Lashley, K. S. (1929). *Brain mechanisms and intelligence: a quantitative study of injuries to the brain*. University of Chicago Press.

(1950). In search of the engram. *Symposium Society for Experimental Biology, 4*, 454–482.

(1951). The problem of serial ordering in behavior. In L. A. Jeffress (ed.), *Cerebral mechanisms of behavior*, pp. 112–136. New York: Wiley.

Lawrence, D. H. (1971). Two studies of visual search for targets with controlled rates of presentation. *Perception and Psychophysics, 10*, 85–89.

Laycock, T. (1860). *Mind and brain*. New York. Appleton.

Lazarus, R. S. (1966). *Psychological stress and the coping process*. New York: McGraw-Hill.

(1982). Thoughts on the relations between emotion and cognition. *American Psychologist, 37*, 1019–1024.

(1984). On the primacy of cognition. *American Psychologist, 39*, 124–129.

Lazarus, R. S., and Alfert, E. (1964). Short-circuiting of threat by experimentally altering cognitive appraisal. *Journal of Abnormal and Social Psychology, 69*, 195–205.

Lazarus, R. S., and Opton, E. M., Jr. (1966). The study of psychological stress: A summary of theoretical formulations and experimental findings. In I. C. D. Spielberger (ed.), *Anxiety and behavior*. New York: Academic Press.

LeDoux, J. E. (1982). Neurorevolutionary mechanisms of cerebral asymmetry in man. *Brain, Behavior and Evolution, 20*, 197–213.

(1984). Cognition and emotion: processing functions and neural systems. In M. S. Gazzaniga (ed.), *Handbook of cognitive neurosciences*, New York: Plenum.

(1985). Mind, brain, language. In D. Oakley (ed.), *Brain and mind*. London: Methuen.

LeDoux, J. E., Barclay, L., and Premack, A. (1980). The brain and cognitive sciences: conference report. *Annals of Neurology, 4*, 391–398.

LeDoux, J. E., Sakaguchi, A., and Reis, D. J. (1984). Subcortical efferent projections of the medulla geniculate nucleus mediate emotional responses conditioned to acoustic stimuli. *J. Neuroscience, 4*, 683–698.

LeDoux, J. E., Wilson, D., and Gazzaniga, M. S. (1979). Beyond commissurotomy: clues to consciousness. In M. S. Gazzaniga (ed.), *Handbook of behavioral neurobiology: Neuropsychology*, Vol. 2, pp. 543–554.

Leiby, C. C., Bender, D. B., and Butter, C. M. (1982). Localization and detection of visual stimuli in monkeys with pulvinar lesion. *Experimental Brain Research, 48*, 449–454.

Lennie, P. (1980). Perceptual signs of parallel pathways. *Philosophical Transactions of the Royal Society of London, B, 290*, 23–37.

Leshner, A. I., Brookshire, K. H., and Stewart, C. N. (1971). The effects of adrenal demedullation on conditioned fear. *Hormones and Behavior, 2*, 43–48.

Lesser, I. M. (1981). A review of the alexithymia concept. *Psychosomatic Medicine, 43*, 531–543.

Lettvin, J. Y., Maturana, H. R., McCullogh, W. S., and Pitts, W. H. (1959). What the frog's eye tells the frog's brain. *Proceedings of the Institute of Radio Engineers, 47*, 1940–1951.

Leventhal, A. G. (1979). Evidence that the different classes of relay cells of the cat's lateral geniculate nucleus terminate in different layers of the striate cortex. *Experimental Brain Research. 37*, 349-372.

Leventhal, A. G., Rodieck, R. W., and Dreher, B. (1981). Retinal ganglion cell classes in old world monkey: morphology and central projections. *Science, 213*, 1139–1142.

Levine, S., and Soliday, S. (1962). An effect of adrenal demedullation on the acquisition of a conditioned avoidance response. *Journal of Comparative and Physiological Psychology, 55*, 214–216.

Lewis, J. L. (1970). Semantic processing of unattended messages using dichotic listening. *Journal of Experimental Psychology, 85*, 225–228.

Linsenmeier, R. A., Frishman, L. J., Jakiela, H. G., and Enroth-Cugell, C. (1982). Receptive field properties of X and Y cells in the cat retina derived from contrast sensitivity measurements. *Vision Research, 22*, 1173–1183.

Linton, M. (1975). Memory for real world events. In D. A. Norman and D. E. Rumelhart (eds.), *Explorations in cognition*. San Francisco: Freeman.

Livingstone, M. S., and Hubel, D. H. (1984). Anatomy and physiology of a color system in the primate visual cortex. *Journal of Neuroscience, 4*, 309–356.

Long, G. M. (1979). The dichoptic viewing paradigm: Do the eyes have it? *Psychological Bulletin, 84*, 391–403.

Lorenz, K. (1966). *On aggression*. New York: Harcourt, Brace, and World.

Lu, S., and Fender, P. H. (1972). The interaction of color and luminance in stereoscopic vision. *Investigative Opthamology, 11*, 482–490.

Lubar, J. F., and Perachio, A. A. (1965). One-way and two-way learning and transfer of an active avoidance response in normal and cingulectomized rats. *Journal of Comparative and Physiological Psychology, 60*, 46–52.

Lucas, M., and Bub, D. (1981). Can practice result in the ability to divide attention between two complex language tasks? *Journal of Experimental Psychology: General, 110*, 495–498.

Lund, J. S., Lund, R. D., Hendrickson, A. E., Bunt, A. H., and Fuchs, A. F. (1975). The origin of efferent pathways from the primary visual cortex, area 17, of the macaque monkey as shown by retrograde transport of horseradish peroxidase. *Journal of Comparative Neurology, 164*, 287–304.

Lund, J. S., Henry, G. H., MacQueen, C. L., and Harvey, A. (1979). Anatomical organization of the primary visual cortex (area 17) of the cat. A comparison with area 17 of the macaque monkey. *Journal of Comparative Neurology, 184*, 599–618.

Luria, A. R. (1982). *Language and cognition*. New York: Wiley.

Lynch, G. and Baudry, M. (1984). The biochemistry of memory: a new and specific hypothesis. *Science, 224*, 1057–1063.

Lynch, J. C. (1980). The functional organization of posterior parietal association cortex. *Behavioral and Brain Sciences, 3*, 485–534.

McCleary, R. A. (1966). Response-modulating functions of the limbic system: initiation and suppression. In E. Stellar and J. M. Sprague (eds.), *Progress in Physiological Psychology*, Vol 1, pp. 209–272, New York: Academic Press.

McClelland, J. L. (1979). On the time relations of mental processes: an examination of systems of processes in cascade. *Psychological Review, 86*, 287–330.

McClelland, J. L., and Miller, J. (1979). Structural factors in figure perception. *Perception and Psychophysics, 26*, 221–229.

McClelland, J. L., and Rumelhart, D. E. (1981). An interactive activation model of context effects in letter perception: Part 1. An account of basic findings. *Psychological Review, 88*, 375–407.

McClurkin, J. W. (1984). Receptive field structure, response properties, and functional organization of the monkey lateral and inferior pulvinar. Doctoral dissertation. University of Oregon.

McClurkin, J. W., and Marrocco, R. T. (1984). Visual cortical input alters spatial tuning in monkey lateral geniculate nucleus cells. *Journal of Physiology, London, 348*, 135–152.

McCollough, C. (1965). Color adaptation of edge-detectors in the human visual system. *Science, 149*, 1115–1116.

McCourt, M. E., and Jacobs, G. H. (1983). Effects of photic environment on the development of spectral response properties of optic nerve fibers in the ground squirrel. *Experimental Brain Research, 49*, 443–452.

McEwen, B. S. (1981). Endocrine effects on the brain and their relation to behavior. In G. J. Siegel, R. W. Alberts, B. Agaroff, and R. Katzman (eds.), *Basic Neurochemistry*, pp. 775–800. Boston: Little, Brown.

McGeoch, J. A., and Irion, A. L. (1952). *The psychology of human learning*. New York: Longmans, Green.

McGuiness, D., and Pribram, K. (1980). The neuropsychology of attention: emotional and motivational controls. In M. C. Wittrock (ed.), *The brain and psychology*. New York: Academic Press.

McIlwain, J. T. (1964). Receptive fields of optic tract axons and lateral geniculate cells: peripheral extent and barbiturate sensitivity. *Journal of Neurophysiology, 27*, 1154–1173.

MacKay, D. G. (1973). Aspects of the theory of comprehension, memory and attention. *Quarterly Journal of Experimental Psychology, 25*, 22–40.

McKoon, G., and Ratcliff, R. (1979). Priming in episodic and semantic memory. *Journal of Verbal Learning and Verbal Behavior, 18*, 463–480.

MacLean, P. D. (1949). Psychomotor disease and the visceral brain. Recent developments bearing on the Papez theory of emotion. *Psychosomatic Medicine, 11*, 338–353.

(1970). The limbic brain in relation to the psychoses. In P. H. Black (ed.), *Physiological correlates of emotion*. New York: Academic Press.

(1973). *A triune concept of the brain and behavior*. Toronto: Toronto University Press.

(1978). A mind of three minds: Educating the triune brain. *Seventy-seventh Yearbook of the National Society for the Study of Education*. Chicago: University of Chicago Press, 308–342.

MacLean, P. D., and Creswell, G. (1970). Anatomical connections of the visual system with limbic cortex of monkey. *Journal of Comparative Neurology, 138*, 265–278.

McQueen, E. W. (1917). The distribution of attention. *The British Journal of Psychology. Monograph Supplements*, Vol. II, Cambridge.

Maffei, L., and Fiorentini, A. (1976). The unresponsive regions of visual cortical receptive fields. *Vision Research, 16*, 1131–1139.

(1977). Spatial frequency rows in the striate visual cortex. *Vision Research, 17*, 257–264.

Malmo, R. B. (1975). *On emotions, needs, and our archaic brain*. New York: Holt, Rinehart & Winston.

Mancia, G., and Zanchetti, A. (1981). Hypothalamic control of autonomic functions. In J. P. Morgane and J. Panskeep (eds.), *Handbook of the hypothalamus: behavioral functions of the hypothalamus*, Vol. 3, pp. 147–202. New York: Dekker

Mandler, G. (1968). Association and organization: facts, fancies, and theories. In T. R. Dixon and D. L. Horton (eds.), *Verbal behavior and general behavior theory*. Englewood Cliffs, N.J.: Prentice-Hall.

(1980). Recognizing: the judgement of previous occurrence. *Psychological Review, 87*, 252–271.

(1975). *Mind and emotion*. New York: Wiley.

(1984). *Mind and body: psychology of emotion and stress*. New York: Norton.

Maranon, G. (1924). Contribution à l'étude de l'action émotive de l'adrenaline. *Revue Française d'Endocrinologie, 2*, 301–325.

Marcel, A. J. (1983a). Conscious and unconscious perception: experiments on visual masking and word recognition. *Cognitive Psychology, 15*, 197–237.

(1983b). Conscious and unconscious perception: an approach to the relations between phenomenal experience and perceptual processes. *Cognitive Psychology, 15*, 238–300.

Marg, E., and Adams, J. E. (1970). Evidence for a neurological zoom system from angular changes in some receptive fields of single neurons with changes in fixation distance in the human visual cortex. *Experientia, 26*, 270–271.

Mark, V. H. and Ervin, F. R. (1970). *Violence and the brain*. New York: Harper & Row.

Markowitsch, H. J., and Pritzel, H. (1976). Learning and the prefrontal cortex of the cat: anatomico-behavioral interrelations. *Physiological Psychology, 4*, 247–261.

Marr, D. (1982). *Vision: a computational investigation into the human representation and processing of visual information*. San Francisco: Freeman.

Marr, D., and Poggio, T. (1976). Cooperative computation of stereo disparity. *Science, 194*, 283–287.

Marrocco, R. T., and Li, R. H. (1977). Monkey superior colliculus: properties of single cells and their afferent inputs. *Journal of Neurophysiology, 40*, 844–860.

Marrocco, R. T., McClurkin, J. W., and Young, R. A. (1981). Spatial properties of superior colliculus cells projecting to the inferior pulvinar and parabigeminal nucleus of the monkey. *Brain Research, 222*, 150–154.

(1982a) Spatial summation and conduction latency classification of cells in the lateral geniculate nucleus of macaques. *Journal of Neuroscience, 2*, 1275–1291.

(1982b). Modulation of LGN cell responsiveness by visual activation of the corticogeniculate pathway. *Journal of Neuroscience, 2*, 256–263.

Marshall, J. F., Turner, B. H., and Teitelbaum, P. (1971). Sensory neglect produced by lateral hypothalamic damage. *Science, 174*, 523–525.

Marslen-Wilson, W. D., and Tyler, L. K. (1980). The temporal structure of spoken language understanding. *Cognition, 8*, 1–71.

Martin, M. (1979). Local and global processing: the role of sparsity. *Memory and Cognition, 7*, 476–484.

Martinez, J. L. (1982). Conditioning: modification by peripheral mechanisms. In C. D. Woody (ed.) *Conditioning: representation of involved neural functions*. New York: Plenum.

Matthews, R. C. (1977). Semantic judgments as encoding operations: the effects of attention to particular semantic categories on the usefulness of interitem relations in recall. *Journal of Experimental Psychology: Human Learning and Memory. 3*, 160–173.

Maunsell, J. H. R., and van Essen, D. C. (1983a). Functional properties of neurons in middle temporal visual area of the macaque monkey. I. Selectivity for stimulus direction, speed, and orientation. *Journal of Neurophysiology, 49*, 1127–1147.

Maunsell, J. H. R., and van Essen, D. C. (1983b). Functional properties of neurons in middle temporal visual area of the macaque monkey. II. Binocular interactions and sensitivity to binocular disparity. *Journal of Neurophysiology, 49*, 1148–1167.

Means, L. W., Walker, D. W., and Isaacson, R. L. (1970). Facilitated single alternation go, no-go acquisition following hippocampectomy in the rat. *Journal of Comparative and Physiological Psychology, 72*, 278–285.

Means, L. W., Leander, J. D., and Isaacson, R. L. (1971). The effect of hippocampectomy on alternation behavior and response to novelty. *Physiology and Behavior, 6*, 17–22.

Mesulam, M. M., van Hoesen, G., Pandya, D. N., and Geschwind, N. (1977). Limbic and sensory connections of the IPL in the rhesus monkey. *Brain Research, 136*, 393–414.

Micco, D. J., and Schwartz, M. (1971). Effects of hippocampal lesions upon the development of Pavlovian internal inhibition. *Journal of Comparative and Physiological Psychology, 76*, 371–177.

Michael, C. R. (1978a). Color vision mechanisms in monkey striate cortex: dual-opponent cells with concentric receptive fields. *Journal of Neurophysiology, 41*, 572–588.

(1978b). Color vision mechanisms in monkey striate cortex: simple cells with dual opponent-color receptive fields. *Journal of Neurophysiology, 41*, 1233–1249.

(1978c). Color-sensitive complex cells in monkey striate cortex. *Journal of Neurophysiology, 41*, 1250–1266.

(1979) Color sensitive hypercomplex cells in monkey striate cortex. *Journal of Neurophysiology, 42*, 726–744.

(1981). Columnar organization of color cells in monkey striate cortex. *Journal of Neurophysiology, 46*, 587–604.

Miezen, F., McGuiness, E., and Allman, J. M. (1982). Antagonistic direction specific mechanisms in area MT in the owl monkey. *Neuroscience Abstracts, 8*, 681.

Mikami, A., Ito, S., and Kubota, K. (1982). Modification of neuron activities of the dorsolateral prefrontal cortex during extrafoveal attention. *Behavioral Brain Research, 5*, 219–223.

Mileunsnic, R., Rose, S.P.R., and Tillson, P. (1980). Passive avoidance learning results in region-specific changes in concentration and incorporation into colchicine-binding proteins in the chick forebrain. *Journal of Neurochemistry, 34*, 1007–1015.

Milgram, S. (1963). Behavioral study of obedience. *Journal of Abnormal and Social Psychology, 67*, 371–378.

Miller, G. A., and Isard, S. (1963). Some perceptual consequences of linguistic rules. *Journal of Verbal Learning and Verbal Behavior, 2*, 217–228.

Miller, G. A. (1956). The magical number seven plus or minus two: some limits on our capacity for processing information. *Psychological Review, 63*, 81–96.

Miller, G. A., and Gazzaniga, M. S. (1984). The cognitive sciences. In M. S. Gazzaniga (ed.), *Handbook of cognitive neuroscience*. New York: Plenum.

Miller, G. A., Galanter, E., and Pribram, K. H. (1960). *Plans and structure of behavior*, New York: Holt, Rinehart & Winston.

Miller, J. (1981). Global precedence in attention and decision. *Journal of Experimental Psychology: Human Perception and Performance, 7*, 1161–1174.

Miller, J. M., Sutton, D., Pfingst, B., Ryan, A., and Beaton, R. (1972). Single cell activity in the auditory cortex of rhesus monkeys: behavioral dependency. *Science, 177*, 449–451.

Miller, N. E. (1969). Learning of visceral and glandular responses. *Science, 163*, 434–445.

Milner, B. (1962). Les troubles de la mémoire accompagnant des lésions hippocampiques bilatérales. In *Physiologie de l'hippocampe*. Paris: Centre National de la Recherche Scientifique.

Milner, B., Corkin, S., and Teuber, H. L. (1968). Further analysis of the hippocampal amnesic syndrome: 14 year follow-up study of H. M. *Neuropsychologia, 6*, 215–234.

Minsky, M. (1975). A framework for representing knowledge. In P. H. Winston (ed.), *The psychology of computer vision*. New York: McGraw-Hill.

Mishkin, M. (1978). Memory in monkeys severely impaired by combined but not separate removal of amygdala and hippocampus. *Nature (London), 273*, 297–298.

(1979). Analogous neural models for tactual and visual learning. *Neuropsychologia, 17*, 139–151.

Mishkin, M., and Aggleton, J. (1981). Multiple functional contributions of the amygdala in the monkey. In Y. Ben-Ari (ed.) *The amygdaloid complex*. Amsterdam: Elsevier.

Mishkin, M., Malamut, B., and Bachevalier, J. (1984). In J. L. McGaugh, G. Lynch, and N. M. Weinberger (eds.), *Neurobiology of learning and memory*. New York: Guilford Press.

Montagu, A. (ed.) (1968). *Man and aggression*. Oxford University Press.

Moran, T., and Desimone, R. (1985). Selective attention gates: visual processing in the extrastriate cortex. *Science, 229*, 782–784.

Moray, N. (1959). Attention in dichotic listening: Affective cues and the influence of instructions. *Quarterly Journal of Experimental Psychology, 11*, 56–60.

(1979). *Mental workload: its theory and measurement*. New York: Plenum.

Moray, N., and O'Brien, T. (1967). Signal-detection theory applied to selective listening. *Journal of the Accoustical Society of America, 42*, 765–772.

Moray, N., Fritter, M., Ostry, D., Faveau, D., and Nagy, V. (1976). Attention to pure tones. *Quarterly Journal of Experimental Psychology, 28*, 271–283.

Morris, C. D., Bransford, J. D., and Franks, J. J. (1977). Levels of processing versus transfer appropriate processing. *Journal of Verbal Learning and Verbal Behavior, 16*, 519–533.

Morris, P. E., Gruneberg, M. M., Sykes, R. N., and Merrick, A. (1981). Football knowledge and the acquisition of new results. *British Journal of Psychology, 72*, 479–483.

Morrow, L., Vatunski, P. B., Kim, Y., and Boller, F. (1981). Arousal responses to emotional stimuli and laterality of lesion. *Neuropsychologica, 19*, 65–71.

Morrone, C. M., Burr, D. C., and Maffei, L. (1982). Functional implications of cross-orientation inhibition of cortical visual cells. I. Neurophysiological evidence. *Proceedings of the Royal Society of London, B., 216*, 335–354.

Morton, J. (1969). Interaction of information in word recognition. *Psychological Review, 76*, 165–178.

(1979). Facilitation in word recognition: experiments causing change in the logogen models. In P. A. Kolers, M. E. Wrolstad, and H. Bouma (eds.), *Processing of visible language*, Vol. 1. New York: Plenum.

Moruzzi, G. and Magoun, H. W. (1949). Brainstem reticular formation and activation of the EEG. *Electroencephalography and Clinical Neurophysiology, 1*, 455–473.

Moscovitch, M. (1982). Multiple dissociations of functions in amnesia. In L. S. Cermak (ed.), *Human memory and amnesia*. Hillsdale, N.J.: Erlbaum.

(1984). The sufficient conditions for demonstrating preserved memory in amnesia: A task analysis. In N. Butters and L. R. Squire (eds.), *The neuropsychology of memory*. New York: Guilford Press.

Mountcastle, V. B., Andersen, R. A., and Motter, B. C. (1981). The influence of attentive fixation upon the excitability of the light-sensitive neurons of the posterior parietal cortex. *Journal of Neuroscience, 1*, 1218–1235.

Mountcastle, V. B., Talbot, W. H., and Yin, T. C. T. (1977). Parietal lobe mechanisms for directed visual attention. *Journal of Neurophysiology, 40*, 362–389.

Movshon, J. A. (1975). The velocity tuning of single units in cat striate cortex. *Journal of Physiology, London, 249*, 445–468.

Movshon, J. A., Thompson, I. D., and Tolhurst, D. J. (1978a). Spatial summation in the receptive fields of simple cells in the cat's striate cortex. *Journal of Physiology, London, 283*, 53–77.

(1978b). Receptive field organization of complex cells in the cat's striate cortex. *Journal of Physiology, London, 283*, 79–99.

Mowrer, O. H. (1960). *Learning theory and behavior*. New York: Wiley.

Moyer, K. E. (1958). Effect of adrenalectomy on anxiety motivated behavior. *Journal of Genetic Psychology, 92*, 11–16.

Moyer, K. E., and Bunnell, B. N. (1958). Effect of injected adrenalin in an avoidance response in the rat. *Journal of Genetic Psychology, 92*, 247–251.

Murakami, M., and Shimoda, Y. (1977). Identification of amacrine and ganglion cells in the carp retina. *Journal of Physiology, London, 264*, 801–818.

Murdock, B. B., Jr. (1982). A theory for the storage and retrieval of item and associative information. *Psychological Review, 89*, 609–626.

Myers, R. E., and Sperry, R. W. (1953). Interocular transfer of a visual form discrimination habit in cats after section of the optic chiasm and corpus callosum *Anatomical Record, 175*, 351 352.

Näätänen, R. (1975). Selective attention and evoked potentials in humans – a critical review. *Journal of Biological Psychology, 2*, 237–307.

Näätänen, R., and Michie, P. T. (1979). Early selective-attention effects on the evoked potential: a critical review and reinterpretation. *Journal of Biological Psychology, 8*, 81–136.

Nauta, W. J. (1972). The central viscero-sensory system: a general survey. In C. H. Hockman (ed.), *Limbic system mechanisms and autonomic function*. Springfield, Ill.: Thomas.

Nauta, W. J., and Haymaker, W. (1969). Hypothalamic nuclei and fiber connections. In W. Haymaker, E. Anderson, and W. J. Nauta (eds.), *The hypothalamus*. Springfield, Ill.: Thomas.

Navon, D. (1977). Forest before trees: the precedence of global features in visual perception. *Cognitive Psychology, 9*, 353–383.

Navon, D. (1984). Resources – A theoretical soupstone. *Psychological Review, 91*, 216–234.

Navon, D. and Gopher, D. (1979). On the economy of the human processing system. *Psychological Review, 86*, 214–255.

Neff, W. D. (1961). Neural mechanisms of auditory discrimination. In W. A. Rosenblith (ed.), *Sensory communication*, pp. 259–278. Cambridge, Mass.: MIT Press.

Neisser, U. (1963). The multiplicity of thought. *British Journal of Psychology, 54*, 1–14.

(1964). Visual Search. *Scientific American, 210*, 94–101.

(1967). *Cognitive psychology*. New York: Appleton-Century-Crofts.

(1976). *Cognition and reality: principles and implications of cognitive psychology*, San Francisco: Freeman.

(1978). Memory: What are the important questions? In M. M. Grunlberg, P. E. Morris, and R. N. Sykes (eds.), *Practical aspects of memory*. London: Academic Press.

(1981). John Dean's memory: a case study. *Cognition, 9*, 1–22.

(1982). *Memory observed*. San Francisco: Freeman.

Neisser, U., and Becklen, R. (1975). Selective looking: attending to visually specified events. *Cognitive Psychology, 7*, 480–494.

Neisser, U., Hirst, W., and Spelke, B. (1981). Limited capacity theories and the motion of automaticity. *Journal of Experimental Psychology: General, 10*, 499–500.

Neisser, U., Novick, R., and Lazar, R. (1963). Searching for ten targets simultaneously. *Perceptual and Motor Skills, 17*, 955–961.

Nelson, R. J. (1984). Responsiveness of monkey primary somatosensory cortical neurons to peripheral stimulation depends on "motor-set." *Brain Research, 304*, 143–148.

Nelson, T. O. (1977). Repetition and depth of processing. *Journal of Verbal Learning and Verbal Behavior, 16*, 151–171.

Nelson, T. O., and Narens, L. (1980). A new technique for investigating the feeling of knowing. *Acta Psychologica, 19*, 69–80.

Nelson, T. O., Leonesia, R. J., Shimamura, A. P., Landwehr, R. F., and Narens,

L. (1982). Overlearning and the feeling of knowing. *Journal of Experimental Psychology: Learning, Memory, and Cognition, 8*, 279–288.

Nemiah, J. C., and Sifneos, P. E. (1970). Psychosomatic illness: a problem in communication. *Psychotherapy and Psychosomatics, 18*, 154–160.

Newell, A. (1972). In A. W. Melton and E. Martin (eds.), *Coding processes in human memory*, Washington, D.C.: Winston.

Newell, A. (1973). You can't play 20 questions with nature and win. In W. G. Chase (ed.), *Visual information processing*. New York: Academic Press.

Newsome, W. T., Wurtz, R. H., Dürsteller, M. R., and Mikami, A. (1983). Deficits in pursuit eye movements after chemical lesions of motion-related visual areas in the superior temporal sulcus of the macaque monkey. *Neuroscience Abstracts, 9*, 154.

Newstead, S. E., and Dennis, I. (1979). Lexical and grammatical processing of unshadowed messages: A reexamination of the Mackay effect. *Quarterly Journal of Experimental Psychology, 31*, 447–488.

Noble, C. E., and McNeely, D. A. (1957). The role of meaningfulness (m) in paired-associate verbal learning. *Journal of Experimental Psychology, 53*, 16–22.

Noda, H., and Adey, R. (1974). Excitability changes in cat lateral geniculate cells during saccadic eye movements. *Science, 183*, 543–545.

Norita, M., and Kawamura, K. (1980). Subcortical afferents to the monkey amygdala. *Brain Research, 190*, 225–230.

Norman, D. A. (1968). Towards a theory of memory and attention. *Psychological Review, 75*, 522–536.

(1969). Memory while shadowing. *Quarterly Journal of Experimental Psychology, 21*, 85–93.

(1981). *Perspectives on cognitive science*. Hillsdale, N.J.: Erlbaum.

Norman, D. A., and Bobrow, D. G. (1975). On data-limited and resource-limited processes. *Cognitive Psychology, 7*, 44–64.

(1976). On the analysis of performance-operating characteristics. *Psychological Review, 83*, 508–510.

Nothdurft, H. C., and Lee, B. B. (1982). Responses to coloured patterns in the macaque lateral geniculate nucleus: pattern processing in single neurons. *Experimental Brain Research, 48*, 43–54.

Nottebohm, F. (1977). Assymetries in neural control of vocalization in the canary. In S. Harnad, R. W. Doty, L. Goldstein, J. Jaynes, and G. Krauthamer (eds.), *Lateralization in the nervous system*, pp. 23–44. New York: Academic Press.

Oakley, D. A. (1981). Brain mechanisms of mammalian memory. *British Medical Bulletin, 37*, 175–180.

(1983). The varieties of memory: a phylogenetic approach. In A. Mayes (ed.), *Memory in animals and humans*, New York: Van Nostrand Reinhold.

Oatman, L. C. (1971). Role of visual attention on auditory evoked potentials in unanesthetized cats. *Experimental Neurology, 32*, 341–356.

(1976). Effects of visual attention on the intensity of auditory evoked potentials. *Experimental Neurology, 51*, 41–53.

Ogren, M. P., and Hendrickson, A. E. (1979). The structural organization of the inferior and lateral subdivisions of the macaca monkey pulvinar. *Journal of Comparative Neurology, 188*, 147–178.

O'Keefe, J., and Bouma, H. (1969). Complex sensory properties of certain amygdala units in freely moving cat. *Experimental Neurology, 23*, 384–398.

O'Keefe, J., and Nadel, L. (1978). *The hippocampus as a cognitive map*. Oxford University Press.

Olds, J. E., and Milner, P. (1954). Positive reinforcement produced by electrical stimulation of the septal area and other regions of the rat brain. *Journal of Comparative Physiology and Psychology, 47*, 419–427.

Olds, J. (1977). *Drives and reinforcement*. New York: Raven.

Olton, D. S. (1984a). Comparative analysis of episodic memory. *The Behavioral and Brain Sciences, 7*, 250–251.

(1984b). *Learning and memory: neuropsychological and ethological approaches to its classification*, Paper presented to Conference on Human and Animal Memory, Umea, Sweden.

Olton, D. S., Becker, J. T., and Handelmann, G. E. (1979). Hippocampus, space, and memory. *The Behavioral and Brain Sciences, 2*, 313–365.

Olton, D. S., Walker, J. A., and Gage, F. H. (1978). Hippocampal connections and spatial discrimination. *Brain Research, 139*, 295–308.

Orban, G. A., Kennedy, H., and Maes, H. (1981). Response to movement of neurons in areas 17 and 18 of the cat: velocity sensitivity. *Journal of Neurophysiology, 45*, 1043–1058.

Orem, J., Schlag-Rey, M., and Schlag, J. (1973). Unilateral visual neglect and thalamic intralaminar lesions in the cat. *Experimental Neurology, 40*, 734–797.

Orona, E., and Gabriel, M. (1983). Unit activity of the prefrontal cortex and the mediodorsal thalamic nucleus during acquisition of discriminative avoidance behavior in rabbits. *Brain Research, 263*, 295–312.

Ostry, D., Moray, N., and Marks, G. (1976). Attention, practice, and semantic targets. *Journal of Experimental Psychology: Human Perception and Performance, 2*, 326–336.

Overton, D. A. (1968). Dissociated learning in drug states (state-dependent learning). In *Psychopharmacology, a review of progress 1957–1967*.

(1972). State-dependent learning produced by alcohol and its relevance to alcoholism. In B. Kissin and H. Begleiter (eds.) *The biology of alcoholism*, Vol. 2, *Physiology and behavior*, 193–217. New York: Plenum.

Paivio, A., Smythe, P. C., and Yuille, J. C. (1968). Imagery versus meaningfulness of nouns in paired-associate learning. *Canadian Journal of Psychology, 22*, 427–441.

Palmer, L. A., and Rosenquist, A. C. (1974). Visual receptive fields of single striate cortical units projecting to the superior colliculus in the cat. *Brain Research, 67*, 27–42.

Palmero, D. (1971). Is a scientific revolution taking place in psychology? *Science Studies, 7*, 135–155.

Pankseep, J. (1981). Hypothalamic integration of behavior. In P. J. Morgane and J. Pankseep (eds.), *Handbook of the hypothalamus: behavioral studies of the hypothalamus*. Vol. 3, pp. 289–431. New York: Dekker.

Papez, J. W. (1937). A proposed mechanism of emotion. *Archives of Neurology and Psychiatry, 38*, 725–743.

Pare, W. P., and Cullen, J. W. (1971). Adrenal influences on the aversive threshold and CBR acquisition. *Hormones and Behavior, 2*, 139–147.

Pascoe, J. P., and Kapp, B. S. (1983). Electrophysiological characteristics of amygdaloid central nucleus neurons in the awake rabbit. *Society for Neuroscience Abstracts, 9*, 222.

Pavlov, I. P. (1927). *Conditioned reflexes*. Oxford University Press.

Peichl, L., and Wässle, H. (1983). The structural correlate of the receptive field centre of alpha ganglion cells in the cat retina. *Journal of Physiology, London, 341*, 309–324.

Penfield, W. (1958). Centrencephalic integrating system. *Brain, 81*, 231–234.

Perry. R. B. (1927). General theory of value. Cited in W. B. Cannon, The James–Lange theory of emotion: a critical examination and an alternative theory. *American Journal of Psychology, 39*, 106–124.

Petersen, S. E., and Robinson, D. L. (1983). Two types of behavioral enhancement of visual responses in the pulvinar of alert rhesus monkeys. *Investigative Ophthalmology and Visual Science, 24*, 106.

Petersen, S. E., Morris, J. D., and Robinson, D. L. (1984). Modulation of attentional behavior by injection of GABA-related drugs into the pulvinar of macaque. *Society for Neuroscience Abstracts, 10*, 475.

Petersen, S. E., Robinson, D. L., and Keys, W. (1982). A physiological comparison of the lateral pulvinar and area 7 in the behaving macaque. *Society for Neuroscience Abstracts, 8*, 681.

Piaget, J. (1971). Piaget's theory. In P. Mussen (ed.), *Handbook of child development*, Vol. 1. New York: Wiley.

Pickard, G. E., and Silverman, A. J. (1981). Direct retinal projections to the hypothalamus piriform cortex, and accessory optic nuclei in the golden hamster as demonstrated by a sensitive anterograde horseradish peroxidase technique. *Journal of Comparative Neurology, 196*, 155–172.

Pinker, S. (1984). Visual cognition: an introduction. *Cognition, 18*, 1–63.

Poggio, G. F., and Fischer, B. (1977). Binocular interaction and depth sensitivity in striate and prestriate cortex of behaving rhesus monkey. *Journal of Neurophysiology, 40*, 1392–1405.

Plutchick, R. (1980). *Emotion: a psychoevolutionary synthesis*. New York: Harper & Row.

Poletti, C. E., Kliot, M., and Boytin, G. (1984). Metabolic influences of the hippocampus on hypothalamus, preoptic and basal forebrain is exerted through amygdalofugal pathways. *Neuroscience Letters, 45*, 211–216.

Pollen, D. A., and Ronner, S. F. (1981). Phase relationship between adjacent simple cells in the visual cortex. *Science, 212*, 1409–1411.

(1982). Spatial computation performed by simple and complex cells in the visual cortex of the cat. *Vision Research, 22*, 101–118.

Pollen, D. A., Lee, J. R., and Taylor, J. H. (1971). How does the striate cortex begin the reconstruction of the visual world? *Science, 173*, 74–77.

Polyak, S. (1957). *The vertebrate visual system*. University of Chicago Press.

Pomerantz, J. (1981). Perceptual organization in information processing. In M. Kubovy and J. Pomerantz (eds.), *Perceptual organization*. Hillsdale, N.J.: Erlbaum.

Pomerantz, J., and Sager, L. (1975). Asymmetric integrality with dimensions of visual pattern. *Perception and Psychophysics, 18*, 460–466.

Pomerantz, J. R. (1983). Global and local precedence: Selective attention in form and motion perception. *Journal of Experimental Psychology: General, 112*, 516–540.

Popper, K. R., and Eccles, J. C. (1977). *The self and its brain*. Berlin: Springer.

Poranen, A., and Hyvärinen, J. (1982). Effects of attention on multi-unit responses to vibration in the somatosensory regions of the monkey's brain. *Electroencephalography and Clinical Neurophysiology, 53*, 525–537.

Posner, M. I. (1980). Orienting of attention. *Quarterly Journal of Experimental Psychology, 32*, 3–25.

(1984). Neural control of the direction of covert visual orienting. University of Oregon Technical Report 84–4, August, 1984.

Posner, M. I., and Klein, R. (1973). On the functions of consciousness. In S. Kornblum (ed.), *Attention and performance*, Vol. IV. New York: Academic Press.

Posner, M. I., and Snyder, C. R. R. (1975). Attention and cognitive control. In R. Solso (ed.), *Information processing and cognition: The Loyola Symposium.* Potomac, Md.: Erlbaum.

Posner, M. I., Cohen, Y., and Rafal, R. D. (1982). Neural systems control of spatial orienting. *Philosophical Transactions of the Royal Society, London B, 298,* 187–198.

Posner, M., Nissen, M. J., and Ogden, W. C. (1978). Attended and unattended processing modes: The role of set for spatial location. In H. L. Pick and I. J. Saltzman (eds.), *Modes of perceiving and processing information.* Hillsdale, N.J.: Erlbaum.

Posner, M. I., Pea, R., and Volpe, B. (1982). Cognitive-neuroscience: developments toward a science of synthesis. In J. Mehler, E. C. T. Walker, and M. Garrett (eds.), *Perspectives on mental representation.* Hillsdale, N.J.: Erlbaum.

Posner, M. I., Walker, J. A., Friedrich, F. J., and Rafal, R. D. (1984) Effects of parietal injury on covert orienting of visual attention. *Journal of Neuroscience, 4,* 1863–1874.

Potter, M. C. (1975). Meaning in visual search. *Science, 187,* 965–966.

(1976). Short-term conceptual memory for pictures. *Journal of Experimental Psychology: Human Learning and Memory, 2,* 509–522.

Powell, D. A., Mankowski, J., and Buchanan, S. (1978). Concomitant heart rate and corenoretinal potential conditioning in the rabbit (*Onycytologus ciniculus*): effect of caudate lesions. *Physiology of Behavior, 20,* 143–150.

Previc, F. H., and Harter, M. F. (1982). Electrophysiological and behavioral indicants of selective attention to multifeature gratings. *Perception and Psychophysics, 32,* 465–472.

Pribram, K., and McGuinness, D. (1975). Arousal, attention, and effort in the control of attention. *Psychological Review, 82,* 116–149.

Puff, C. R. (ed.). (1982). *Handbook of research methods in human memory and cognition.* New York: Academic Press.

Pylyshyn, Z. W. (1981). The imagery debate: analogue media versus tacit knowledge. *Psychological Review, 87,* 16–45.

Raaijmakers, J. G. W., and Shiffrin, R. N. (1981). Search of associative memory. *Psychological Review, 88,* 93–135.

Radna, R. J., and MacLean, P. D. (1981). Vagal elicitation of respiratory-type and other unit responses in basal limbic structures of squirrel monkeys. *Brain Research, 213,* 45–61.

Ragsdale, D., Sparenborg, S. P., Stolar, N., Mills, S., and Gabriel, M. (1984). Lesions in the cingulate cortex and neuronal activity in the anterior thalamus during conditioning in rabbits. *Society for Neuroscience Abstracts, 10,* 127.

Ramachandran, V. S., and Gregory, R. L. (1978). Does color provide an input to human motion perception? *Nature (London), 275,* 55–56.

Ramon y Cajal, S. (1909). *Histologie du système nerveux de l'homme et des vertébrés.* Paris: Maloine.

Ratcliff, R. (1978). A theory of memory retrieval. *Psychological Review, 85,* 59–108.

(1979). Group reaction time distributions and an analysis of distribution statistics. *Psychological Bulletin, 86,* 446–461.

Reddy, D. R., Erman, L. D., Fennell, R. D., and Neely, R. B. (1973). The HEARSAY speech understanding system: an example of the recognition process. *Proceedings of the Third International Conference on Artificial Intelligence,* Stanford, Calif.

Reeves, A. G., and Hagaman, W. D. (1971). Behavioral and EEG asymmetry following unilateral lesions of the forebrain and midbrain in cats. *Electroencephalography and Clinical Neurophysiology, 30,* 83–86.

Regan, D., Beverley, K., and Cynader, M. (1979). The visual perception of motion in depth. *Scientific American, 241*, 136–151.

Reicher, G. M. (1969). Perceptual recognition as a function of meaningfulness of stimulus material. *Journal of Experimental Psychology, 81*, 274–280.

Reis, D. J. (1974). Central neurotransmitters and aggression. *Research Publication of the Association for Research in Nervous and Mental Diseases, 52*, 119–148.

(1980). The nucleus tractus solitarii (NTS) and experimental neurogenic hypertension. In M. J. Hughes and C. D. Barnes (eds.) *Neural control of circulation.* New York: Academic Press.

Remington, R. (1980). Visual attention, detection, and the control of eye movements. *Journal of Experimental Psychology: Human Perception and Performance, 6*, 724–744.

Rescorla, R. A., and Solomon, R. L. (1967). Two-process learning theory: Relationships between Pavlovian conditioning and instrumental learning, *Psychological Review, 74*, 151–182.

Richard, D., Gioanni, Y., Kitsikis, A., and Buser, P. (1975). A study of geniculate unit activity during cryogenic blockade of the primary visual cortex in the cat. *Experimental Brain Research, 22*, 235–242.

Richardo, J. A., and Koh, E. T. (1978). Anatomical evidence of direct projections from the nucleus of the solitary tract to the hypothalamus, amygdala, and other forebrain structures in the rat. *Brain Research, 153*, 1–26.

Richmond, B. J., and Sato, T. (1982). Visual responses of inferior temporal neurons are modified by attention to different stimulus dimensions. *Society for Neuroscience Abstracts, 3*, 812.

Richmond, B. J., Wurtz, R. H., and Sato, T. (1983). Visual responses of inferior temporal neurons in awake rhesus monkey. *Journal of Neurophysiology, 50*, 1415–1432.

Rizzolatti, G., Matelli, M., and Pavesi, G. (1983). Deficits in attention and movement following the removal of postarcuate (area 6) and prearcuate (area 8) cortex in macaque monkeys. *Brain, 106*, 655–673.

Rizzolatti, G., Scandolara, C., Gentilucci, M., and Camarda, R. (1981). Response properties and behavioral modulation of "mouth" neurons of the postarcuate cortex (area 6) in macaque monkeys. *Brain Research, 225*, 421–424.

Robinson, D. L., and Petersen, S. E. (1984). Posterior parietal cortex of the awake monkey: visual responses and their modulation by behavior. In F. Reinoso-Suarez and C. Ajmone-Marsan (eds.), *Cortical integration: basic, archicortical, and cortical association levels of neural integration*, Vol. II, IBRO Monograph Series. New York: Raven.

Robinson, D. L., Baizer, J. S., and Dow, B. M. (1980). Behavioral enhancement of visual responses of prestriate neurons of the rhesus monkey. *Investigative Ophthalmology and Visual Science, 19*, 1120–1123.

Robinson, D. L., and Wurtz, R. H. (1976). Use of extraretinal signal by monkey superior colliculus to distinguish real from self-induced stimulus movement. *Journal of Neurophysiology, 39*, 852–870.

Robinson, D. L., Goldberg, M. E., and Stanton, G. B. (1978). Parietal association cortex in the primate: sensory mechanisms and behavioral modulations. *Journal of Neurophysiology, 41*, 910–932.

Robinson, D. L., Morris, J. D., and Petersen, S. E. (1984). Cued visual behavior and the pulvinar of the awake macaque. *Investigative Ophthalmology and Visual Science, 25*, 33.

Robinson, J. A. (1976). Sampling autobiographical memory. *Cognitive Psychology, 8*, 578–595.

Robson, J. G. (1975). Receptive fields: neural representation of the spatial and intensive attributes of the visual image. In E. C. Carterette and M. P. Friedman (eds.), *Handbook of perception*. New York: Academic Press.

Rock, I. (1975) *An introduction to perception*. New York: Macmillan.

(1983). *The logic of perception*. Cambridge, Mass.: MIT Press.

Rodieck, R. W. (1973). *The vertebrate retina*. San Francisco: Freeman.

(1979). Visual pathways. *Annual Review of Neuroscience, 2*, 193–225.

Rodieck, R. W., and Stone, J. (1965). Analysis of receptive fields of cat retinal ganglion cells. *Journal of Neurophysiology, 28*, 819–832.

Roediger, H. L., III. (1980). Memory metaphors in cognitive psychology. *Memory & Cognition, 8*, 231–236.

Rogers, L. J., and Anson, J. N. (1979). Lateralisation of function in the chick forebrain. *Pharmacology, Biochemistry and Behavior, 10*, 679–686.

Rogers, T. B., Kuiper, N. A., and Kirker, W. S. (1977). Self-reference and the encoding of personal information. *Journal of Personality and Social Psychology, 35*, 677–688.

Roland, P.E. (1982). Cortical regulation of selective attention in man. A regional cerebral blood flow study. *Journal of Neurophysiology, 48*, 1059–1078.

Rollins, H. A., and Hendricks, R. (1980). *Journal of Experimental Psychology: Human Perception and Performance, 6*, 99–109.

Rolls, E. T. (1976). The neurophysiological basis of brain-stimulation reward. In A. Wanguier and E. T. Rolls (eds.), *Brain-stimulation reward*. Amsterdam: North Holland.

Rolls, E. T., Perrett, D., Thorpe, S. J., Puerto, A., Roper-Hall, A., and Maddison, S. (1979). Responses of neurons in area 7 of the parietal cortex to objects of different significance. *Brain Research, 169*, 194–198.

Rorty, R. (1979). *Philosophy and the mirror of nature*. Princeton University Press.

Rose, D. (1977). Responses of single units in cat visual cortex to moving bars of light as a function of bar length. *Journal of Physiology, London, 271*, 1–17.

Rose, S.P.R. (1981). What should a biochemistry of learning and memory be about? *Neuroscience, 6*, 811–821.

Rose, S.P.R., and Harding, S. (1984). Training increases tritium-labeled fucose incorporation in chick brain only if followed by memory storage. *Neuroscience, 12*, 663–667.

Rose, S.P.R., Gibbs, M.E., and Hambley, J. (1980). Transient increase in forebrain muscarinic cholinergic receptor binding following passive avoidance learning in the young chick. *Neuroscience, 5*, 169–172.

Ross, C. A., Ruggiero, D. A., Park, D. H., Joh, T. H., Sved, A. F., and Fernandez-Pardal, J. (1984). Tonic vasomotor control by the rostral ventrolateral medulla: effect of electrical or chemical stimulation of the area containing C1 adrenalin-neurons on arterial pressure, heart rate and plasma catecholamines and vasopressin. *Journal of Neuroscience, 4*, 474–494.

Rossi, G., and Rosadini, G. (1967). Experimental analysis of cerebral dominance in man. In C. H. Millikan and F. L. Darley (eds.). *Brain mechanisms underlying speech and language*. New York: Grune & Stratton.

Routtenberg. A. (1971). Stimulus processing and response execution: a neurobehavioral theory. *Physiology and Behavior, 6*, 589–596.

(1982). Memory formation as a posttranslational modification of brain proteins. In Ed. C. Ajmone-Marsan and H. Matthies (eds.), *Neuronal plasticity and memory formation*, pp. 17–24. New York: Raven.

(1984). Brain phosphoproteins kinase C and protein F1: Protagonists of plasticity in particular pathways. In G. Lynch, J. McGaugh, and N. Weinberger (eds.), *Neurobiology of learning and memory*. New York: Guilford.

Routtenberg, A., Lovinger, D. M., and Steward, O. (1985). Synaptic plasticity predicted by the phosphorylation state of a 47RD protein. *Behavioral and Neural*

Biology, 43, 3–11.

Rowe, M. H., and Stone, J. (1977). Naming of neurons. *Brain, Behavior, and Evolution, 14*, 185–216.

Rumelhart, D. E., and McClelland, J. L. (1982). An interactive activation model of context effects in letter perception: Part 2. The contextual enhancement effect and some tests and extensions of the model. *Psychological Review, 89*, 60–94.

Russell, B., (1903) *The principles of mathematics*. London: Allen & Unwin.

(1912). *Problems of philosophy*. Oxford University Press.

(1921). *The analysis of mind*. London: Allen & Unwin.

(1927/1961). *An outline of philosophy*. Cleveland: World. Reprinted in 1961.

Ryan, A., and Miller, J. (1977). Effects of behavioral performance on single-unit firing patterns in inferior colliculus of the rhesus monkey. *Journal of Neurophysiology, 40*, 943–956.

Ryle, G. (1949). *The concept of mind*. London: Hutchinson.

Sackeim, H. A., Gur, R. C., and Saucy, M. C. (1978). Emotions are expressed more intensely on the left side of the face. *Science, 202*, 434–435.

Safer, M. A. (1981). Sex and hemisphere differences in access to codes for processing emotional expressions and faces. *Journal of Experimental Psychology: General, 110*, 86–100.

Safer, M. A., and Leventhal, H. (1977). Ear differences in evaluating emotional tones of voice and verbal content. *Journal of Experimental Psychology: Human Perception and Performance, 3*, 75–82.

Saito, H. (1983a). Morphology of physiologically identified X-, Y-, and W-type retinal ganglion cells of the cat. *Journal of Comparative Neurology. 1983, 221*, 279–288.

(1983b). Pharmacological and morphological differences between X- and Y-type ganglion cells in the cat's retina. *Vision Research. 23*, 1299–1308.

Sakaguchi, A., LeDoux, J. E., Sved, A., and Reis, D. J. (1984). Strain differences in fear between spontaneously hypertensive and normotensive rats is mediated by adrenal cortical hormones. *Neuroscience Letters, 46*, 53–58.

Sakata, H., Shibutani, H., and Kwano, K. (1983). Functional properties of visual tracking neurons in the posterior parietal association cortex of the monkey. *Journal of Neurophysiology, 49*, 1364–1380.

Samuels, A. G. (1981a). Phonemic restoration: insights from a new methodology. *Journal of Experimental Psychology: General, 110*, 474–494.

(1981b). The role of bottom-up confirmation in the phonemic restoration illusion. *Journal of Experimental Psychology: Human Perception and Performance, 7*, 1124–1131.

Saper, C. B. (1982). Convergence of autonomic and limbic connections in the insular cortex of the rat. *Journal of Comparative Neurology, 210*, 163–173.

Saper, C. B., Lowey, A. D., Swanson, L. W., and Cowan, W. M. (1976). Direct hypothalmo-autonomic projections. *Brain Research, 117*, 305–312.

Scarborough, D. L., Cortese, C., and Scarborough, H. S. (1977). Frequency and repetition effects in lexical memory. *Journal of Experimental Psychology: Human Perception and Performance, 3*, 1–17.

Scarborough, D. L., Gerard, L., and Cortese, C. (1979). Accessing lexical memory: the transfer of word repetition effects across tasks and modality. *Memory & Cognition, 7*, 3–12.

Schachter, S. (1964). The interaction of cognitive and physiological determinants of emotion state. In L. Berkowitz (ed.), *Advances in experimental social psychology*, Vol. 1. New York: Academic Press.

(1975). Cognition and peripheralist-centralist controversies in motivation and emotion. In M. S. Gazzaniga and C. Blakemore (eds.), *Handbook of Psychobiology*, pp. 529–564. New York: Academic Press.

Schachter, S., and Singer, J. (1962). Cognitive, social, and physiological determinants of emotional state. *Psychological Review, 69*, 379–399.

Schacter, D. L. (1982). *Stranger behind the engram: theories of memory and the psychology of science*. Hillsdale, N.J.: Erlbaum.

(1983a). Feeling of knowing in episodic memory. *Journal of Experimental Psychology: Learning, Memory, and Cognition, 9*, 39–54.

(1983b). Amnesia observed: remembering and forgetting in a natural environment. *Journal of Abnormal Psychology, 92*, 236–242.

(1984). Toward the multidisciplinary study of memory: ontogeny, phylogeny, and pathology of memory systems. In N. Butters and L. R. Squire (eds.), *The neuropsychology of memory*. New York: Guilford.

(1985a). Priming of old and new knowledge in amnesic patients and normal subjects. *Annals of the New York Academy of Sciences, 444*, 41–53.

(1985b). Multiple forms of memory in humans and animals. In J. McGaugh, G. Lynch, and N. Weinberger (eds.), *Proceedings of the Second Conference on the Neurobiology of Learning and Memory*. New York: Guilford.

Schacter, D. L., and Crovitz, H. F. (1977). Memory function after closed head injury: a review of the quantitative research. *Cortex, 13*, 150–176.

Schacter, D. L., and Graf, P. (In press). Preserved learning in amnesic patients: Perspectives from research on direct priming. *Journal of Clinical and Experimental Neuropsychology*.

Schacter, D. L., and Moscovitch, M. (1984). Infants, amnesics, and dissociable memory systems. In M. Moscovitch (ed.), *Infant memory*. New York: Plenum.

Schacter, D. L., and Tulving, E. (1982a). Amnesia and memory research. In L. S. Cermak (ed.), *Human memory and amnesia*. Hillsdale, N.J.: Erlbaum.

(1982b). Memory, amnesia, and the episodic/semantic distinction. In R. L. Isaacson and N. E. Spear (eds.), *The expression of knowledge*. New York: Plenum.

Schacter, D. L., and Worling, J. (1985). Attribute information and the feeling of knowing. *Canadian Journal of Psychology, 39*, 467–475.

Schacter, D. L., Eich, J. E., and Tulving, E. (1978). Richard Semon's theory of memory. *Journal of Verbal Learning and Verbal Behavior, 17*, 721–743.

Schacter, D. L., Harbluk, J. A., and McLachlan, D. R. (1984). Retrieval without recollection: an experimental analysis of source amnesia. *Journal of Verbal Learning and Verbal Behavior, 23*, 593–611.

Schacter, D. L., Wang, P. L., Tulving, E., and Freedman, M. (1982). Functional retrograde amnesia: a quantitative case study. *Neuropsychologia, 20*, 523–532.

Schallert, T., Whishaw, I., Ramirez, V., and Teitelbaum, P. (1978). Compulsive abnormal walking caused by anticholinergics in akinetic, 6-hydroxydopamine-treated rats. *Science, 199*, (4336), 1461–1463.

Schein, S. J., Desimone, R., and de Monastcrio, F. M. (1983). Spectral properties of area V4 cells of macaque monkey. *Investigative Ophthalmology and Visual Science, 24*, 107.

Schein, S. J., Marrocco, R. T., and de Monasterio, F. M. (1982). Is there a high concentration of color-selective cells in area V4 of monkey visual cortex? *Journal of Neurophysiology, 47*, 193–213.

Scherer, K. (1984). Les émotions: fonctions et composantes (Emotions: functions and components). *Cahiers de Psychologie Cognitive, 4*, 9–39.

Schiller, P. H., and Malpeli, J. G. (1977). The effect of striate cortex cooling on area 18 cells in the monkey. *Brain Research, 126*, 366–369.

Schiller, P. H., Finlay, B. L., and Volman, S. F. (1976a). Quantitative studies of single cell properties in monkey striate cortex. I. Spatio-temporal organization of receptive fields. *Journal of Neurophysiology, 39*, 1288–1319.

(1976b). Quantitative studies of single-cell properties in monkey visual cortex. III. Spatial frequency. *Journal of Neurophysiology, 39*, 1334–1351.

Schmaltz, L. W., and Theios, J. (1972). Acquisition and extinction of a classically conditioned response in hippocampectomized rabbits (*Oryctolagus cuniculus.*) *Journal of Comparative and Physiological Psychology, 79,* 328–333.

Schneider, W., and Fisk, A. D. (1982). Concurrent automatic and controlled visual search: Can processing occur without resource cost? *Journal of Experimental Psychology: Learning, Memory, and Cognition, 8,* 261–278.

Schneider, W. E., and Shiffrin, R. M.. (1977). Controlled and automatic human information processing: I. Detection, search and attention. *Psychological Review, 84,* 1–66.

Schulman, G. L., Remington, R. W., and McLean, J. P. (1979). Moving attention through space. *Journal of Experimental Psychology: Human Perception and Performance, 5,* 522–526.

Schultzberg, M. (1983). The peripheral nervous system. In P. C. Emson (ed.), *Chemical neuroanatomy,* New York: Raven.

Schustack, M. W., and Anderson, J. R. (1979). Effect of analogy to prior knowledge on memory for new information. *Journal of Verbal Learning and Verbal Behavior, 18,* 565–583.

Schwaber, J. S., Kapp, B. S. and Higgins, G. (1978). The origin and extent of direct amygdala projections to the region of the dorsal motor nucleus of the vagus and the nucleus of the solitary tract. *Neuroscience Letters, 20,* 15–20.

Schwartz, E. L., Desimone, R., Albright, T. O., and Gross, C. G. (1985). Shape recognition and inferior temporal neurons. *Proceedings of the National Academy of Science, 80,* 5776–5778.

Schwartz, G. E., Davidson, R. J. and Maer, F. (1975). Right hemisphere lateralization for emotion in the human brain: interactions with cognition. *Science, 190,* 286–288.

Schwent, V. L., Hillyard, S. A., and Galambos, R. (1976a). Selective attention and the auditory vertex potential. I. Effects of stimulus delivery rate. *Electroencephalography and Clinical Neurophysiology, 40,* 604–614.

(1976b). Selective attention and the auditory vertex potential. II. Effects of signal intensity and masking noise. *Electroencephalography and Clinical Neurophysiology, 40,* 615–622.

Scoville, W. B., and Milner, B. (1957). Loss of recent memory after bilateral hippocampal lesions. *Journal of Neurology, Neurosurgery, and Psychiatry, 20,* 11–21.

Searle, J. (1978). *Brain and mind: Ciba foundation symposium 69.* Amsterdam: Excerpta Medica.

Seibel, R. (1963). Discrimination reaction time for a 1.023-alternative task. *Journal of Experimental Psychology, 66,* 215–226.

Seidenberg, M. S., Tanenhaus, M. K., Leiman, J. M., and Bienkowski, M. (1982). Automatic access to the meanings of ambiguous words in context: some limitations of knowledge-based processing. *Cognitive Psychology, 14,* 489–537.

Sekuler, R., Wilson, H. R., and Owsley, C. (1984). Structural modeling of spatial vision. *Vision Research, 24,* 689–700.

Selye, H. (1950). *The physiology and pathology of exposure to stress.* Montreal: Acta. (1978). *The stress of life.* New York: McGraw-Hill.

Shaffer, L. H. (1975). Multiple attention in continuous verbal tasks. In P. Rabbit and S. Dornic (eds.), *Attention and performance* Vol. V. New York: Academic Press.

Shallice, T. (1979). Neuropsychological research and the fractionation of memory systems. In L.-G. Nilsson (ed.), *Perspectives on memory research.* Hillsdale, N.J.: Erlbaum.

Shavit, Y., Lewis, J. W., Terman, G. W., Gale, R. P., and Liebeskind, J. C. (1983). The effects of stress and morphine on immune function in rats. *Society for Neuroscience Abstracts, 9,* 117.

Shaw, R. E., and Pittenger, J. (1977). Perceiving the face of change in changing faces: implications for a theory of object perception. In R. E. Shaw and J. Bransford (eds.), *Perceiving, acting and knowing*. Hillsdale, N.J.: Erlbaum.

Shearer, S. L., and Tucker, D. M. (1981). Differential cognitive contributions of the cerebral hemispheres in the modulation of emotional arousal. *Cognitive Theory and Research, 5*, 85–93.

Sherk, H. (1978). Area 18 cell responses in cat during reversible inactivation of area 17. *Journal of Neurophysiology, 41*, 204–215.

Sherman, S. M., Wilson, J. R., Kaas, J. H., and Webb, S. V. (1976). X- and Y-cells in the dorsal lateral geniculate nucleus of the owl monkey (*Aotes trivirgatus*). *Science, 192*, 475–477.

Sherrington, C. S. (1900). Experiments on the value of vascular and visceral factors for the genesis of emotion. *Proceedings of the Royal Society, 66*, 390–403.

Shiffrin, R. M., and Dumais, S. T. (1981). The development of automatism. In J. R. Anderson (ed.), *Cognitive skills and their acquisition*. Hillsdale, N.J.: Erlbaum.

Shiffrin, R. M., and Schneider, W. (1977). Controlled and automatic human information processing: II. Perceptual learning, automatic attending, and a general theory. *Psychological Review, 84*, 128–190.

(1984). Automatic and controlled processing revisited. *Psychological Review, 91*, 269–276.

Siegel, A., and Edinger, H. (1981). Neural control of aggression and rage behavior. In P. J. Morane and J. Pankseep (eds.), *Handbook of the hypothalamus: behavioral studies of the hypothalamus*, Vol. 3, pp. 203–232. New York: Dekker.

Siegel, A., and Flynn, J. P. (1968). Differential effects of electrical stimulation and lesions of the hippocampus and adjacent regions upon attack behaviors in cats. *Brain Research, 7*, 252–267.

Siegel, A., and Skog, D. (1970). Effects of electrical stimulation of the septum upon attack behavior elicited from the hypothalamus in the cat. *Brain Research, 23*, 371–380.

Sillito, A. M. (1975). The contribution of inhibitory mechanisms to the receptive field properties of neurons in the striate cortex of the cat. *Journal of Physiology, London, 250*, 305–329.

Sillito, A. M., and Versiani, V. (1977). The contribution of excitatory and inhibitory inputs to the length of preference of hypercomplex cells in layers II and III of the cat's striate cortex. *Journal of Physiology, London, 273*, 775–790.

Singer, W., and Bedworth, N. (1973). Inhibitory interaction between X and Y units in the cat lateral geniculate nucleus. *Brain Research, 49*, 291–307.

Singer, W., and Creutzfeldt, O. D. (1970). Reciprocal lateral inhibition of on- and off-center neurons in the lateral geniculate body of the cat. *Experimental Brain Research, 10*, 311–330.

Slaughter, M. M., and Miller, R. F. (1982). 2-Amino-4-phosphonobutyric acid: a new pharmacological agent for retina research. *Science, 211*, 182–185.

Smith, G. J. W., Spence, D. P., and Klein, G. S. (1959). Subliminal effects of verbal stimuli. *Journal of Abnormal and Social Psychology, 59*, 167–77.

Smith, G. P. (1974). Adrenal hormones and emotional behavior. In E. Stallar and J. M. Sprague (eds.), *Progress in physiological psychology*, Vol. 5, pp. 299–352. New York: Academic Press.

Smith, O. A. (1974). Reflex and central mechanisms involved in the control of the heart and circulation. *Annual Review of Physiology, 36*, 93–123.

Smith, O. A., and DeVito, J. L. (1984). Central neural integration for the control of autonomic responses associated with emotion. *Annual Review of Neuroscience, 7*, 43–65.

Smith, O. A., Astley, C. A., DeVito, J. L., Stein, J. M., and Walsh, R. E. (1980). Functional analysis of hypothalamic control of the cardiovascular responses accompanying emotional behavior. *Federation Proceedings, 29*, 2487–2494.

Sokoloff, L., Reivich, M., Kennedy, C., De Rosiers, M. H., Pattak, C. S., Pettigrew, K. C., Sakurada, O., and Shinohara, M. (1977). The [¹⁴C]-deoxyglucose method for measurement of local cerebral glucose utilization: theory, procedure, and normal values in the conscious and anesthetized albino rat. *Journal of Neurochemistry, 28*, 897–916.

Solomon, P. R. (1977). Role of the hippocampus in blocking and conditional inhibition of the rabbit's nictitating membrane response. *Journal of Comparative and Physiological Psychology, 91*, 407–417.

Solomon, P. R., and Moore, J. W. (1975). Latent inhibition and stimulus generalization of the classically conditioned nictitating membrane response in rabbits (*Oryctolagus cuniculus*) following dorsal hippocampal ablations. *Journal of Comparative and Physiological Psychology, 89*, 1192–1203.

Solomon, R. L., and Corbit, J. P. (1974). An opponent-process theory of motivation. *Psychological Review, 81*, 119–145.

Solomons, L., and Stein, G. (1896). Normal motor automatism. *Psychological Review, 3*, 492–512.

Sourkes, T. L. (1981). Psychopharmacology. In G. J. Seigel, R. W. Albers, B. W. Agranoff, and R. D. Katzman (eds.), *Basic neurochemistry*, pp. 737–758. Boston: Little, Brown.

Spatz, W. B., Tigges, J., and Tigges, M. (1970). Subcortical projections, cortical associations, and some intrinsic interlaminar connections in the striate cortex of the squirrel monkey. *Journal of Comparative Neurology, 140*, 155–174.

Spear, N., Smith, G. J., Bryan, R. G., Gordon, W. C., Timmons, R., and Chiszar, D. A. (1980). Contextual influences on the interaction between conflicting memories in the rat. *Animal Learning & Behavior, 8*, 273–281.

Spear, N. E. (1978). *The processing of memories*, Hillsdale, N.J.: Erlbaum.

Spelke, E., Hirst, W., and Neisser, U. (1976). Skills of divided attention. *Cognition, 4*, 215–230.

Sperling, G. (1960). The information available in brief visual presentations. *Psychological Monographs, 74*.

(1984). A unified theory of attention and signal detection. In R. Parasuramen and D. R. Davies (eds.), *Varieties of attention*, pp. 103–182. Orlando, Fla.: Academic Press.

Sperry, R. W. (1968). A modified concept of consciousness. *Psychological Review, 76*, 532–536.

Sperry. R. W., Gazzaniga, M. S., and Bogen, J. E. (1969). Interhemispheric relationships: the neocortical commissures; syndromes of hemisphere disconnection. In P. J. Vinken and G. W. Bruyn (eds.), *Handbook of clinical neurology*, Vol. 4. Amsterdam: North Holland.

Sperry, R. W., Zaidel, E., and Zaidel, D. (1979). Self-recognition and social awareness in the disconnected minor hemisphere. *Neuropsychologia, 17*, 153–166.

Spiesman, J. C., Lazarus, R. S., Mordkoff, A. M., and Davison, L. A. (1964) Experimental reduction of stress based on ego-defense theory. *Journal of Abnormal and Social Psychology, 68*, 367–380.

Spillich, G. J., Vesonder, G. T., Chiesi, H. L., and Voss, J. F. (1979). Text processing of domain-related information for individuals with high and low domain knowledge. *Journal of Verbal Learning and Verbal Behavior, 18*, 279–290.

Spoehr, K. T., and Lehmkuhle, S. W. (1982). *Visual information processing*. San Francisco: Freeman.

Sprague, J. M., and Meikle, T. H. (1965). The role of the superior colliculus in visually guided behavior. *Experimental Neurology, 11*, 115–146.

Sprague, J. M., Chambers, W. W., and Stellar, E. (1961). Attentive, affective and adaptive behavior in the cat. *Science, 133*, 165–173.

Sprague, J. M., Hughes, H. C., and Berlucchi, G. (1981). Cortical mechanisms in pattern and form perception. In O. Pompeiano and C. Ajmone-Marsan (eds.), *Brain mechanisms and perceptual awareness*. New York: Raven.

Squire, L. R. (1982). The neuropsychology of human memory. *Annual Review of Neuroscience, 5*, 241–273.

Squire, L. R., and Cohen, N. J. (1984). Human memory and amnesia. In J. McGaugh, G. Lynch, and N. Weinberger (eds.), *Proceedings of the conference on the neurobiology of learning and memory*. New York: Guilford.

Squire, L. R., and Zola-Morgan, S. (1983). The neurology of memory: The case for correspondence between the findings for human and nonhuman primates. In J. A. Deutsch & D. Deutsch (eds.), *The physiological basis of memory*, pp. 199–268. New York: Academic Press.

Stanford, L. R., Friedlander, M. J., and Sherman, S. M. (1983). Morphological and physiological properties of geniculate W-cells of the cat: a comparison with X- and Y-cells. *Journal of Neurophysiology, 50*, 582–608.

Starr, A., and Phillips, L. (1970). Verbal and motor memory in the amnesic syndrome. *Neuropsychologia, 8*, 75.

Sternberg, S. (1969). Memory scanning: Mental processes revealed by reaction-time experiments. *American Scientist, 57*, 421–457.

Stewart, M. G., Rose, S.P.R., King, T. S., Gabbott, P.L.A., and Bourne, R. (1984). Hemispheric asymmetry of synapses in chick medial hyperstriatum ventrale following passive avoidance training: a stereological investigation. *Developmental Brain Research, 12*, 261–269.

Stone, J., and Dreher, B. (1973). Projection of X- and Y-cells of the cat's lateral geniculate nucleus to areas 17 and 18 of visual cortex. *Journal of Neurophysiology, 36*, 551–567.

(1982). Parallel processing of information in the visual pathways. *Trends in Neuroscience, 5*, 441–446.

Stone, J., Dreher, B., and Leventhal, A. G. (1979). Hierarchical and parallel mechanisms in the organization of visual cortex. *Brain Research Reviews, 1*, 345–394.

Stone, J., and Fukuda, Y. (1974). Properties of cat retinal ganglion cells: a comparison of W-cells with X- and Y-cells. *Journal of Neurophysiology, 37*, 722–748.

Suga, N., Niwa, H., and Taniguchi, I. (1983). Representation of biosonar information in the auditory cortex of the mustached bat, with emphasis on representation of target velocity information. In J. P. Ewert, R. Capranica, and D. J. Ingle (eds.), *Advances in vertebrate neuroethology*. New York: Plenum.

Sumal, K. K., Blessing, W. W., Joh, T. H., Reis, D. J., and Pickel, V. M. (1982). Ultrastructrual evidence that vagal afferents terminate on catecholaminergic neurons in the nucleus tractus solitarius. *Neuroscience Abstracts, 8*, 429.

Sur, M., and Sherman, S. M. (1982). Linear and nonlinear W-cells in C-laminae of the cat's lateral geniculate nucleus. *Journal of Neurophysiology, 47*, 869–884.

Sverko, B. (1977). Individual differences in time sharing performance. Technical Report ARL–77–4/AFSOR–77–4, Aviation Research Laboratory, University of Illinois, Urbana-Champaign.

Sverko, B., Jerneic, A., and Kulenovic, A. (1983). A contribution to the investigation of time-sharing ability. *Ergonomics, 26*, (No. 2), 151–160.

Swanson, L. W. (1981). The hippocampus and the concept of the limbic system. In W. Seifert (ed.), *Neurobiology of the hippocampus*, London: Academic Press.

Swanson, L. W., Cowan, W. M., and Jones, E. G. (1974). An autoradiographic study of the efferent connections of the ventral lateral geniculate nucleus of the albino rat and cat. *Journal of Comparative Neurology, 156*, 143–163.

Swanson, L. W., Teyler, T. J., and Thompson, R. F. (1982). *Hippocampal long-term potentiation: mechanisms and implications for memory*, Neurosciences Research Program Bulletin 20. Cambridge, Mass.: MIT Press.

Swinney, D. (1979). Lexical access during sentence comprehension: (Re)consideration of content effects. *Journal of Verbal Learning and Verbal Behavior, 18*, 645–660.

Talman, W. T., Granata, A. R., and Reis, D. J. (1984). Glutamatergic mechanisms in the nucleus tractus solitarius in blood pressure control. *Federation Proceedings, 43*, 39–44.

Tavris, C. (1974). A conversation with Stanley Milgram. *Psychology Today, 8*, 71–80.

Taylor, D. A. (1976). Stage analysis of reaction time. *Psychological Bulletin, 83*, 161–191.

Teasdale, J. D., and Fogarty, S. J. (1979). Differential effects of included mood on retrieval of pleasant and unpleasant events from episodic memory. *Journal of Abnormal Psychology, 88*, 248–257.

Terzian, H. (1964). Behavioural and EEG effects of intracarotid sodium amytal injections. *Acta Neurochirurgica (Wien), 12*, 230–239.

Terzian, H., and Ceccotto, C. (1959). Su un nuova metoda per la determinazione e lo studio della dominanza emisferica. *Giornale di Psichiat. Neuropat., 87*, 889–924.

Thompson, R., Chetta, H., and LeDoux, J. E., (1974). Brightness discrimination loss after lesions of the corpus striatum in the white rat. *Bulletin of the Psychonomic Society, 4*, 78–80.

Thompson, R., Lesse, H., and Rich, I. (1963). Pretectal lesions in rats and cats. *Journal of Comparative Neurology, 121*, 161–171.

Thompson, R. F., McCormick, D. A., Lavond, D. G., Clark, G. A., Kettner, R. E., and Mauk, M. D. (1983). The engram found? Initial localization of the memory trace for a basic form of associative learning. *Progress in Psychobiology and Physiological Psychology, 10*, 167–196.

Thomson, D. M., and Tulving, E. (1970). Associative encoding and retrieval: weak and strong cues. *Journal of Experimental Psychology, 86*, 255–262.

Till, R. E. (1977). Sentence memory prompted with inferential recall cues. *Journal of Experimental Psychology: Human Learning and Memory, 3*, 129–141.

Todd, J. T., and Mingolla, E. (1983). Perception of surface curvature and direction of illumination from patterns of shading. *Journal of Experimental Psychology: Human Perception and Performance, 9*, 583–595.

Tolhurst, D. J. (1973). Separate channels for the analysis of the shape and the movement of a moving visual stimulus. *Journal of Physiology, London, 231*, 385–402.

Tolhurst, D. J., and Thompson, I. D. (1981). On the variety of spatial frequency selectivities shown by neurons in area 17 of the cat. *Proceedings of the Royal Society of London, B, 213*, 183–199.

Tomkins, S. (1962). *Affect, imagery, and consciousness: The positive affects*, Vol. 1. New York: Springer.

(1963). *Affect, imagery, and consciousness: The negative affects*, Vol. 2. New York: Springer.

(1982). Affect theory. In P. Ekman (ed.), *Emotion in the human face* (2nd ed.). Cambridge University Press.

Tootell, R. B., Silverman, M., and De Valois, R. L. (1981). Spatial frequency columns in primary visual cortex. *Science, 214*, 813–815.

Treisman, A. M. (1960). Contextual cues in selective listening. *Quarterly Journal of Experimental Psychology, 12*, 242–248.

(1964). Selective attention in man. *British Medical Bulletin, 20*, 12–16.

(1969). Strategies and models of selective attention. *Psychological Review, 76,* 282–299.

(1979). The psychological reality of levels of processing. In L. S. Cermak, and F.I.M. Craik (eds.). *Levels of processing and human memory.* Hillsdale, N.J.: Erlbaum.

(1985). Properties, Parts, and Objects. In K. Boff, L. Kaufman, and J. Thomas (eds.), *Handbook of perception and human performance.* New York: Wiley.

Treisman, A., and Davies, A. (1973). Divided attention between eye and ear. In S. Kornblum (ed.), *Attention and performance,* Vol. IV. New York: Academic Press.

Treisman, A. M., and Geffen, G. (1968). Selective attention and cerebral dominance in perceiving and responding to speech messages. *Quarterly Journal of Experimental Psychology, 20,* 139–151.

Treisman, A., and Gelade, G. (1980). A feature integration theory of attention. *Cognitive Psychology, 12,* 97–136.

Treisman, A., and Paterson, R. (1984). Emergent features, attention, and object perception. *Journal of Experimental Psychology: Human Perception and Performance, 10,* 12–31.

Treisman, A., and Schmidt, H. (1982). Illusory conjunctions in the perception of objects. *Cognitive Psychology, 14,* 107–141.

Treisman, A., Nauires, R., and Green, J. (1974). Semantic processing in dichotic listening: a replication. *Memory and Cognition, 2,* 641–646.

Tretter, F., Cynader, M., and Singer, W. (1975). Cat parastriate cortex: a primary or secondary visual area? *Journal of Neurophysiology, 38,* 1099–1113.

Tucker, D., and Williamson, P. (1984). Asymmetric neural control systems in human self-regulation. *Psychological Review, 91,* 185–215.

Tucker, D. M. (1981). Lateral brain function, emotion, and conceptualization. *Psychological Bulletin, 89,* 19–46.

Tucker, D. M., and Newman, J. P. (1981). Lateral brain function and the cognitive inhibition of emotional arousal. *Cognitive Therapy and Research, 5,* 197–202.

Tucker, D. M., Roth, R. S., Arneson, B. A., and Buckingham, V. (1977). Right hemisphere activation during stress. *Neuropsychologist, 15,* 697–700.

Tucker, D. M., Watson, R. T., and Heilman, K. M. (1977). Discrimination and evocation of affectively intoned speech in patients with right parietal disease. *Neurology, 27,* 947–950.

Tulving, E. (1962). Subjective organization in free recall of "unrelated" words. *Psychological Review, 69,* 344–354.

(1972). Episodic and semantic memory. in E. Tulving and W. Donaldson (eds.), *Organization of memory.* New York: Academic Press.

(1976). Ecphoric processes in recognition and recall. In J. Brown (ed.), *Recall and recognition.* London: Wiley.

(1979). Memory research: What kind of progress? In L.-G. Nilsson (ed.), *Perspectives on memory research.* Hillsdale, N.J.: Erlbaum.

(1983). *Elements of episodic memory.* New York: Oxford University Press.

(1984). Multiple learning and memory systems. In K. M. J. Lagerspetz and P. Niemi (eds.), *Psychology in the 1990's.* Amsterdam: Elsevier Science.

Tulving, E., and Madigan, S. A. (1970). Memory and verbal learning. *Annual Review of Psychology, 21,* 437–484.

Tulving, E., and Thomson, D. M. (1973). Encoding specificity and retrieval processes in episodic memory. *Psychological Review, 80,* 352–373.

Tulving, E., Schacter, D. L., and Stark, H. A. (1982). Priming effects in word-fragment completion are independent of recognition memory. *Journal of Experimental Psychology: Learning, Memory, and Cognition, 8,* 336–342.

424   References

Turner, B. H., Mishkin, M., and Knapp, M. (1980). Organization of the amygdalopetal projections from modality-specific cortical association areas in the monkey. *Journal of Comparative Neurology, 191*, 515–543.

Underwood, B. J. (1969). Attributes of memory. *Psychological Review, 76*, 559–573.

Underwood, B. J., Runquist, W. N., and Schwartz, R. W., (1959). Response learning in paired-associate lists as a function of intra-list similarity. *Journal of Experimental Psychology, 58*, 70–78.

Underwood, G. (1974). Moray vs. the rest: the effect of extended shadowing practice. *Quarterly Journal of Experimental Psychology, 26*, 368–372.

Underwood, G., and Moray, N. (1971). Shadowing and monitoring for selective attention. *Quarterly Journal of Experimental Psychology, 23*, 284–295.

Ungerleider, L. G., and Christensen, C. A. (1977). Pulvinar lesions in monkeys produce abnormal eye movements during visual discrimination training. *Brain Research, 136*, 189–196.

   (1979). Pulvinar lesions in monkeys produce abnormal scanning of a complex visual array. *Neuropsychologia, 17*, 493–501.

Ungerleider, L. G., and Mishkin, M. (1982). Two cortical visual systems. In D. J. Ingle, M. A. Goodale, and R. J. W. Mansfield (eds.), *Analysis of visual behavior*. Cambridge, Mass.: MIT Press.

Ungerstedt, U. (1974). Brain dopamine neurons and behavior. In F. O. Schmitt and F. G. Worden (eds.), *The neurosciences. Third Study Program*, pp. 695–703. Cambridge, Mass.: MIT Press.

Unvas, B. (1960). Central cardiovascular control. *Handbook of Physiology, 2*, 1131.

Ursin, H., and Kaada, B. R. (1960). Functional localization within the amygdaloid complex in the cat. *Electroencephalography and Clinical Neurophysiology, 12*, 1–20.

Ursin, H., Linck, P., and McCleary, R. A. (1969). Spatial differentiation of avoidance deficit following septal and cingulate lesions. *Journal of Comparative and Physiological Psychology, 68*, 74–79.

Uttal, U. R. (1981). *A taxonomy of visual processes*. Hillsdale, N.J.: Erlbaum.

Vanderwolf, C. H., and Robinson, T. E. (1981). Reticulo-cortical activity and behavior: a critique of arousal theory and a new synthesis. *The Behavioral and Brain Sciences, 4*, 459–514.

van Essen, D. C. (1979). Visual areas of the mammalian cerebral cortex. *Annual Review of Neuroscience, 2*, 227–263.

van Essen, D. C., and Maunsell, J. H. R. (1983). Hierarchical organization and functional streams in the visual cortex. *Trends in Neuroscience, 6*, 370–375.

van Essen, D. C., Maunsell, J. H. R., and Bixby, J. L. (1981). The middle temporal area in the macaque: myeloarchitecture, connections, functional properties, and topographic organization. *Journal of Comparative Neurology, 199*, 293–326.

van Hoesen, G. W., Pandya, D. N., and Butters, N. (1972). Cortical afferents to the entorhinal cortex of the rhesus monkey. *Science, 175*, 1471–1473.

Van Voorhis, S., and Hillyard, S. A. (1977). Visual evoked potentials and selective attention to points in space. *Perceptual Psychophysiology, 22*, 54–62.

Vogt, B. A., and Peters, A. (1981). Form and distribution of neurons in rat cingulate cortex: areas 32, 24, and 29. *Journal of Comparative Neurology, 195*, 603–625, *200*, 461.

Vogt, B. A., Rosene, D. L., and Peters, A. (1981). Synaptic termination of thalamic and callosal afferents in cingulate cortex. *Journal of Comparative Neurology, 201*, 265–283.

Volpe, B. T., LeDoux, J., and Gazzaniga, M. S. (1979). Information processing of visual stimuli in an "extinguished" field. *Nature (London), 228*, 722–724.

Voss, J. F., Tyler, S. W., and Bisanz, G. L. (1982). Prose comprehension and memory. In C. R. Puff (ed.), *Handbook of research methods in human memory and cognition*. New York: Academic Press.

Wagner, A. R. (1979). Habituation and memory. In A. Dickinson and R. A. Boakes (eds.), *Mechanisms of learning and motivation*, pp. 53–82. Hillsdale, N.J.: Erlbaum.

Walk, R. D. (1965). The study of visual depth and distance perception in animals, In D. S. Lehrman, R. A. Hinde, and E. Shaw (eds.), *Advances in the study of behavior*. New York: Academic Press.

Walker, S. F. (1980). Lateralization of function in the vertebrate brain: a review. *British Journal of Psychology, 71*, 329–367.

Wallach, H. (1948). Brightness constancy and the nature of achromatic colors. *Journal of Experimental Psychology, 38*, 310–324.

Walters, D., Biederman, I., and Weisstein, N. (1983). The combination of spatial frequency and orientation is not effortlessly perceived. *Investigative Opthalmology and Visual Science, Supplement, 24*, 238.

Waltz, D. (1975) Understanding line drawings of scenes with shadows. In P. H. Winston (ed.), *The psychology of computer vision*. New York: McGraw-Hill.

Wanquier, A., and Rolls, E. T. (1976). *Brain-stimulation reward*. Amsterdam: North-Holland.

Wardlaw, K. A., and Kroll, N. E. (1976). Autonomic responses to shock-associated words in a non-attended message: A failure to replicate. *Journal of Experimental Psychology: Human Perception and Performance, 2;*, 357–360.

Warren, J. M., Warren, H. B., and Akert, K. (1961). Learning by cats with lesions in the prestriate cortex. *Journal of Comparative and Physiological Psychology, 54*, 629–632.

Warren, R. M. (1970). Perceptual restoration of missing speech sounds. *Science, 167*, 392–393.

Warrington, E. K., and Sanders, H. I. (1971). The fate of old memories. *Quarterly Journal of Experimental Psychology, 23*, 432–442.

Warrington, E. K., and Weiskrantz, L. (1970). Amnesia: Consolidation or retrieval? *Nature (London), 228*, 628–630.

(1973). An analysis of short-term and long-term memory defects in man. In J. A. Deutsch (ed.) *The physiological basis of memory*, pp. 365–395. New York: Academic Press.

(1974). The effects of prior learning on subsequent retention in amnesic patients. *Neuropsychologia, 12*, 419–428.

(1982). Amnesia: a disconnection syndrome? *Neuropsychologia, 20*, 233–248.

Wässle, H., Boycott, B. B., and Illing, R.-B. (1981). Morphology and mosaic of on- and off-beta cells in the cat retina and some functional considerations. *Proceedings of the Royal Society of London, B, 212*, 177–195.

Wässle, H., Peichl, L., and Boycott, B. B. (1981). Morphology and topography of on- and off-alpha cells in the cat retina. *Proceedings of the Royal Society of London, B., 212*, 157–175.

Wassman, M., and Flynn, J. P. (1962). Directed attack elicited from the hypothalamus. *Archives of Neurology, 6*, 220–227.

Watson, A. B., Barlow, H. B., and Robson, J. G. (1983). What does the eye see best? *Nature (London), 302*, 419–422.

Watson, R. T., and Heilman, K. M. (1979). Thalamic neglect. *Neurology, 29*, 690–694.

Watson, R. T., Heilman, K. M., Cauthen, J. C., and King, F. A. (1973). Neglect after cingulectomy. *Neurology, 23*, 1003–1007.

Watson, R. T., Heilman, K. M., Miller, B. D., and King, F. A. (1974). Neglect after mesencephalic reticular formation lesions. *Neurology, 24*, 294–298.

Watson, R. T., Miller, B. D., and Heilman, K. M. (1978). Nonsensory neglect. *Annals of Neurology, 3*, 505–508.

Weber, J. T., Huerta, M. F., Kaas, J. H., and Harting, J. K. (1983). The projections of the lateral geniculate nucleus of the squirrel monkey: studies of the interlaminar zones and the S layers. *Journal of Comparative Neurology, 213*, 135–145.

Weingartner, H., Miller, H., and Murphy, D. L. (1977). Mood-state-dependent retrieval of verbal associations. *Journal of Abnormal Psychology*, 276–284.

Weinstein, E. A., Kahn, R. L., and Slote, W. H. (1955). Withdrawal, inattention, and pain asymbolia. *Archives of Neurology and Psychiatry, 74*, 235–248.

Weiskrantz, L. (1956). Behavioral changes associated with ablation of the amygdaloid complex in monkeys. *Journal of Comparative Physiology and Psychology, 49*, 381–391.

——— (1970). Verbal learning and retention by amnesic patients using partial information. *Psychonomic Science, 20*, 210.

Weiskrantz, L. (1978). A comparison of hippocampal pathology in man and other animals. *Functions of the septo-hippocampal system (CIBA Foundation Symposium No. 58)*. Amsterdam: Elsevier.

Weiss, J. M. (1972). The influence of physiological variables on stress induced pathology. *Physiology, emotion, and psychosomatic illness. (CIBA Foundation Symposium No. 8)*. Amsterdam: Elsevier.

Weiss, J. M., McEwen, B. S., Silva, T., and Kalkent, M. (1970). Pituitary-adrenal alterations and fear responding. *American Journal of Physiology, 218*, 864–868.

Weisstein, N., and Harris, C. S. (1974). Visual detection of line segments: an object superiority effect. *Science, 186*, 752–755.

Weisz, D. J., Clark, G. A., and Thompson, R. F. (1984). Increased responsivity of dentate granule cells during nictitating membrane response conditioning in rabbit. *Behavioural Brain Research, 12*, 145–154.

Welch, K., and Stuteville, P. (1958). Experimental production of unilateral neglect in monkeys. *Brain, 81*, 341–347.

Welford, A. T. (1968). *Fundamentals of skill*, London: Methuen.

Weller, R. E., and Kaas, J. H. (1981). Cortical and subcortical connections of visual cortex in primates. In C. N. Woolsey (ed.), *Cortical sensory organization*, Vol. 2. Chicago: Humana.

Werblin, F. S. (1970). Response of retinal cells to moving spots: intracellular recording in *Necturus maculosus. Journal of Neurophysiology, 33*, 342–350.

Werblin, F. S., and Dowling, J. E. (1969). Organization of the retina of the mudpuppy, *Necturus maculosus*. II. Intracellular recording. *Journal of Neurophysiology, 32*, 339–355.

Wheeler, D. (1970). Processes in word recognition. *Cognitive Psychology, 1*, 59–85.

White, R. W. (1959). Motivation reconsidered: the concept of competence. *Psychological Review, 66*, 297–333.

Whitty, C. W. M., and Zangwill, O. L. (eds.). (1977). *Amnesia*. London: Butterworth.

Wickelgren, W. A. (1977). *Learning and memory*. Englewood Cliffs, N.J.: Prentice-Hall.

——— (1979). Chunking and consolidation: a theoretical synthesis of semantic networks, configuring in conditioning, S-R versus cognitive learning, normal forgetting, the amnesic syndrome, and the hippocampal arousal system. *Psychological Review, 86*, 44–60.

Wickens, C. (1980). The structure of attentional processes. In R. S. Nickerson (ed.), *Attention and performance*, Vol. VIII. Hillsdale, N.J.: Erlbaum.

——— (1984). Processing resources in attention. In R. Parasuraman and D. R. Davies (eds.), *Varieties of attention*. Orlando, Fla.: Academic Press.

Wickens, C. D., and Vidulich, M. (1982). S-C-R compatibility and dual task performance in two complex information processing tasks: Threat evaluation and fault diagnosis. Engineering-Psychology Research Laboratory, University of Illinois at Urbana-Champaign, Technical Report EPL–82–3/ONR–82–3, December, 1982.

Wickens, C., Kramer, A., Vanasse, L., and Donchin, E. (1983). Performance of concurrent tasks: a physiological analysis of the reciprocity of information processing resources. *Science, 221,* 1080–1082.

Wiesel, T. N., and Hubel, D. H. (1966). Spatial and chromatic interactions in the lateral geniculate body of the rhesus monkey. *Journal of Neurophysiology, 29,* 1115–1156.

Wiesendanger, R., and Wiesendanger, M. (1982). The corticopontine system in the rat. II. The projection pattern. *Journal of Comparative Neurology, 208,* 227–238.

Williams, A., and Weisstein, N. (1978). Line segments are perceived better in a coherent context than alone: an object-line effect in visual perception. *Memory and Cognition, 6,* 85–90.

Wiliams, M. D., and Hollan, J. D. (1981). The process of retrieval from very long-term memory. *Cognitive Science, 5,* 87–119.

Williams, M. D., and Santos-Williams, S. (1980). Method for exploring retrieval processes using verbal protocols. In R. Nickerson (ed.), *Attention and performance,* Vol. VIII. Hillsdale, N.J.: Erlbaum.

Wilson, E. O. (1975). *Sociobiology: the new synthesis.* Cambridge, Mass.: Belknap.

Wilson, H. R., and Bergen, J. R. (1979). A four mechanism model for threshold spatial vision. *Vision Research, 19,* 19–32.

Wilson, J. R., and Hendrickson, A. E. (1981). Neuronal and synaptic structure of the dorsal lateral geniculate nucleus in normal and monocularly deprived macaca monkeys. *Journal of Comparative Neurology, 197,* 517–539.

Wilson, M. (1957). Effects of circumscribed cortical lesions upon somaesthetic and visual discrimination in the monkey. *Journal of Comparative and Physiological Psychology, 52,* 630–635.

Wilson, P. D., Rowe, M. H., and Stone, J. (1976). Properties of relay cells in the cat's lateral geniculate nucleus. A comparison of W-cells with X- and Y-cells. *Journal of Neurophysiology, 39,* 1193–1209.

Wilson, W. R. (1979). Feeling more than we can know: exposure effects without learning. *Journal of Personality and Social Psychology, 37,* 811–821.

Winnick, W. A., and Daniel, S. A. (1970). Two kinds of response priming in tachistoscopic recognition. *Journal of Experimental Psychology, 84,* 74081.

Winocur, G., and Breckenridge, C. B. (1973). Cue dependent behavior of hippocampally damaged rats in a complex maze. *Journal of Comparative and Physiological Psychology, 82,* 512–522.

Winocur, G., and Olds, J. (1978). Effects of context manipulation on memory and reversal learning in rats with hippocampal lesions. *Journal of Comparative and Physiological Psychology, 92,* 312–321.

Winograd, E., and Killinger, W. A. (1983). Relating age at encoding in early childhood to adult recall: Development of flashbulb memories. *Journal of Experimental Psychology: General, 112,* 413–422.

Woodworth, R. S., and Schlosberg, H. (1954). *Experimental psychology.* New York: Holt.

Woodworth, R. S., and Sherrington, C. S. (1904). A pseudoaffective reflex and its spinal path. *Journal of Physiology (London), 31,* 234–243.

Woody, C. D. (1982). Neurophysiologic correlates of latent facilitation. In. C. D. Woody, (ed.), *Advances in behavioral biology,* Vol. 26, *Conditioning: Representation of involved neural functions,* pp. 233–248, New York: Plenum.

Wurtz, R. H. (1969). Response of striate cortex neurons to stimuli during rapid eye movements in the monkey. *Journal of Neurophysiology, 32*, 975–986.

Wurtz, R. H., and Goldberg, M. E. (1972). Activity of superior colliculus in behaving monkey: III. Cells discharging before eye movements. *Journal of Neurophysiology, 35*, 575–586.

Wurtz, R. H., and Mohler, C. W. (1976a). Organization of monkey superior colliculus: Enhanced visual response of superficial layer cells. *Journal of Neurophysiology, 39*, 745–765.

(1976b). Enhancement of visual responses in monkey striate cortex and frontal eye fields. *Journal of Neurophysiology, 39*, 766–772.

Wurtz, R. H., Goldberg, M. E., and Robinson, D. L. (1980). Behavioral modulation of visual responses in the monkey: stimulus selection for attention and movement. In J. M. Sprague and A. N. Epstein (eds.), *Progress in psychobiology and physiological psychology*, Vol. 9. New York: Academic Press.

Wynne, L. C., and Solomon, R. L. (1955). Traumatic avoidance learning: acquisition and extinction in dogs deprived of normal peripheral autonomic function. *Genetic Psychology Monograph, 52*, 241–284.

Wyss, J. M., and Sripanidkulchai, K. (1984). The topography of the mesencephalic and pontine projections from the cingulate cortex of the rat. *Brain Research, 293*, 1–15.

Yin, T. C. T., and Mountcastle, V. B. (1977). Visual input to the visuomotor mechanisms of the monkey's parietal lobe. *Science, 197*, 1381–1383.

Zajonc, R. B. (1980). Feeling and thinking: preferences need no inferences. *American Psychologist, 35*, 151–175.

(1984). On the primacy of affect. *American Psychologist, 39*, 117–123.

Zanchetti, A., and Bartorelli, C. (1977). Central nervous mechanisms in arterial hypertension: experimental and clinical evidence. In S. Julius and M. D. Esler (eds.), *Hypertension*. Springfield, Ill.: Thomas.

Zeki, S. M. (1974). Functional organization of a visual area in the posterior bank of the superior temporal sulcus of the rhesus monkey. *Journal of Physiology, London, 236*, 549–573.

(1976). The functional organization of projections from striate to prestriate cortex in the rhesus monkey. *Cold Spring Harbor Symposia on Quantitative Biology, 15*, 591–600.

(1977). Colour coding in the superior temporal sulcus of the rhesus monkey visual cortex. *Proceedings of the Royal Society of London, B., 197*, 195–223.

(1978). Uniformity and diversity of structure and function in rhesus monkey prestriate visual cortex. *Journal of Physiology, London, 277*, 273–290.

(1980). The representation of colours in the cerebral cortex. *Nature (London), 284*, 412–418.

(1983). The distribution of wavelength and orientation selective cells in different areas of monkey visual cortex. *Proceedings of the Royal Society of London, B, 217*, 449–470.

Zihl, J., and von Cramon, D. (1979). The contribution of the "second" visual system to directed visual attention in man. *Brain, 102*, 835–856.

Zihl, J., von Cramon, D., and Mai, N. (1983). Selective disturbance of movement vision after bilateral brain damage. *Brain, 106*, 313–340.

Zola-Morgan, S. M., and Oberg, R. G. E. (1980). Recall of life experiences in an alcoholic Korsakoff patient: a naturalistic approach. *Neuropsychologia, 18*, 549–557.

Zurif, E. B., and Blumstein, S. E. (1978). Language and the brain. In M. Halle, J. Bresnan, and G. A. Miller (eds.), *Linguistic theory and psychological reality*. Cambridge, Mass.: MIT Press.

Author index

429

Subject index

accessory optic system, 40
accommodation, 277
acetylcholine (ACh), 58, 309–11, 312
acetylcholinesterase (AChE), 41
action potentials, 35, 145
adaptation, 276–7, 279–81, 286–7, 297, 300, 319, 341, 359
adenosine monophosphate (AMP), cyclic, 222–3
adrenal cortex, 280, 319
adrenal medulla, 280, 288, 311–12, 315, 318, 323, 324, 325, 338
adrenergic transmission, 311, 336
adrenocorticotropic hormone (ACTH), 280, 281, 319, 326
affective processing, 356, 357, 374
affect(s), 3, 293–4, 300; primary, 277, 291–2
afferents, 45, 46, 81, 307, 328, 332, 348; exteroceptive, 307, 323; visceral, 323, 332, 336, 347
aggression, 282, 299, 318, 338, 339, 358
alarm reaction, 280, 319
alexithymia, 284
algorithms, 91, 93
aliments, 277–8
alternation behavior, 232
Alzheimer's disease, 207
amacrine cells, 38–9
amino acids, 37, 45
γ-aminobutyric acid (GABA), 161–2
amnesia (amnèsics), 231, 234, 259, 264, 369, 376, 377
amobarbitol, 294
amplitude, 55, 56, 57
amygdala, 237, 254, 331, 332, 333, 338–9, 341–2, 343, 352, 361, 362, 363; reward sites, 347; viscerosensory processing, 347–8
analytic cognition, 292–4
anger, 277, 291–2, 320–1, 365
animal consciousness, 335, 351, 364–5, 366–7
animal learning, 257, 272
animal models, 255, 319, 354

anterior thalamus, 254, 329, 333
anterior ventral nucleus (AVN), 242–5, 246–53, 258–60
anxiety, 185, 295, 299, 317
aphasia, 293
Aplysia (mollusc), 222–4, 228, 375–6
appraisal (emotion), 289–90, 356, 362–4, 366
approach behavior, 50
arousal, 217, 233, 296–7, 314, 315, 321, 322; attention, 142, 143, 144, 151, 153, 155–6, 165, 169, 177, 180–1; autonomic, 290, 315–19, 357; central pathways controlling cardiovascular system in, 338–40; nonspecific, 353; single cell studies, 149–55
arousal system(s), 157, 159
arousal theory, 287
artificial intelligence models, 213
aspartate, 41, 58
assimilation, 277–8
atropine, 311
attack behavior, 328, 338–9
attention, 2–3, 170, 189, 217, 362, 363–4, 371; cognitive psychology/neurobiology interaction, 171, 374–5; definitions, 142–4, 180–2; and detection, 182–3, 186; feature-integration theory of, 11–15; function of, 114, 120; mechanisms of, 101; models of, 112–13, 138, 370, 374; neurobiological view of psychology of, 172–9; neurobiology of, 142–71; neurobiology/psychology differences in, 85, 87, 89–90; in perception, 8, 44, 83, 97, 98, 100–1; processes in, 158; psychological view of neurobiology of, 180–6; psychology of, 105–41; serial, 86; *see also* spatial attention
attentional shifts, 118–20, 140, 142–3, 151, 152, 158, 159, 161–2, 165, 171, 182
attentional system, 142, 158–9, 177–8
attenuation model (information processing), 110, 183
attribute-specific areas (ASA), 68–75
attribution theory, 198, 278

438